Knowledge Management in the Intelligence Enterprise

For a complete listing of the *Artech House Information Warfare Library*,
turn to the back of this book.

Knowledge Management in the Intelligence Enterprise

Edward Waltz

AH

Artech House
Boston • London
www.artechhouse.com

Library of Congress Cataloging-in-Publication Data
Waltz, Edward.
   Knowledge management in the intelligence enterprise/Edward Waltz.
      p. cm.—(Artech House information warfare library)
   Includes bibliographical references and index.
   ISBN 1-58053-494-5 (alk. paper)
   1. Military intelligence.   2. Knowledge management.   3. Military intelligence—
United States.   4. Knowledge management—United States.   I. Title.   II. Series.

   UB250.W33 2003
   658.4'7—dc21

                                                                    2003041892

British Library Cataloguing in Publication Data
Waltz, Edward
   Knowledge management in the intelligence enterprise. — (Artech House information
   warfare library)
   1. Intelligence services   2. Knowledge management   3. Business intelligence
   I. Title
   327.1'2

   ISBN 1-58053-494-5

Cover design by Gary Ragaglia

© 2003 ARTECH HOUSE, INC.
685 Canton Street
Norwood, MA 02062

International Standard Book Number: 1-58053-494-5
Library of Congress Catalog Card Number: 2003041892

10 9 8 7 6 5 4 3 2 1

Dedicated to my daughter, Debbie

*From her wheelchair, she has shown me
what the Scriptures describe as those believers who are
clothed with compassion, kindness, humility,
gentleness, and patience (Colossians 3:12).*

Contents

Preface

In this rapidly changing and volatile world, the expectations required of those in the intelligence discipline are high—knowledge of the hidden and foreknowledge of the unpredictable. The consumers of intelligence—national policymakers, military planners, warfighters, law enforcement, business leaders—all expect accurate and timely information about the their areas of interest and threats to their security. They want strategic analyses, indications and warnings, and tactical details. This book is about the application of knowledge management (KM) principles to the practice of intelligence to fulfill those consumers' expectations.

I began this manuscript shortly before the September 11, 2001, terrorist attack on the United States. Throughout the period that I was writing this manuscript, the nation was exposed to an unprecedented review of the U.S. intelligence organizations and processes; intelligence has entered our nation's everyday vocabulary. Unfortunately, too many have reduced intelligence to a simple metaphor of "connecting the dots." This process, it seems, appears all too simple after the fact—once you have seen the picture and you can ignore irrelevant, contradictory, and missing dots. Real-world intelligence is not a puzzle of connecting dots; it is the hard daily work of planning operations, focusing the collection of data, and then processing the collected data for deep analysis to produce a flow of knowledge for dissemination to a wide range of consumers. From a torrent of data, real-world intelligence produces a steady stream of reliable and actionable knowledge. Intelligence organizations have performed and refined this process to deliver knowledge long before the term *knowledge management* became popular; today they are applying new collaborative methods and technologies to hone their tradecraft. This book focuses on those methods and technologies.

While the focus of this book is intelligence, it is also an outgrowth of a 2-day military KM seminar that I teach in the United States to describe the methods to integrate people, processes, and technologies into knowledge-creating enterprises. I have benefited from my interaction with a wide range of government and industry participants in the seminar, and the structure of this book reflects their interests in a balance between the abstract, organizational, practical, and technology aspects of KM.

The book progresses from an introduction to KM applied to intelligence (Chapters 1 and 2) to the principles and processes of KM (Chapter 3). The characteristics of collaborative knowledge-based intelligence organizations are described (Chapter 4) before detailing its principle craft of analysis and synthesis (Chapter 5 introduces the principles and Chapter 6 illustrates the practice). The wide range of technology tools to support analytic thinking and allow analysts to interact with information is explained (Chapter 7) before describing the automated tools that perform all-source fusion and mining (Chapter 8). The organizational, systems, and technology concepts throughout the book are brought together in a representative intelligence enterprise (Chapter 9) to illustrate the process of architecture design for a small intelligence cell. An overview of core, enabling, and emerging KM technologies in this area is provided in conclusion (Chapter 10).

I am thankful to God for the understanding of my family in allowing me the time to enter into my writing projects; I spent our vacation this year at a northern Michigan cottage writing (while my daughter read and my wife completed puzzles). Their understanding runs deep and wide. I am also grateful for the ever-present encouragement and support of Mark Lazaroff, Woody Spring, and Dr. Christopher Davis, my leadership at Veridian, and for the insight and encouragement for my research provided by Dr. Susan Durham. Thanks to Jeff Malone of the Australian Department of Defence for his thorough review of the manuscript; I incorporated virtually every one of excellent his comments to make this book more accurate to the trade and more readable for the intelligence user. I am particularly indebted to my friends and colleagues, Mike Bennett, Val Johnson, Tom Tulenko, and Dr. Russ Vane for performing independent reviews of sections of the manuscript to help me sharpen the finer points (and avoid some real blunders). And, of course, I am grateful to have learned much from a large number of devoted and uniquely talented individuals across the intelligence community (IC); they are friends and colleagues whom I cannot name. These analysts and operations officers, developers, and systems operators have taught me the tradecraft, toned my optimism on technology solutions, and guided my ideas toward real-world intelligence applications.

1

Knowledge Management and Intelligence

This is a book about the management of knowledge to produce and deliver a special kind of knowledge: intelligence—that knowledge that is deemed most critical for decision making both in the nation-state and in business. In each case, intelligence is required to develop policy and strategy and for implementation in operations and tactics. The users of intelligence range from those who make broad policy decisions to those who make day-to-day operational decisions. Thus, the breadth of this product we call intelligence is as wide as the enterprise it serves, with users ranging from executive decision makers to every individual in the enterprise, including its partners, suppliers, and customers.

First, we must define the key terms of this text that refer to the application of technology, operations, and people to the creation of knowledge:

- *Knowledge management* refers to the organizational disciplines, processes, and information technologies used to acquire, create, reveal, and deliver knowledge that allows an enterprise to accomplish its mission (achieve its strategic or business objectives). The components of knowledge management are the people, their operations (practices and processes), and the information technology (IT) that move and transform data, information, and knowledge. All three of these components make up the entity we call *the enterprise.*

- *Intelligence* refers to a special kind of knowledge necessary to accomplish a mission—the kind of strategic knowledge that reveals critical threats and opportunities that may jeopardize or assure mission accomplishment. Intelligence often reveals hidden secrets or conveys a deep understanding that is covered by complexity, deliberate denial, or outright deception. The intelligence process has been described as the process of the *discovery of secrets by secret means.* In business and in national security, secrecy is a process of protection for one party; discovery of the secret is the object of competition or security for the competitor or adversary. The need for security in the presence of competition, crisis, and conflict drives the need for intelligence. While a range of definitions of intelligence exist, perhaps the most succinct is that offered by the U.S. Central Intelligence Agency (CIA): "Reduced to its simplest terms, intelligence is knowledge and foreknowledge of the world around us—the prelude to decision and action by U.S. policymakers" [1]. These classical components of intelligence, knowledge, and foreknowledge provide the insight and warning that leaders need for decision making to provide security for the business or nation-state [2].

- *The intelligence enterprise* encompasses the integrated entity of people, processes, and technologies that collects and analyzes intelligence data to synthesize intelligence products for decision-making consumers.

Indeed, intelligence (whether national or business) has always involved the management (acquisition, analysis, synthesis, and delivery) of knowledge. In this book, we emphasize the application of knowledge management operations to refer to the organizational culture, automated processes, and enterprise architecture that enables the automated management of data, information, and knowledge to complement human analysis and decision making. At least three driving factors continue to make this increasing need for automation necessary. These factors include:

- *Breadth of data to be considered.* The effect of globalization in politics, nation-state collaboration (in both cooperative trade and coalition warfare), economics, and communication has increased the breadth of intelligence analysis to include a wide range of influences related to security and stability. While intelligence has traditionally focused on relatively narrow collection of data by trusted sources, a floodgate of open sources of data has opened, making available information on virtually any topic. However, these new avenues come with the attendant uncertainty in sources, methods, and reliability.

- *Depth of knowledge to be understood.* Driven by the complexity of operations on a global scope, national policies and business strategies involve the consideration of many interactive variables. This complexity requires models that allow alternative policies and strategies to be evaluated. These models require accurate data about the environment (e.g., markets, nation-state economies, or military orders of battle) in general and focused problems in particular (e.g., market niches, specific companies, or military targets).

- *Speed required for decision making.* The pace of operations, in national policymaking, military warfare, and business operations is ever increasing, placing demands for the immediate availability of intelligence about the dynamic world or marketplace to make nation-state policy and business strategy decisions.

Throughout this book, we distinguish between three levels of abstraction of knowledge, each of which may be referred to as *intelligence* in forms that range from unprocessed reporting to finished intelligence products [3]:

1. *Data.* Individual observations, measurements, and primitive messages form the lowest level. Human communication, text messages, electronic queries, or scientific instruments that sense phenomena are the major sources of data. The terms *raw intelligence* and *evidence* (data that is determined to be relevant) are frequently used to refer to elements of data.

2. *Information.* Organized sets of data are referred to as information. The organization process may include sorting, classifying, or indexing and linking data to place data elements in relational context for subsequent searching and analysis.

3. Knowledge. Information once analyzed, understood, and explained is knowledge or foreknowledge (predictions or forecasts). In the context of this book, this level of understanding is referred to as the *intelligence product.* Understanding of information provides a degree of comprehension of both the static and dynamic relationships of the objects of data and the ability to model structure and past (and future) behavior of those objects. Knowledge includes both static content and dynamic processes.

These abstractions are often organized in a *cognitive hierarchy,* which includes a level above knowledge: human wisdom. In this text, we consider wisdom to be a uniquely human cognitive capability—the ability to correctly apply knowledge to achieve an objective. This book describes the use of IT to support the creation of knowledge but considers wisdom to be a human capacity out of the realm of automation and computation. IT can enable humans to gain experience through training, simulation, and enhanced understanding of real-life events; this way, technology can contribute to a human's growth in wisdom [4].

1.1 Knowledge in a Changing World

This strategic knowledge we call intelligence has long been recognized as a precious and critical commodity for national leaders. Sixth century B.C. military strategist Sun Tzu is often quoted for his recognition of the importance of intelligence in military strategy. On the use of spies, he acknowledged the necessity of knowledge of the adversary:

> The means by which enlightened rulers and sagacious generals moved and conquered others, that their achievements surpassed the masses, was advance knowledge.
>
> Advance knowledge cannot be gained from ghosts and spirits, inferred from phenomena, or projected from the measures of Heaven, but must be gained from men for it [i.e., advance knowledge] is the knowledge of the enemy's true situation [5].

Sun Tzu's treatise also defined five categories of spies [6], their tasks, and the objects of their intelligence collection and covert operations. More than seven centuries before Sun Tzu, the Hebrew leader Moses commissioned and documented an intelligence operation to explore the foreign land of Canaan. That classic account clearly describes the phases of the intelligence cycle, which proceeds from definition of the *requirement* for knowledge through *planning, tasking, collection,* and *analysis* to the *dissemination* of that knowledge. He first detailed the intelligence *requirements* by describing the eight essential elements of information to be collected, and he described the *plan* to covertly enter and reconnoiter the denied area:

> When Moses sent [12 intelligence officers] to explore Canaan, he said, "Go up through the Negev and on into the hill country. See what the land is like and whether the people who live there are strong or weak, few or many.
> What kind of land do they live in?
> Is it good or bad?
> What kind of towns do they live in?
> Are they unwalled or fortified?
> How is the soil?
> Is it fertile or poor?
> Are there trees on it or not?
> Do your best to bring back some of the fruit of the land. [It was the season for the first ripe grapes.]
> (Numbers 13:17-20, NIV) [7].

A 12-man reconnaissance team was *tasked,* and it carried out a 40-day *collection* mission studying (and no doubt mapping) the land and collecting crop samples. The team traveled nearly 200 miles north from the desert of Zin (modern Gaza) observing fortified cites and natural resources. Upon return, the intelligence observations were delivered and the data *analysis* and report *synthesis* phase began as the leaders considered the implications of the data (Numbers 13:26-33). As all too often is the case in intelligence, the interpretation of the data and judgments about the implications for the Hebrew people were in severe dispute. In the account of this analysis, the dispute over the interpretation of the data and the estimated results once *disseminated* to the leaders and the nation at large led to a major national crisis (see Numbers 14–15).

The analysis of intelligence data has always been as significant as the collection, because analysis of the data and synthesis of a report creates *meaning* from the often-scant samples of data about the subject of interest. Before becoming the first U.S. president, George Washington commissioned intelligence-collection operations when he was a general officer of the revolutionary army. He recognized the crucial importance of analysis. In a letter of appreciation to James Lovell in April 1782, Washington specifically noted the value of all-source intelligence analysis and synthesis that integrates disparate components of evidence:

I THANK YOU FOR THE TROUBLE you have taken in forwarding the intelligence which was inclosed in your Letter of the 11th of March. It is by comparing a variety of information, we are frequently enabled to investigate facts, which were so intricate or hidden, that no single clue could have led to the knowledge of them in this point of view, intelligence becomes interesting which but from its connection and collateral circumstances, would not be important [8].

While each of these leaders acknowledged the value of applied intelligence, their processes of requirements articulation, planning, collection, analysis-synthesis, and dissemination were entirely manual. Since the days of Washington, intelligence has undergone transformation even as the consumers of intelligence—those who maintain national security and wage warfare, and those who create wealth—have been in transformation. Political, military, and business thinkers have widely analyzed the revolutionary changes in the nation-state, the military, and business as a result of information technologies.

The most popular and widely cited general view of the transformation attributable to IT is the thesis introduced by Alvin and Heidi Toffler that defines three great waves of civilization based on the changing means of maintaining power, creating wealth, and waging war [9]. The thesis can be summarized in four essential points. First, history can be described in terms of three distinct periods (phases or *waves*) during which mankind's activity—both production and destruction—have changed in quantum transitions. In the conduct of both commerce and warfare, the necessary resources and core competencies radically shifted at the transition between waves. Second, each distinct wave is characterized by its means of wealth production—and a central resource at the core of the production mechanism. Third, technology is the cause of the rapid transitions, because as new technologies are introduced, the entire basis for wealth (production) and power (the potential for economic strength and destruction) change. These changes between waves also bring the potential to rapidly change the world order. Finally, each new wave has partitioned the nation-states of the world into categories, each characterized by their maturity (e.g., an information-age society is characterized as *third wave*). The world is now trisected into nations in each of the three wave categories.

Table 1.1 summarizes the three waves identified by the Tofflers, transitioning from the agricultural to the information age. The agricultural wave was characterized by peasant-based crop production, dependent upon the central resource of land ownership. The industrial age rapidly shifted the balance of world power, as raw materials for mass production became the central resource. Mass production, and the comparable ability to wage mass destruction, transferred power to the nation-states with industrial technology.

The last decades of the twentieth century brought the transition to a new information age, in which the Tofflers asserted:

- Information (the raw material of knowledge) is the central resource for wealth production and military power.

- Wealth production is based on the ownership of information—the creation of knowledge and delivery of custom products based on that knowledge.

- Conflicts are based on geo-information competitions—over ideologies and economies.

The intelligence discipline has always faced a competition for information—critical information about competitors and adversaries. Table 1.1 also distinguishes the significant transitions in the focus of intelligence throughout the Tofflers' waves of civilization. Throughout the agricultural age, intelligence collection remained centered on human observation and interaction, or human intelligence (HUMINT), as cited earlier in the accounts of Moses, Sun Tzu, and General Washington. This *human collection-centric* means was dependent upon *physical human access* and covert means to communicate information from collectors to decision makers.

The industrial age introduced increasingly complex remote sensing instruments and stand-off collection platform technologies, ranging from early telescopes and hot air balloons to post–World War II radars and more recent satellite platforms. These sensors and platforms combined to provide

Table 1.1
Three Waves of Civilization and the Transitions in Wealth Creation, Warfare, and Intelligence

Age	Agricultural	Industrial	Information	
Approx. Period	**Until–1700**	**1700–2000**	**2000–Future**	
Wealth Creation: Power and Business	Method: peasant-based crop production Central resource: land	Method: mass production of goods Central resource: raw materials	Method: customized production of knowledge services Central resource: knowledge	
Nation-State Warfare, Conflict, and Competition	Object of conflicts: land Infantry warfare: attrition of infantry (target human bodies)	Objects of conflict: regional economies, access to materials Mechanized warfare: mass destruction of weapons (target mechanized weapons)	Objects of conflicts: global economies, ideologies Information warfare: attrition of will and capability, precision targeting, speed and agility, management of perception (target the human mind)	
Focus of Intelligence	Human collection centric (covert access)	Technical sensing centric (remote access)	Network centric (network access)	Knowledge-centric (perceptual access)
Intelligence Examples	Moses, Sun Tzu, General George Washington	World War II: radio, radar, cryptography; use of air platforms Cold War: space reconnaissance	Post–Gulf War: emphasis on network-centric warfare, battle-field digitization, rapid targeting and data dissemination	Future emphasis on human congition, decision-making and influence

revolutionary, powerful intelligence-collection capabilities. Intelligence consumers increased their dependence on these sources to complement and validate their traditional HUMINT sources. Aggressors' orders of battle were essentially hidden until radar, electro-optics, and radio receivers were refined throughout the Cold War to provide remote sensing of large weapons and production facilities, both for monitoring treaties and providing indications and warnings of large-scale attacks. Revolutionary space capabilities introduced by electronic sensors and spaceborne platforms in the 1960s and 1970s moved intelligence toward a mature *sensor-centric* emphasis. In the Gulf War, these sensor assets benefited the United States–led coalition on the battlefield, providing unprecedented surveillance and targeting. In that sensor-centric world of the early

1990s, information superiority required *sensing coverage* and the key technologies were *global sensors*.

But the Gulf War also pointed out a weakness in the ability to reap the benefits of global sensing—the difficulties in developing collaboration between intelligence and operational communities and the inability to rapidly disseminate knowledge to the warfighter [10]. Since the war, as remote sensing and global communications has proliferated and become available to all, the information competition has shifted from coverage to *speed of access and dissemination*. The U.S. defense community has developed a *network-centric* approach to intelligence and warfare that utilizes the power of networked information to enhance the speed of command and the efficiency of operations [11]. Sensors are linked to shooters, commanders efficiently coordinate agile forces, and engagements are based on prediction and preemption. The keys to achieving information superiority in this network-centric model are *network breadth* (or *connectivity*) and bandwidth; the key technology is *information networking*.

Winning future intelligence competitions, where the conflict space is global and extends across the physical, symbolic, and cognitive realms, will require yet a new strategy. The future emphasis will become dependent on maintaining a *knowledge-centric* advantage. This is because we are moving into a world environment where no single player will maintain the sole and significant margin in global sources of information or in the ability to network information. Global sensing and networking capabilities will become a commodity with most global competitors at parity. Like the open chess game where everyone sees all the pieces, the world will be an open chessboard of readily available information accessible by all intelligence competitors. The ability to win will depend upon the ability to select and convert raw data into accurate decision-making knowledge. Intelligence superiority will be defined by the ability to make decisions most quickly and effectively—with the same information available to virtually all parties. The key enabling technology in the next century will become *processing* and *cognitive power* to rapidly and accurately convert data into comprehensive explanations of reality—sufficient to make rapid and complex decisions.

Consider several of the key premises about the significance of knowledge in this information age that are bringing the importance of intelligence to the forefront. First, knowledge has become the central resource for competitive advantage, displacing raw materials, natural resources, capital, and labor. This resource is central to both wealth creation and warfare waging. Second, the management of this abstract resource is quite complex; it is more difficult (than material resources) to value and audit, more difficult to create and exchange, and much more difficult to protect. Third, the processes for producing knowledge from raw data are as diverse as the manufacturing processes for physical materials, yet are implemented in the same virtual manufacturing plant—the computer. Because of these factors, the management of knowledge to produce strategic intelligence has become a necessary and critical function within nations-states and business enterprises—requiring changes in culture, processes, and infrastructure to compete.

According to Gary Hamel in *Leading the Revolution* [12], the business revolution of the twenty-first century is characterized by complex, nonlinear behaviors (in technology, the competition, and the highly interconnected global marketplace) that demand continuous innovation for competitive wealth creation. Similarly, those who envision a revolution in military affairs (RMA) see identical challenges to the business of national security.

The rapid transition over the past 3 decades from industrial age linearity has progressed in three stages (Figure 1.1):

- The focus on continuous improvement in the 1970s focused on innovation to improve products and services. Management focused on improving production capital assets. In the military, this included the refinement of weapons (precision guided munitions, data links, stand-off surveillance, etc.) using closed loop command and control.

- The 1980s and 1990s brought greater awareness of the value of intellectual capital, and attention was turned to enhancing processes, through business process re-engineering (BPR). These process refinements were

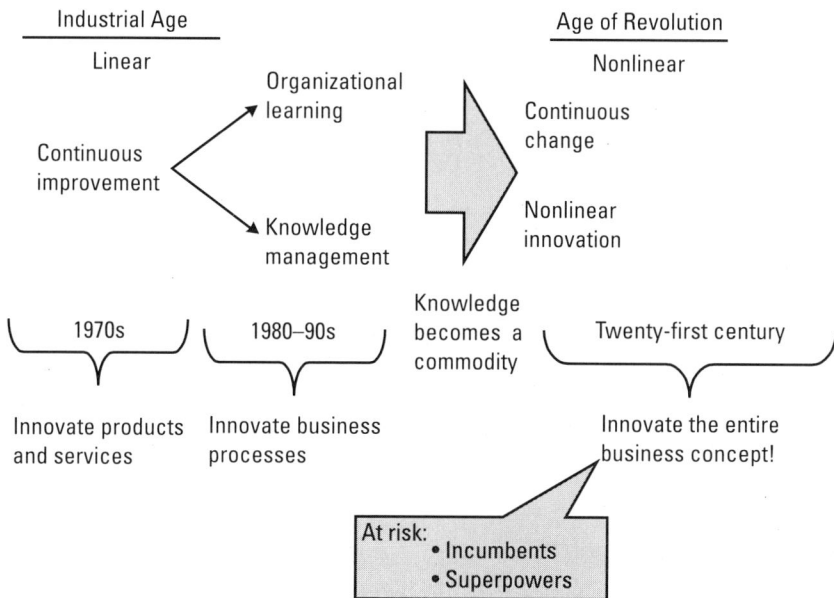

Figure 1.1 Transformations to the nonlinearity of revolution.

accompanied by the development of learning organizations, and the introduction of knowledge management (KM) infrastructures and practices.

- Now, with rapidly emerging information technologies, the complexities of globalization and diverse national interests (and threats), businesses and militaries must both adopt radically new and innovative agendas to enable continuous change in their *entire operating concept.* Innovation and agility are the watchwords for organizations that will remain competitive in Hamel's age of nonlinear revolution.

According to Hamel:

> Business concept innovation will be *the* defining competitive advantage in the age of revolution. Business concept innovation is the capacity to reconceive existing business models in ways that create new value for customers, rude surprises for competitors, and new wealth for investors. Business concept innovation is the only way for newcomers to succeed in the face of enormous resource disadvantages, and the only way for incumbents to renew their lease on success [13].

In this view, those at greatest risk in this new nonlinear environment are incumbents (in business) and superpowers (in national security). The U.S. emphasis on RMA to become innovative and agile is observed in the investments to address asymmetric threats and information warfare. And the exploration of a new network-centric doctrine illustrates the move to restructure the military to an adaptive warfighting *organism* that emphasizes networked collaborative knowledge rather than a command hierarchy that emphasizes control of weaponry [14].

1.2 Categories of Intelligence

The U.S. IC defines intelligence as a temporal knowledge product that is the result of collection, analysis, and production:

> Reduced to its simplest terms, intelligence is knowledge and foreknowledge of the world around us. The prelude to decision and action by U.S. policymakers. Intelligence organizations provide this information in a fashion that helps consumers, either civilian leaders or military commanders, to consider alternative options and outcomes. The intelligence process involves the painstaking and generally tedious collection of facts, their analysis, quick and clear evaluations, production of intelligence assessments, and their timely

dissemination to consumers. Above all, the analytical process must be rigorous, timely, and relevant to policy needs and concerns [15].

A functional taxonomy (Figure 1.2) based on the type of analysis and the temporal distinction of knowledge and foreknowledge (warning, prediction, and forecast) distinguishes two primary categories of analysis and five subcategories of intelligence products [16]:

- *Descriptive analyses* provide little or no evaluation or interpretation of collected data; rather, they enumerate collected data in a fashion that organizes and structures the data so the consumer can perform subsequent interpretation. Descriptive analytic tasks include the enumeration and organization of such topics as census data, production, geospatial data (maps), organizational data, public records (e.g., telephone books, government officials), and weather. Descriptive analysis tasks include compiling, organizing, structuring, indexing, and crosschecking.

- *Inferential analyses* require the analysis of collected relevant data sets (evidence) to infer and synthesize explanations that describe the *meaning* of the underlying data. We can distinguish four different focuses of inferential analysis:

  1. Analyses that explain past events (How did this happen? Who did it?);

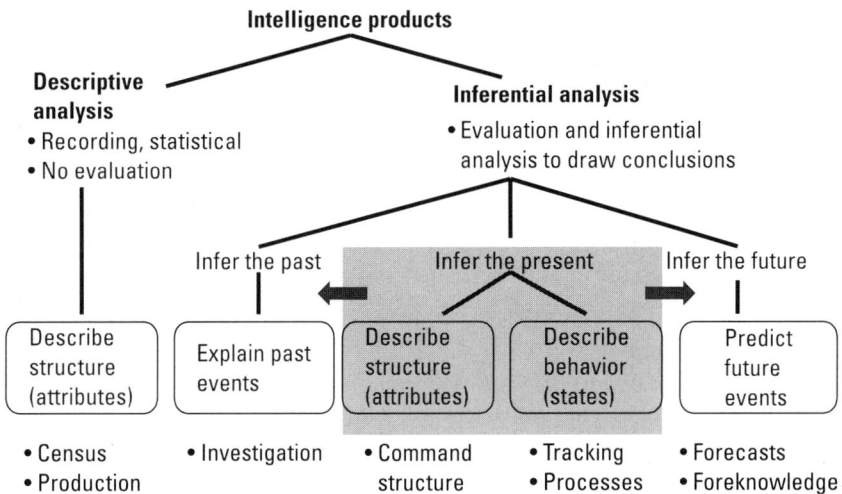

Figure 1.2 Taxonomy of intelligence products by analytic methods.

2. Analyses that explain the structure of current structure (What is the organization? What is the order of battle?);

3. Analyses that explain current behaviors and states (What is the competitor's research and development process? What is the status of development?);

4. Foreknowledge analyses that forecast future attributes and states (What is the expected population and gross national product growth over the next 5 years? When will force strength exceed that of a country's neighbors? When will a competitor release a new product?).

In further chapters, we will expand this basic taxonomy in greater detail to describe the many analytic techniques that may be applied to inferential analysis. Indeed, the focus of this book is on the issues of inferential analysis, though the KM processes presented provide the foundation capabilities for both descriptive and inferential analysis.

1.3 The Intelligence Disciplines and Applications

While the taxonomy of intelligence products by analytic methods is fundamental, the more common distinctions of intelligence are by discipline or consumer. In this section, we compare and distinguish between those applications and consumers: national and military, business and competitive intelligence. Throughout the book, the principles of KM for all intelligence applications will be treated in a general manner; it is important to carefully describe these four distinct uses of intelligence up front. The KM processes and information technologies used in all cases are identical (some say, "bits are bits," implying that all digital data at the bit level is identical), but the content and mission objectives of these four intelligence disciplines are unique and distinct.

Consider first the top-level similarities (Table 1.2) between users, security concerns, intelligence functions, and intelligence consumers in the nation-state and in business. Nation-state security interests deal with sovereignty; ideological, political, and economic stability; and threats to those areas of national interest. Intelligence serves national leadership and military needs by providing strategic policymaking knowledge, warnings of foreign threats to national security interests (economic, military, or political) and tactical knowledge to support day-to-day operations and crisis responses. Nation-state intelligence also serves a public function by collecting and consolidating open sources of foreign information for analysis and publication by the government on topics of foreign relations, trade, treaties, economies, humanitarian efforts, environmental concerns, and other foreign and global interests to the public and businesses at large.

Table 1.2
Nation-State and Business Uses of Intelligence

| User | Security Concerns | Intelligence Functions | | | Intelligence Consumers |
		Strategic	Indications and Warning (I&W)	Operational and Tactical	
Nation-state	Sovereignty Political, economic stability Treaties, alliances Threats to defined national interests	Global political, economic, and military analysis; threat analysis	Threat event warning	Diplomatic support; crisis support; military targeting	National leaders; military; public
Business	Competitiveness Growth Real and Intellectual property Business alliances Threats to market position	Market analysis; competitor analysis	Market discontinuities and trends	Marketing and sales support; supply-chain management; customer relations management	Leaders and management; operations; employees

Businesses seek to maintain competitiveness in a marketplace and similarly require intelligence to provide awareness of the threats to its economic stability and growth, intellectual property, and position in the marketplace. Intelligence therefore plays a critical role in strategic business planning, as well as more tactical roles in supporting marketing, sales, and customer-supplier relationship management. Similar to the threat-warning intelligence function to the nation-state, business intelligence is chartered with the critical task of foreseeing and alerting management of marketplace discontinuities [17]. The consumers of business intelligence range from corporate leadership to employees who access supply-chain data, and even to customers who access information to support purchase decisions.

While these nation-state and business uses and functions of intelligence are quite analogous, the distinctions between national-military and business-competitor intelligence are sharp. These distinctions are based on the scope of the objects (targets or subjects) of intelligence addressed by each of the four disciplines (Figure 1.3). The objects of intelligence fall in three broad categories: own resources and position, the neutral environment in which all participants interact, and potential security threats to the nation-state or business. Notice

Objects (Targets) of Intelligence			
	Own situation (own)	The environment (neutral)	Threats to security (adversary)
Nation-state	• Own military force dispositions • Called friendly force information (FFI)	• Global affairs, foreign relations, treaties, economics, politics • Transnationals, • Battlespace factors, constraints: terrain, weather, lines of communication, etc.	• Adversary nations • Transnational and NGO threats • Adversary military threats • Infrastructure threats
Business	• Business operations • Customer relations mgmt • Sales force automation • Supply chain mgmt • E-Commerce	• The Market (customers, products and services, etc.) • Market factors (economy, season, sales area, etc.) • Market dynamics	• Competitive landscape • Competitor market position • Competitor operations, products

National intelligence

Military intelligence

Business intelligence

Competitor intelligence

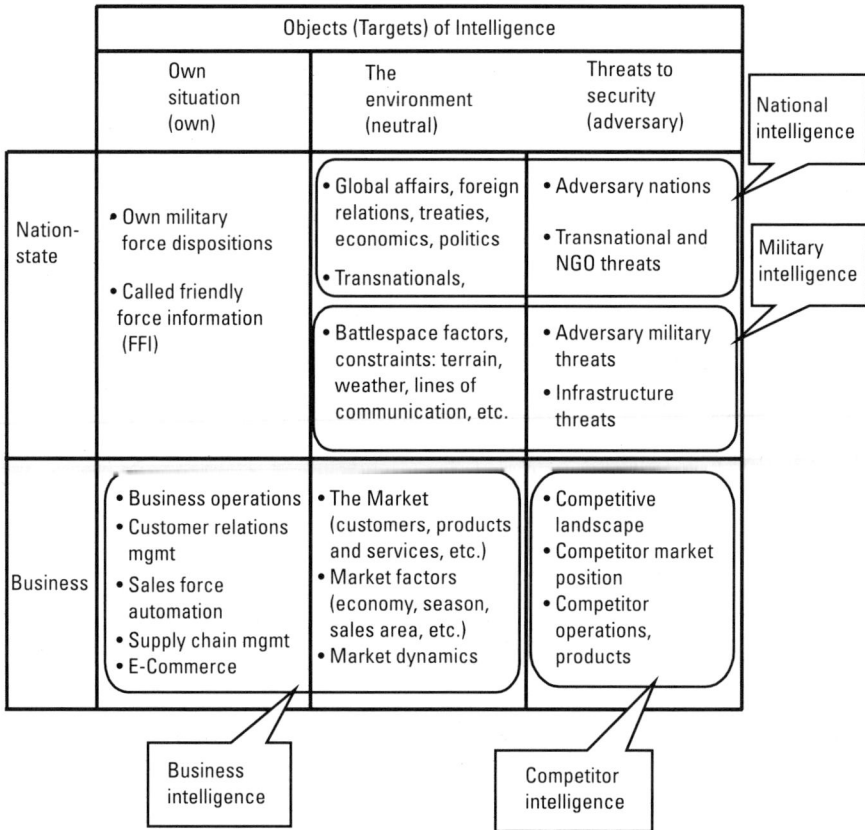

Figure 1.3 The subjects of the four categories of intelligence disciplines.

that each of the four disciplines, as defined by their users, partition the target subjects in ways that are not analogous:

1. National intelligence focuses on the understanding of the global environment (political, economic, natural environmental, science, and technology areas) and its important participants (foreign nation-states and their political organizations, nongovernmental organizations [NGOs], and influential individuals).

2. Military intelligence (MI) refers to the intelligence processes that focus on understanding foreign military threats to provide threat assessments, I&W, weapons targeting, and damage assessments (in time of conflict).

3. Business intelligence (BI) refers to the acquisition, organization, analysis, and reporting of internal and external factors to enable

decision makers to make faster, more accurate, and effective decisions to meet business objectives. The general market focus of BI is often called competitive intelligence, a term not to be confused with competitor intelligence (CI).

4. CI is a subdivision of business intelligence that concerns the current and proposed business activities of competitors [18]. It uses legal and ethical means to collect and analyze data to focus narrowly on the competitive landscape and targets specific competitors (which, it should be noted, can also become strategic partners, acquisition targets, or future owners) and their roles in the marketplace.

It is important to note that this book marks a sharp distinction between business (private sector) and nation-state (public sector) intelligence activities, though they are not necessarily distinct in all countries. In the United States, public and private sector intelligence activities have been officially separated. Debate has centered on the importance and value of maintaining a separation of (public sector) national intelligence products from (private sector) businesses [19]. A European Parliament study has enumerated concern over the potential for national intelligence sources to be used for nation-state economic advantages by providing competitive intelligence directly to national business interests [20]. The United States has acknowledged a policy of applying national intelligence to protect U.S. business interests from fraud and illegal activities, but not for the purposes of providing competitive advantage [21].

1.3.1 National and Military Intelligence

National intelligence refers to the strategic knowledge obtained for the leadership of nation-states to maintain national security. National intelligence is focused on national security—providing strategic warning of imminent threats, knowledge on the broad spectrum of threats to national interests, and foreknowledge regarding future threats that may emerge as technologies, economies, and the global environment changes. National intelligence also supports national leaders in such areas as foreign policymaking, assessments of global economies, and validation of treaty compliance by foreign countries.

The term intelligence refers to both a process and its product. The U.S. Department of Defense (DoD) provides the following product definitions that are rich in description of the processes involved in producing the product [22]:

1. The product resulting from the collection, processing, integration, analysis, evaluation, and interpretation of available information concerning foreign countries or areas;

2. Information and knowledge about an adversary obtained through observation, investigation, analysis, or understanding.

Michael Herman accurately emphasizes the essential components of the intelligence process [23]: "The Western intelligence system is two things. It is partly the collection of information by special means; and partly the subsequent study of particular subjects, using all available information from all sources. The two activities form a sequential process." While we introduce the subject of collection in Chapter 2, this book is about the "subsequent study" process that includes analysis (decomposition of the data to its essential parts) and synthesis (construction of essential parts of data to infer knowledge about the subject).

From a military perspective, intelligence is the enabler to achieve military dominance. Martin Libicki has provided a practical definition of *information dominance*, and the role of intelligence coupled with command and control and information warfare:

Information dominance may be defined as superiority in the generation, manipulation, and use of information sufficient to afford its possessors *military* dominance. It has three sources:

- Command and control that permits everyone to know *where* they (and their cohorts) are in the battlespace, and enables them to execute operations *when* and as quickly as necessary.
- Intelligence that ranges from knowing the enemy's dispositions to knowing the location of enemy assets in real-time with sufficient precision for a one-shot kill.
- Information warfare that confounds enemy information systems at various points (sensors, communications, processing, and command), while protecting one's own [24].

This superiority in the information domain is the enabling concept in the U.S. DoD's initial Joint Vision 2010 and the updated JV 2020 [25]. The superiority is achieved by gaining superior intelligence and protecting information assets while fiercely degrading the enemy's information assets. The goal of such superiority is not the attrition of physical military assets or troops—it is the attrition of the quality, speed, and utility of the adversary's decision-making ability.

The military has acknowledged the similarity, from a knowledge perspective, between the commercial business environment and military missions. Applying a commercial business model, the U.S. Navy offered the following description of its acquisition knowledge environment [26]: "A *knowledge environment* is an organizations (business) environment that enhances its capability to deliver on its mission (competitive advantage) by enabling it to build and leverage it intellectual capital."

1.3.2 Business and Competitive Intelligence

The focus of business intelligence is on understanding all aspects of a business enterprise: internal operations and the external environment, which includes customers and competitors (the marketplace), partners, and suppliers. The external environmental also includes independent variables that can impact the business, depending on the business (e.g., technology, the weather, government policy actions, financial markets). All of these are the objects of business intelligence in the broadest definition. But the term *business intelligence* is also used in a narrower sense to focus on only the internals of the business, while the term *competitor intelligence* refers to those aspects of intelligence that focus on the externals that influence competitiveness: competitors.

A taxonomy of the business intelligence terminology (Table 1.3) distinguishes business intelligence proper from competitive intelligence by the objects of their study. Neutral external factors are often included in the definitions of both categories of intelligence.

Each of the components of business intelligence has distinct areas of focus and uses in maintaining the efficiency, agility, and security of the business; all are required to provide active strategic direction to the business. In large companies with active business intelligence operations, all three components are essential parts of the strategic planning process, and all contribute to strategic decision making.

1.4 The Intelligence Enterprise

The intelligence enterprise includes the collection of people, knowledge (both internal tacit and explicitly codified), infrastructure, and information processes that deliver critical knowledge (intelligence) to the consumers. This enables them to make accurate, timely, and wise decisions to accomplish the mission of the enterprise. This definition describes the enterprise as a process—devoted to achieving an objective for its stakeholders and users. The enterprise process includes the production, buying, selling, exchange, and promotion of an item, substance, service, or system. The definition is similar to that adopted by Daimler-Chrysler's extended virtual enterprise, which encompasses its suppliers:

> A DaimlerChrysler coordinated, goal-driven process that unifies and extends the business relationships of suppliers and supplier tiers in order to reduce cycle time, minimize systems cost and achieve perfect quality [27].

This all-encompassing definition brings the challenge of describing the full enterprise, its operations, and component parts. Later in Chapter 9, we introduce the DoD three-view architecture [28] description, which defines three

Table 1.3
Taxonomy of the Components of Business Intelligence

	Business Intelligence: Acquisition, organization, analysis, and reporting of internal and external factors to enable decision makers to make faster, more accurate and effective decisions to meet business objectives.		
	Business Intelligence		**Competitive Intelligence**
Focus of Intelligence	**Internal**	**External: Neutral Factors**	**External: Competitive Factors**
Objects (Targets) of Intelligence	Business operations Supply chain Customer relations management Buyers and suppliers	Customer structure, preferences, behaviors Financial environment Regulatory climate Marketplace environment • Segmentation • Market drivers Buying patterns	Competitors Strategic partner candidates
Objective	Efficiency	Agility in the marketplace	Security
Uses by Intelligence Consumers	Business process performance analysis, refinement, and reengineering	Market dynamics modeling and forecasting Market positioning Learning customer trends Identifying threats, technology, regulation	Identifying competitor threats Tracking and forecasting competitor actions Identifying, qualifying strategic partner candidates
	Strategic Business Planning Process		

interrelated perspectives or architectural descriptions that define the operational, system, and technical aspects of an enterprise [29]. The *operational architecture* is a people- or organization-oriented description of the operational elements, intelligence business processes, assigned tasks, and information and work flows required to accomplish or support the intelligence function. It defines the type of information, the frequency of exchange, and the tasks that are supported by these information exchanges. The *systems architecture* is a description of the systems and interconnections providing for or supporting intelligence functions. The system architecture defines the physical connection, location, and identification of the key nodes, circuits, networks, and users, and specifies system and component performance parameters. The *technical architecture* is the minimal set of rules (i.e., standards, protocols, interfaces, and services) governing the arrangement, interaction, and interdependence of the elements of the system. These three views of the enterprise (Figure 1.4) describe three layers of people-oriented operations, system structure, and procedures (protocols) that must be defined in order to implement an intelligence enterprise.

The operational layer is the highest (most abstract) description of the concept of operations (CONOPS), human collaboration, and disciplines of the knowledge organization. The technical architecture layer describes the most detailed perspective, noting specific technical components and their operations, protocols, and technologies. In the middle is the system architecture layer, which defines the network structure of nodes and interconnections. The performance of these layers is quantified by the typical kinds of metrics depicted in the figure. The intelligence supply chain that describes the flow of data into knowledge to create consumer value is measured by the value it provides to intelligence consumers. Measures of human intellectual capital and organizational knowledge describe the intrinsic value of the organization. The distributed computing architecture is measured by a variety of performance-level metrics that characterize the system capability in terms of information volume, capacity, and delivery rates. The technical physical (or hardware) and abstract

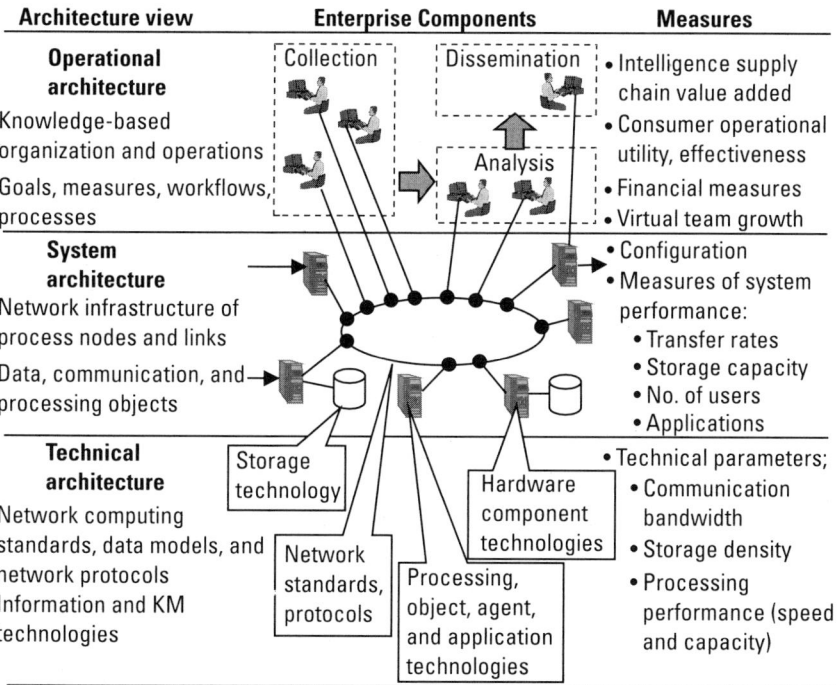

Figure 1.4 Enterprise information architecture elements.

(or software) elements of the enterprise are described by engineering dimensional performance parameters (e.g., bandwidth, storage density, and processing gain).

Throughout this book, we introduce the KM principles and practice that allow intelligence officers, enterprise architects, and engineers to implement these abstract models into a working intelligence enterprise of people and their processes, systems, and technology.

1.5   The State of the Art and the State of the Intelligence Tradecraft

The subject of intelligence analysis remained largely classified through the 1980s, but the 1990s brought the end of the Cold War and, thus, open publication of the fundamental operations of intelligence and the analytic methods employed by businesses and nation-states. In that same period, the rise of commercial information sources and systems produced the new disciplines of open source intelligence (OSINT) and business/competitor intelligence. In each of these areas, a wealth of resources is available for tracking the rapidly changing technology state of the art as well as the state of the intelligence tradecraft.

1.5.1   National and Military Intelligence

Numerous sources of information provide management, legal, and technical insight for national and military intelligence professionals with interests in analysis and KM (rather than intelligence operations, collection, or covert action). These sources include:

- *Studies in Intelligence*—Published by the U.S. CIA Center for the Study of Intelligence and the Sherman Kent School of Intelligence, unclassified versions are published on the school's Web site (http://odci. gov.csi), along with periodically issued monographs on technical topics related to intelligence analysis and tradecraft.

- *International Journal of Intelligence and Counterintelligence*—This quarterly journal covers the breadth of intelligence interests within law enforcement, business, nation-state policymaking, and foreign affairs.

- *Intelligence and National Security*—A quarterly international journal published by Frank Cass & Co. Ltd., London, this journal covers broad intelligence topics ranging from policy, operations, users, analysis, and products to historical accounts and analyses.

- *Defense Intelligence Journal*—This is a quarterly journal published by the U.S. Defense Intelligence Agency's Joint Military Intelligence College.

- *American Intelligence Journal*—Published by the National Military Intelligence Association (NMIA), this journal covers operational, organizational, and technical topics of interest to national and military intelligence officers.

- *Military Intelligence Professional Bulletin*—This is a quarterly bulletin of the U.S. Army Intelligence Center (Ft. Huachuca) that is available on-line and provides information to military intelligence officers on studies of past events, operations, processes, military systems, and emerging research and development.

- *Jane's Intelligence Review*—This monthly magazine provides open source analyses of international military organizations, NGOs that threaten or wage war, conflicts, and security issues.

In addition to these specific sources, intelligence topics are frequently covered in national policy-related publications such as *Foreign Affairs* and *Washington Monthly*, and in technical publications such as *Aviation Week and Space Technology*.

1.5.2 Business and Competitive Intelligence

Several sources focus on the specific areas of business and competitive intelligence with attention to the management, ethical, and technical aspects of collection, analysis, and valuation of products.

- *Competitive Intelligence Magazine*—This is a CI source for general applications-related articles on CI, published bimonthly by John Wiley & Sons with the Society for Competitive Intelligence (SCIP).

- *Competitive Intelligence Review*—This quarterly journal, also published by John Wiley with the SCIP, contains best-practice case studies as well as technical and research articles.

- *Management International Review*—This is a quarterly refereed journal that covers the advancement and dissemination of international applied research in the fields of management and business. It is published by Gabler Verlag, Germany, and is available on-line.

- *Journal of Strategy and Business*—This quarterly journal, published by Booz Allen and Hamilton focuses on strategic business issues, including regular emphasis on both CI and KM topics in business articles.

1.5.3   KM

The developments in the field of KM are covered by a wide range of business, information science, organizational theory, and dedicated KM sources that provide information on this diverse and fast growing area. Among the major sources of current practice in the field are the following:

- *CIO Magazine*—This monthly trade magazine for chief information officers and staff includes articles on KM, best practices, and related leadership topics.

- *Harvard Business Review, Sloan Management Review*—These management journals cover organizational leadership, strategy, learning and change, and the application of supporting ITs.

- *Journal of Knowledge Management*—This is a quarterly academic journal of strategies, tools, techniques, and technologies published by Emerald (UK). In addition, Emerald also publishes quarterly *The Learning Organization—An International Journal.*

- *IEEE Transactions of Knowledge and Data Engineering*—This is an archival journal published bimonthly to inform researchers, developers, managers, strategic planners, users, and others interested in state-of-the-art and state-of-the-practice activities in the knowledge and data engineering area.

- *Knowledge and Process Management*—A John Wiley (UK) journal for executives responsible for leading performance improvement and contributing thought leadership in business. Emphasis areas include KM, organizational learning, core competencies, and process management.

- *American Productivity and Quality Center* (APQC)—THE APQC is a nonprofit organization that provides the tools, information, expertise, and support needed to discover and implement best practices in KM. Its mission is to discover, research, and understand emerging and effective methods of both individual and organizational improvement, to broadly disseminate these findings, and to connect individuals with one another and with the knowledge, resources, and tools they need to successfully manage improvement and change. They maintain an on-line site at www.apqc.org.

- *Data Mining and Knowledge Discovery*—This Kluwer (Netherlands) journal provides technical articles on the theory, techniques, and practice of knowledge extraction from large databases.

- *International Journal on Multi-Sensor, Multi-Source Information Fusion*—This Elsevier Science journal provides technical articles on the theory, techniques, and practice of creating knowledge from diverse multiple sources of data.

1.6 The Organization of This Book

This book is structured to introduce the unique role, requirements, and stakeholders of intelligence (the applications) before introducing the KM processes, technologies, and implementations. The chapter structure (Figure 1.5) therefore moves from applications (Section I) to organizational and functional KM processes for the intelligence application (Section II) and then to implementations (Section III). Beyond the introduction in this chapter, we describe the mission and functions of the intelligence application (Chapter 2) and the KM processes that are applied to intelligence problems (Chapter 3). The socialization aspects of KM that develop a knowledge-based organization of people are described (Chapter 4) before explaining many of the principles of the immanently human process of collaborative intelligence analysis and synthesis (Chapters 5 and 6). Next, the methods of transferring tacit and explicit knowledge to create knowledge, and the practical application of intelligence analysis and synthesis in networks with automated systems are described (Chapter 7). The applications of fully automated explicit knowledge combination (reasoning) capabilities are then introduced (Chapter 8). Finally, intelligence enterprise

I. Intelligence and its applications	1. Introduction: knowledge management and intelligence		
	2. The intelligence enterprise 3. Knowledge management processes		
II. Knowledge management processes	Socialization	Transfer	Combination
	4. The knowledge-based intelligence organization 5. Intelligence analysis and synthesis	6. Implementing analysis and synthesis 7. Knowledge internalization and externalization	8. Explicit knowledge capture and combination
III. Intelligence enterprise	9. The intelligence enterprise 10. Knowledge management technology		

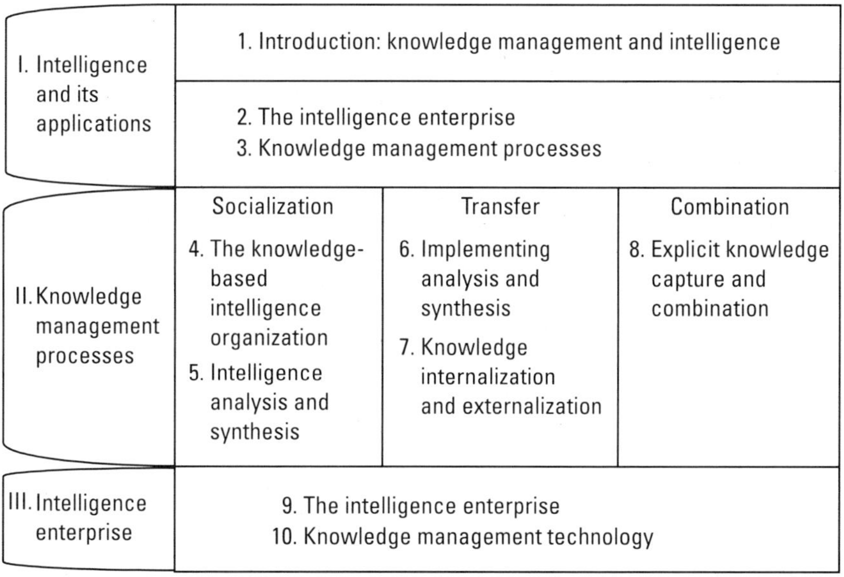

Figure 1.5 Logical organization of this book.

architectures that integrate people, processes, and IT to conduct intelligence operations are described to illustrate how intelligence enterprises are implemented from the principles described in earlier chapters (Chapter 9). Chapter 10 provides a broad survey of key information technologies—many which now enable and more that are emerging to increase the effectiveness of intelligence enterprises in the future.

Endnotes

[1]   "A Consumer's Guide to Intelligence," CIA (Office of Public Affairs), Washington, D.C., 1999, p. vii. For a comprehensive discussion of the range of definitions of intelligence and its central meaning, see: "Wanted: A Definition of 'Intelligence,'" in *Studies in Intelligence*, Vol. 46, No. 3, CIA, Washington D.C., 2002, Unclassified Edition, accessed on-line October 3, 2002, at http://www.cia.gov/csi/studies/vol46no3/index.html.

[2]   The United States distinguishes intelligence proper as the service of obtaining and delivering knowledge to government users (consumers); counterintelligence and covert action are *intelligence-related* operations. In this book, we do not discuss these secondary intelligence-related activities.

[3]   These engineering distinctions are refinements of the common terminology to distinguish three levels of information *content*. General dictionary definitions of information often include data and knowledge as synonyms.

[4]   The Greeks distinguished wisdom (*sophia*) and understanding (*sunesis*) as the principles by which we live and the ability to apply those principles in daily life, respectively.

[5]   Tzu, Sun, *The Art of War*, translated by R. D. Sawyer, Boulder, CO: Westview Press, 1994, p. 231.

[6]   Tzu, Sun, *The Art of War*, translated by R. D. Sawyer, Boulder, CO: Westview Press, 1994, pp. 231–232. Sun Tzu's five categories can be compared to current HUMINT terminology: 1) *local spies* (agents native to the foreign country), 2) *inward spies* (foreign agents who are officials), 3) *converted spies* (double agents—foreign agents turned to one's use), 4) *doomed spies* (one's own expendable agents sent with fabricated intelligence for purposes of deception), and 5) *surviving spies* (defectors or those returning with intelligence).

[7]   *Relevant Information* is comprised of *intelligence* (information about the operational environment, adversaries, and third parties), *friendly force information* (information about own forces), and *essential elements of friendly information* (specific information about friendly forces we seek to deny to an adversary). See *Field Manual 3-0—Operations*, Washington, D.C.: HQ Dept. of U.S. Army, June 2001, Chapter 11: "Information Superiority," accessed on-line at http://www.adtdl.army.mil/cgi-bin/atdl.dll/fm/3-0/toc.htm. The enumeration of intelligence requirements effectively defined the instructions to perform the process defined in U.S. Army doctrine as Intelligence Preparation of the Battlefield. See *Field Manual 34-130—Intelligence Preparation of the Battlefield*, Washington, D.C.: HQ Dept. of U.S. Army, July 1994, accessed on-line at http://www.adtdl.army.mil/cgi-bin/atdl.dll/fm/34-130/toc.htm.

[8]   From "Presidential Reflections on US Intelligence," CIA Center for the Study of Intelligence, accessed on-line November 2001 at http://www.odci.gov/csi/monograph/firstln/washington.html.

[9]   These concepts are described in, for example: Toffler, A., *Third Wave*, New York: Bantam, 1991; Toffler, A., and H. Toffler, *War and Anti-War*, New York: Warner, 1995, and A. Toffler, *Powershift—Knowledge, Wealth and Violence at the Edge of the 21st Century*, New York: Bantam, 1991.

[10]   Keaney, T. A., and E. Cohen, *Gulf War Air Power Survey Summary Report*, Washington D.C.: Government Printing Office, 1993, Chapter 4: "What Was the Role of Intelligence?"

[11]   Cebrowski, A. K. (VADM, USN), and J. J. Garstka, "Network-Centric Warfare: Its Origin and Future," Naval Institute Proceedings, January 1998, accessed on-line November 2001 at http://www.usni.org/Proceedings/Articles98/PROcebrowski.htm. See also Alberts, D. S., et al., *Network Centric Warfare: Developing and Leveraging Information Superiority*, (2nd ed.), Washington, D.C.: C4ISR Cooperative Research Program, August 1999.

[12]   Hamel, G., *Leading the Revolution*, Boston: HBS Press, 2000.

[13]   Hamel, G., *Leading the Revolution*, Boston: HBS Press, 2000, p. 18.

[14]   Edwards, S. J. A., *Swarming on the Battlefield: Past, Present and Future*, Santa Monica, CA: RAND, 2000.

[15]   *Consumer's Guide to Intelligence*, Washington, D.C.: CIA, September 1993, updated February 1994.

[16]   This taxonomy is based on a categorization in the text: Schum, David, *Inference and Analysis for the Intelligence Analyst*, Volumes 1 and 2, Washington D.C.: University Press of America, 1987.

[17]   While national and business intelligence is required to understand and estimate continuous processes in threat and market environments, intelligence analysis in both domains must also consider discontinuities—surprises or unexpected emergent behavior in complex processes. Discontinuities arising from new technologies, cultural shifts, globalization, and other factors can create radical changes in the threats to nation-states as well as to business.

[18]   Definition from "Glossary of Competitive Intelligence Terms," *Competitive Intelligence Review*, Vol. 9, No. 2, April–June 1998, p. 66.

[19]   See, for example the argument posed by Stanley Kober in "WHY SPY?—The Uses and Misuses of Intelligence," Cato Policy Analysis No. 265, CATO Institute, December 12, 1996, accessed on-line at http://www.cato.org/pubs/pas/pa-265.html.

[20]   "Development of Surveillance Technology and Risk of Abuse of Economic Information: An Appraisal of Technologies of Political Control," Working document for the Scientific and Technological Options Assessment Panel, PE 168.184/Int.St./part 1 of 4, European Parliament, Luxembourg, May 1999.

[21]   For a discussion of the disagreements over and implications of using national intelligence organizations in support of private-sector economic intelligence, see Greenberg, M. R., and R. N. Haas, *Making Intelligence Smarter: The Future of U.S. Intelligence*, New York: Council on Foreign Relations, 1996—see section entitled, "Economic Intelligence"—and

Gregory, S., *Economic Intelligence in the Post-Cold War Era: Issues for Reform,* Policy Paper for The Woodrow Wilson School of Public and International Affairs, Princeton University, January 1997.

[22]   Joint Publication 1-02, "DoD Dictionary of Military and Associated Terms."

[23]   Herman, M., *Intelligence Power in Peace and War,* Cambridge, England: Cambridge University Press, 1996, p. 56.

[24]   Libicki, Martin C., "Information Dominance" in *Strategic Forum,* Number 132, Institute for Strategic Studies, National Defense University, Washington D.C., November 1997.

[25]   Joint Vision 2020 (JV 20202), U.S. Joint Chiefs of Staff, Department of Defense, 2000.

[26]   Robertson, E., Department of the Navy Acquisition Reform FY2000, April 19, 2000, accessed on-line at http://www.ace.navy.mil/alrweek2000/briefs/jackson/sld001.htm; this document cites the Computer Sciences Corporation (CSC) corporate knowledge environment.

[27]   DaimlerChrysler Extended Enterprise definition, accessed at http://supplier.chrysler.com/purchasing/extent/index.html.

[28]   An architecture is defined in IEEE 610.12 as the structure of components, their relationships, and the principles and guidelines governing their design and evolution over time.

[29]   *C4ISR Architecture Framework Version 2.0,* Office of the Assistant Secretary of Defense for Command, Control, Communications, and Intelligence, Washington, D.C., November 1997.

Selected Bibliography

The following bibliography identifies major texts in the application areas of national and military intelligence and business and competitive intelligence.

Consumer's Guide to Intelligence, CIA, Washington, D.C., September 1993, updated February 1994.

National and Military Intelligence (texts on intelligence that emphasize the processing, analytic and decision-making support roles of intelligence (rather than collection or covert action).

Berkowitz, B. D., and A. E. Goodman, *Best Truth: Intelligence in the Information Age,* New Haven, CT: Yale University Press, 2000.

Berkowitz, B. D., and A. E. Goodman, *Strategic Intelligence for American National Security,* Princeton, NJ: Princeton University Press, 1989.

Bozeman, A. B., *Strategic Intelligence and Statecraft: Selected Essays,* Washington, D.C.: Brassey's, 1992.

Clark, R., *Intelligence Analysis,* Baltimore: American Lit Press, 1996.

Codevilla, A., *Informing Statecraft: Intelligence for a New Century,* NY: The Free Press, 1992.

Dulles, A., *The Craft of Intelligence* New York: Harper & Row, 1963; London: Weidenfeld & Nicolson, 1963; Boulder, CO: Westview, 1985.

Herman, M., *Intelligence Power in Peace and War,* Cambridge, England: Cambridge University Press, 1996.

Heuer, R., *The Psychology of Intelligence Analysis,* Washington D.C.: CIA Sherwood Kent School of Intelligence Analysis, 1999.

Johnson, L., *Bombs, Bugs, Drugs, and Thugs: Intelligence and America's Quest for Security,* New York: New York University Press, 2000.

Krizan, L., *Intelligence Essentials for Everyone,* Occasional Paper No. 6., Joint Military Intelligence College, Washington, D.C.: Government Printing Office, 1999.

Steele, R. D., *On Intelligence: Spies and Secrecy in an Open World,* Washington D.C.: AFCEA International Press, 2000.

Treverton, G. D., *Reshaping National Intelligence for an Age of Information,* Cambridge, England: Cambridge University Press, 2001.

Competitor Intelligence

Gilad, B., and T. Gilad, *The Business Intelligence System—A New Tool for Competitive Advantage,* New York: American Management Association, 1988.

Fuld, L. M., *The New Competitor Intelligence,* New York: John Wiley & Sons, 1995, and *Competitor Intelligence: How to Get It; How to Use It,* New York: John Wiley, 1985.

Hohhof, B., *Competitive Information Systems Development,* Glastonbury, CT: The Futures Group, 1993.

Porter, M. E., *Competitive Strategy: Techniques for Analyzing Industries and Competitors,* New York: The Free Press, 1980.

Stoffels, J. D., *Strategic Issues Management: A Comprehensive Guide to Environmental Scanning,* New York: Pergamon, 1993.

Tyson, K. W. M., *Competitor Intelligence Manual and Guide: Gathering, Analyzing, and Using Business Intelligence,* Englewood Cliffs, NJ: Prentice Hall) 1990.

Business Intelligence

Harvard Business Review on Knowledge Management, Boston: Harvard Business School Press, 1998.

Dixon, N. M., *Common Knowledge: How Companies Thrive by Sharing What They Know,* Boston: Harvard Business School Press, 2000.

Krogh, G. V., Johan Roos, and Dirk Kleine, *Knowing in Firms: Understanding, Managing and Measuring Knowledge,* London: Sage, 1998.

Liebowitz, J., *Building Organizational Intelligence: A Knowledge Management Primer,* Boca Raton, FL: CRC Press, 1999.

2

The Intelligence Enterprise

Intelligence, the strategic information and knowledge about an adversary and an operational environment obtained through observation, investigation, analysis, or understanding [1], is the product of an enterprise operation that integrates people and processes in a organizational and networked computing environment. The intelligence enterprise exists to produce intelligence goods and service—knowledge and foreknowledge to decision- and policy-making customers. This enterprise is a production organization whose prominent infrastructure is an information supply chain. As in any business, it has a "front office" to manage its relations with customers, with the information supply chain in the "back office." The intellectual capital of this enterprise includes sources, methods, workforce competencies, and the intelligence goods and services produced. As in virtually no other business, the protection of this capital is paramount, and therefore security is integrated into every aspect of the enterprise.

In this chapter we introduce the stakeholders, functions, and operations of the intelligence enterprise. We also examine the future of intelligence and how the global environment is requiring changes in the organization and operations of intelligence.

2.1 The Stakeholders of Nation-State Intelligence

The intelligence enterprise, like any other enterprise providing goods and services, includes a diverse set of stakeholders in the enterprise operation. The business model for any intelligence enterprise, as for any business, must clearly identify the stakeholders who own the business and those who produce and consume its goods and services. The stakeholders in the U.S. IC (Figure 2.1)

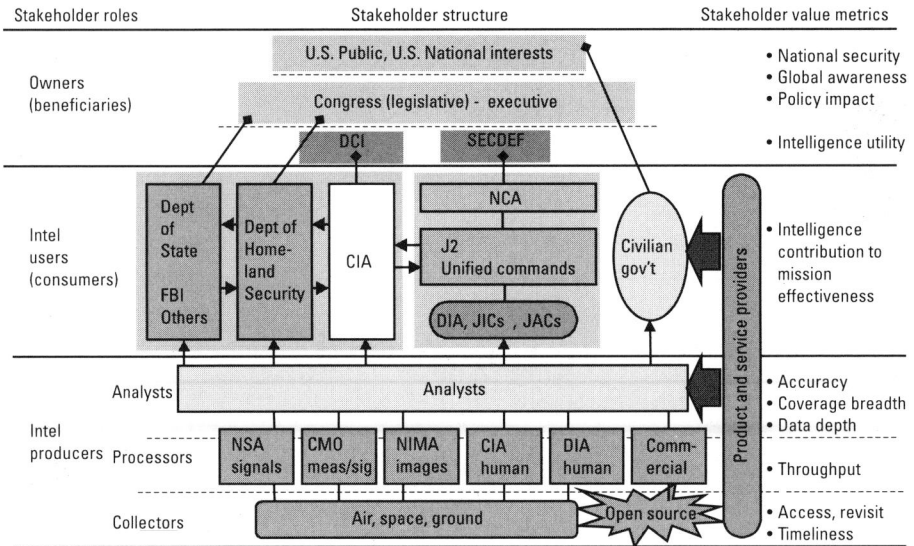

Figure 2.1 Structure and metrics of the stakeholders of the U.S. IC.

illustrate the relationships between key stakeholder roles and the metrics by which these stakeholders value the enterprise:

- The *owners* of the process include the U.S. public and its elected officials, who measure intelligence value in terms of the degree to which national security is maintained. These owners seek awareness and warning of threats to prescribed national interests.

- *Intelligence consumers (customers or users)* include national, military, and civilian user agencies that measure value in terms of intelligence contribution to the mission of each organization, measured in terms of its impact on mission effectiveness.

- *Intelligence producers,* the most direct users of raw intelligence, include the collectors (HUMINT and technical), processor agencies, and analysts. The principal value metrics of these users are performance based: information accuracy, coverage breadth and depth, confidence, and timeliness.

The purpose and value chains for intelligence (Figure 2.2) are defined by the stakeholders to provide a foundation for the development of specific value measures that assess the contribution of business components to the overall enterprise. The corresponding chains in the U.S. IC include:

- *Source*—the source or basis for defining the purpose of intelligence is found in the U.S. Constitution, derivative laws (i.e., the National Security Act of 1947, Central Intelligence Agency Act of 1949, National Security Agency Act of 1959, Foreign Intelligence Surveillance Act of 1978, and Intelligence Organization Act of 1992), and orders of the executive branch [2]. Derived from this are organizational mission documents, such as the Director of Central Intelligence (DCI) Strategic Intent [3], which documents communitywide purpose and vision, as well as derivative guidance documents prepared by intelligence providers.

- *Purpose chain*—the causal chain of purposes (objectives) for which the intelligence enterprise exists. The ultimate purpose is national security, enabled by information (intelligence) superiority that, in turn, is enabled by specific purposes of intelligence providers that will result in information superiority.

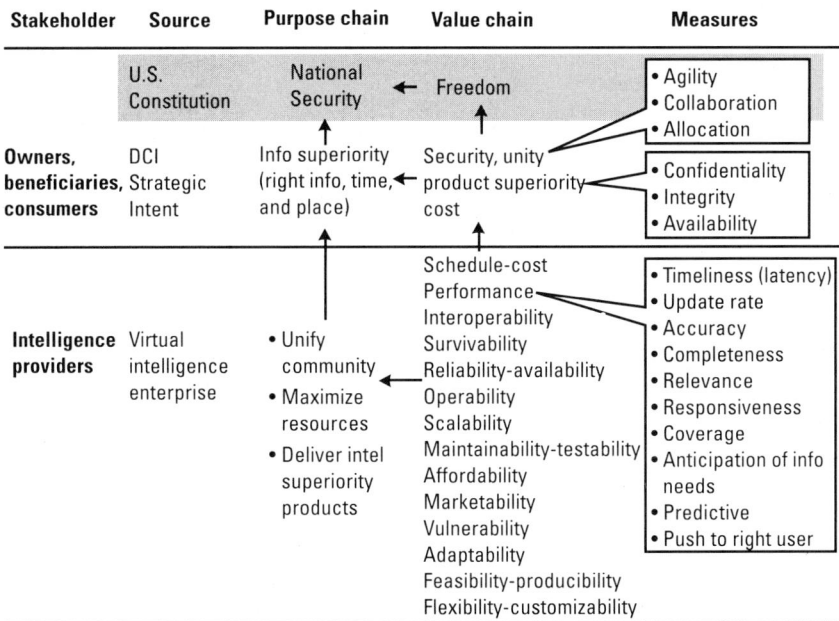

Figure 2.2 Chains of purposes and values for the U.S. IC.

- *Value chain*—the chain of values (goals) by which achievement of the enterprise purpose is measured.

- *Measures*—Specific metrics by which values are quantified and articulated by stakeholders and by which the value of the intelligence enterprise is evaluated.

Three major categories of intelligence products can be distinguished: strategic, military-operational, and military-tactical intelligence. Table 2.1 contrasts the categories, which are complementary and often share the same sources to deliver their intelligence products. The primary difference in the categories is the perspective (long- to short-term projection) and the reporting cycle (annual to near-real-time updates).

Table 2.1
Major Categories of Nation-State Intelligence

Intelligence Category	Focus (Intel Users)	Objects of Analysis	Reporting Cycle
Strategic or National Intelligence	Understanding of current and future status and behavior of foreign nations. Estimates of the state of global activities. Indications and warnings of threats. (National policymakers)	Foreign policy Political posture National stability Socioeconomics Cultural ideologies Science and technology Foreign relationships Military strength, intent	Infrequent (annual, monthly) long-duration estimates and projections (months, years) Long-term analyses (months, years) Frequent status reports (weekly, daily)
Military Operational Intelligence	Understanding of military powers, orders of battle, technology maturity, and future potential. (Military commanders)	Orders of battle Military doctrine Science and technology Command structure Force strength Force status, intent	Continually updated status databases (weekly) Indications and warnings (hours and days) Crisis analysis (daily, hourly)
Military Tactical Intelligence	Real-time understanding of military units, force structure, and active behavior (current and future) on the battlefield. (Warfighters)	Military platforms Military units Force operations Courses of action (past, current, potential future)	Weapon support (real-time: seconds to hours) Situation awareness applications (minutes, hours, days)

In a similar fashion, business and competitive intelligence, introduced in the first chapter, have stakeholders that include customers, shareholders, corporate officers, and employees. Each holds a stake in achieving the enterprise mission; there must exist a purpose and value chain that guides the KM operations. These typically include:

- *Source*—the business charter and mission statement of a business elaborates the market served and the vision for the businesses role in that market.

- *Purpose chain*—the objectives of the business require knowledge about internal operations and the market (BI objectives) as well as competitors (CI).

- *Value chain*—the chain of values (goals) by which achievement of the enterprise purpose is measured.

- *Measures*—Specific metrics by which values are quantified. A balanced set of measures includes vision and strategy, customer, internal, financial, and learning-growth metrics.

2.2 Intelligence Processes and Products

The process that delivers strategic and operational intelligence products is generally depicted in cyclic form (Figure 2.3), with five distinct phases [4]. This cycle, briefly introduced in the first chapter, begins with the need for knowledge by policy or decision makers (consumers) and concludes with the delivery of that knowledge. The need may be a standing requirement, a special request, or an urgent necessity in time of crisis. In every case, the need is the basis for a logical process to deliver the knowledge to the requestor.

1. *Planning and direction.* The process begins as policy and decision makers define, at a high level of abstraction, the knowledge that is required to make policy, strategic, or operational decisions. The requests are parsed into information required, then to data that must be collected to estimate or infer the required answers. Data requirements are used to establish a plan of collection, which details the elements of data needed and the targets (people, places, and things) from which the data may be obtained.

2. *Collection.* Following the plan, human and technical sources of data are tasked to collect the required raw data. The next section introduces the major collection sources, which include both openly available and closed sources that are accessed by both human and technical methods.

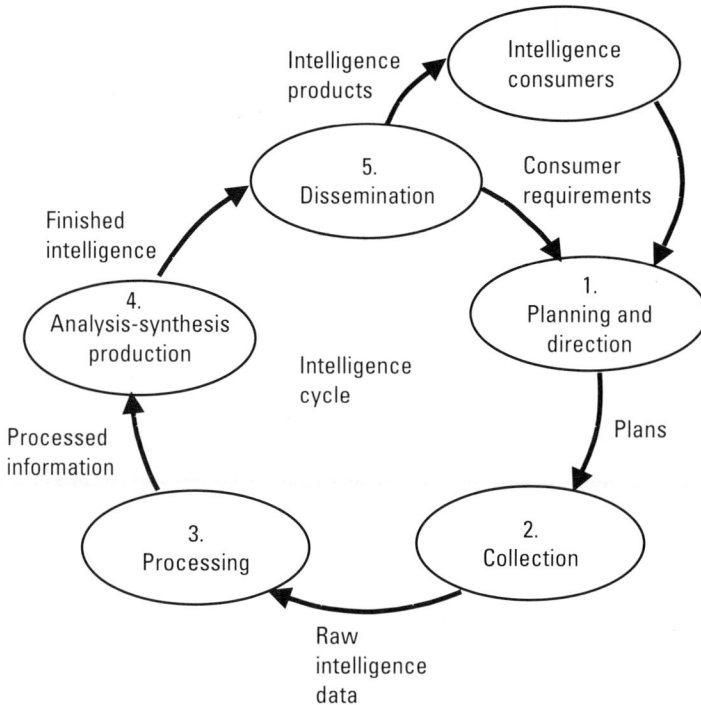

Figure 2.3 The intelligence cycle delivers reports in response to specific requests and queries for knowledge needed to make decisions and set policies.

These *sources and methods* are among the most fragile [5]—and most highly protected—elements of the process. Sensitive and specially compartmented collection capabilities that are particularly fragile exist across all of the collection disciplines.

3. *Processing.* The collected data is processed (e.g., machine translation, foreign language translation, or decryption), indexed, and organized in an information base. Progress on meeting the requirements of the collection plan is monitored and the tasking may be refined on the basis of received data.

4. *All-source analysis-synthesis and production.* The organized information base is processed using estimation and inferential (reasoning) techniques that combine all-source data in an attempt to answer the requestor's questions. The data is analyzed (broken into components and studied) and solutions are synthesized (constructed from the accumulating evidence). The topics or subjects (intelligence *targets*) of study are modeled, and requests for additional collection and processing may be made to acquire sufficient data and achieve a sufficient

level of understanding (or confidence to make a judgment) to answer the consumer's questions.

5. *Dissemination.* Finished intelligence is disseminated to consumers in a variety of formats, ranging from dynamic operating pictures of warfighters' weapon systems to formal reports to policymakers. Three categories of formal strategic and tactical intelligence reports are distinguished by their past, present, and future focus: *current intelligence reports* are news-like reports that describe recent events or indications and warnings, *basic intelligence reports* provide complete descriptions of a specific situation (e.g., order of battle or political situation), and *intelligence estimates* attempt to predict feasible future outcomes as a result of current situation, constraints, and possible influences [6].

The intelligence product is disseminated to the user, providing answers to queries and estimates of accuracy of the product delivered.

Though introduced here in the classic form of a *cycle*, in reality the process operates as a *continuum* of actions with many more feedback (and feedforward) paths that require collaboration between consumers, collectors, and analysts. In Chapters 3 and 4, we illustrate the intensely collaborative processes of collection, processing, and analysis-synthesis in an integrated intelligence enterprise.

2.3 Intelligence Collection Sources and Methods

A taxonomy of intelligence data sources (Table 2.2) includes sources that are openly accessible or closed (e.g., denied areas, secured communications, or clandestine activities). Due to the increasing access to electronic media (i.e., telecommunications, video, and computer networks) and the global expansion of democratic societies, OSINT is becoming an increasingly important source of global data. While OSINT must be screened and cross validated to filter errors, duplications, and deliberate misinformation (as do all sources), it provides an economical source of public information and is a contributor to other sources for cueing, indications, and confirmation [7].

In contrast with open sources, clandestine HUMINT and both open and clandestine technical means of collection provide data on topics and subjects that are protected by denial of access or secrecy [8].

Imagery intelligence (IMINT) provides assessments of resolvable objects from imagery of the Earth. IMINT reveals the location, composition, and characterization of resources, infrastructure, facilities, and lines of communication to perform order of battle estimates, indications and warning, situation assessment, targeting, and battle damage assessment functions. Signals intelligence (SIGINT) monitors electromagnetic signals for electronic data (e.g., radar) and

Table 2.2
Major Intelligence Categories Are Partitioned by Access (Open or Closed)
and Collection Means (Human or Technical)

Access	Source Type	Intelligence Source Category	Representative Sources
Open	Human and technical means	Open source intelligence (OSINT)	Foreign radio and television news sources
			Foreign printed materials: books, magazines, periodicals, journals
			Diplomatic and attaché reporting
			Shortwave radio, telecom, Internet conversations
			Foreign network computer sources
			Gray literature (printed and electronic)
Closed Sources	Human means	Human intelligence (HUMINT)	Reports from agents in foreign nations
			Discussions with personnel in foreign nations
			Reports from defectors from foreign nations
			Messages from friendly third-party sources
	Technical means	Imagery intelligence (IMINT)	Surveillance imagery (static air and space imagery of the earth)
			Surveillance imagery (terrestrial static and video imagery)
		Signals intelligence (SIGINT)	Electromagnetic signals monitoring (ELINT): *externals*—events, activities, relationships, frequency of occurrence, modes, sequences, patterns, signatures—or *internals*—contents of messages
			Radar intelligence (RADINT), including moving target indications (MTIs) tracking data
			Communications traffic monitoring (COMINT) for externals and internals
			Foreign instrumentation signals intelligence (FISINT): telemetry (TELINT), beacons, video links
		Computer network exploitation (CNE)	Network analysis and monitoring
			Network message interception, traffic analysis
			Computer intrusion, penetration, and exploitation
Closed Sources	Technical means	Measurements and signatures intelligence (MASINT)	Technically derived intelligence from all sources (parametric data) to support real-time operations (e.g., electronic support measures, combat identification, and tactical intelligence analysis)
			MASINT exploits physical properties (nuclear, biological, chemical), emitted/reflected energy (radio frequencies, infrared (IR), shock waves, acoustics), mechanical sound, magnetic properties, motion, and materials composition

communications (e.g., voice and data telecommunications) to detect traffic and geolocate individual emitters. The emerging requirement to collect intelligence from digital networks (rather that radiated emissions) is provided by computer network exploitation (CNE). This involves the understanding of network infrastructures, network traffic externals, and data communication internals, as well as access to computer nodes and exploitation of networked computers [9]. Measurements and signatures intelligence (MASINT) is technically derived knowledge from a wide variety of sensors, individual or fused, either to perform special measurements of objects or events of interest or to obtain signatures for use by the other intelligence sources. MASINT is used to characterize the observable phenomena *(observables)* of the environment and objects of surveillance.

U.S. intelligence studies have pointed out specific changes in the use of these sources as the world increases globalization of commerce and access to social, political, economic, and technical information [10–12]:

- The increase in unstructured and transnational threats requires the robust use of clandestine HUMINT sources to complement extensive technical verification means.
- Technical means of collection are required for both broad area coverage and detailed assessment of the remaining denied areas of the world.

Competitive intelligence operations are also conducted in the commercial business world, with growing use of electronic collection sources and open sources available on the Internet. The same principles of strategic intelligence planning, development of the intelligence cycle processes, source development, and analysis apply. Leonard Fuld's *The New Competitor Analysis* details the intelligence processes applied to commercial businesses and the sources available in this domain [13].

2.3.1 HUMINT Collection

HUMINT refers to all information obtained directly from human sources [14]. HUMINT sources may be overt or covert (clandestine); the most common categories include:

- *Clandestine intelligence case officers.* These officers are own-country individuals who operate under a clandestine "cover" to collect intelligence and "control" foreign agents to coordinate collections.
- *Agents.* These are foreign individuals with access to targets of intelligence who conduct clandestine collection operations as representatives

of their controlling intelligence officers. These agents may be recruited or "walk-in" volunteers who act for a variety of ideological, financial, or personal motives.

- *Émigrés, refugees, escapees, and defectors.* The open, overt (yet discrete) programs to interview these recently arrived foreign individuals provide background information on foreign activities as well as occasional information on high-value targets.

- *Third party observers.* Cooperating third parties (e.g., third-party countries and travelers) can also provide a source of access to information.

The HUMINT discipline follows a rigorous process for acquiring, employing, and terminating the use of human assets that follows a seven-step sequence [13]. The sequence followed by case officers includes:

1. *Spotting*—locating, identifying, and securing low-level contact with agent candidates;
2. *Evaluation*—assessment of the potential (i.e., value or risk) of the spotted individual, based on a background investigation;
3. *Recruitment*—securing the commitment from the individual;
4. *Testing*—evaluation of the loyalty of the agent;
5. *Training*—supporting the agent with technical experience and tools;
6. *Handling*—supporting and reinforcing the agent's commitment;
7. *Termination*—completion of the agent assignment by ending the relationship.

HUMINT is dependent upon the reliability of the individual source, and lacks the collection control of technical sensors. Furthermore, the level of security to protect human sources often limits the fusion of HUMINT reports with other sources and the dissemination of wider customer bases. Directed high-risk HUMINT collections are generally viewed as a precious resource to be used for high-value targets to obtain information unobtainable by technical means or to validate hypotheses created by technical collection analysis.

2.3.2 Technical Intelligence Collection

Technical collection is performed by a variety of electronic (e.g., electromechanical, electro-optical, or bioelectronic) sensors placed on platforms in space, the atmosphere, on the ground, and at sea to measure physical phenomena (observables) related to the subjects of interest (intelligence targets). A wide variety of sensor-platform combinations (Table 2.3) collect data that may be used

Table 2.3
Surveillance and Reconnaissance Sources

Platforms	Radar and IFF	IMINT	SIGINT	MASINT
Space Geostationary spacecraft Polar orbital spacecraft Low Earth orbit spacecraft Cooperative spacecraft constellations	Spaceborne radar (MTI or target tracking modes) surveillance	Weather satellites Imaging broad area search and precision imaging	SIGINT ferrets	IR missile warning/tracking Nuclear detection
Air Tactical aircraft Standoff manned Recce aircraft Penetrating high, medium altitude endurance unmanned air vehicles (UAVs)	Airborne warning and control aircraft Fighter aircraft	Synthetic aperture radar (SAR), electro-optical (EO), infrared (IR), and multispectral imaging sensors on manned and unmanned Recce	Airborne SIGINT standoff and penetrating UAVs	IR/EO, laser surveillance aircraft Atmospheric sampling Nonacoustic antisubmarine warfare (ASW) sensors
Ground Attended fixed sites Mobile manned vehicles Man portable sensors Unattended ground sensors in denied areas	Air defense, air surveillance sensors Counterbattery radar Ground surveillance (intrusion) radar	Combat tactical digital cameras Long range IR/EO video IR night vision IR search and track	Ground-based ESM sites and vehicles Unattended electronic support measures (ESM) sensors	Seismic arrays Acoustic arrays IR radiometers
Sea, Undersea Shipboard sensors Submarine sensors Ship/submarine towed sensors Heliborne dipping, air dropped sensors Fixed, autonomous buoys Underwater arrays	Shipboard and submarine air, surface surveillance radar	Ship and submarine long-range IR/EO video IR search and track	Ship, sub, and helo ESM sensors UAV ESM Sensors	Ship, sub towed sonar array Ship, sub hull sonar array Nonacoustic ASW sensors Sonobuoys Dipping sonar

Recce: Reconnaissance

for tactical, operational, or strategic intelligence. The operational utility of these collectors for each intelligence application depends upon several critical factors:

- *Timeliness*—the time from collection of event data to delivery of a tactical targeting cue, operational warnings and alerts, or formal strategic report;

- *Revisit*—the frequency with which a target of interest can be revisited to understand or model (track) dynamic behavior;

- *Accuracy*—the spatial, identity, or kinematic accuracy of estimates and predictions;

- *Stealth*—the degree of secrecy with which the information is gathered and the measure of intrusion required.

2.4 Collection and Process Planning

The technical collection process requires the development of a detailed collection plan, which begins with the decomposition of the subject target into activities, observables, and then collection requirements. From this plan, technical collectors are tasked and data is collected and fused (a *composition*, or reconstruction that is the dual of the *decomposition* process) to derive the desired intelligence about the target.

This methodology is illustrated in Figure 2.4, which uses an illicit drug manufacturing and distribution operation example for analysis. The example follows the common intelligence collection plan that may be established by a local police force (on a small scale) or by a nation-state intelligence agency to understand a global drug cartel [16]. Beginning with the hypothesized model of the targeted drug operation process at the top, the elements of activity that characterize each step in the production and distribution processes are identified. In this oversimplified example, these activities are the six most observable events that are time-sequenced in the process model:

1. Planting of the crops;
2. Harvesting and processing;
3. Transportation of bulk products;
4. Delivery to local distributors;
5. Local covert storage;
6. Bank transfers closely related to delivery.

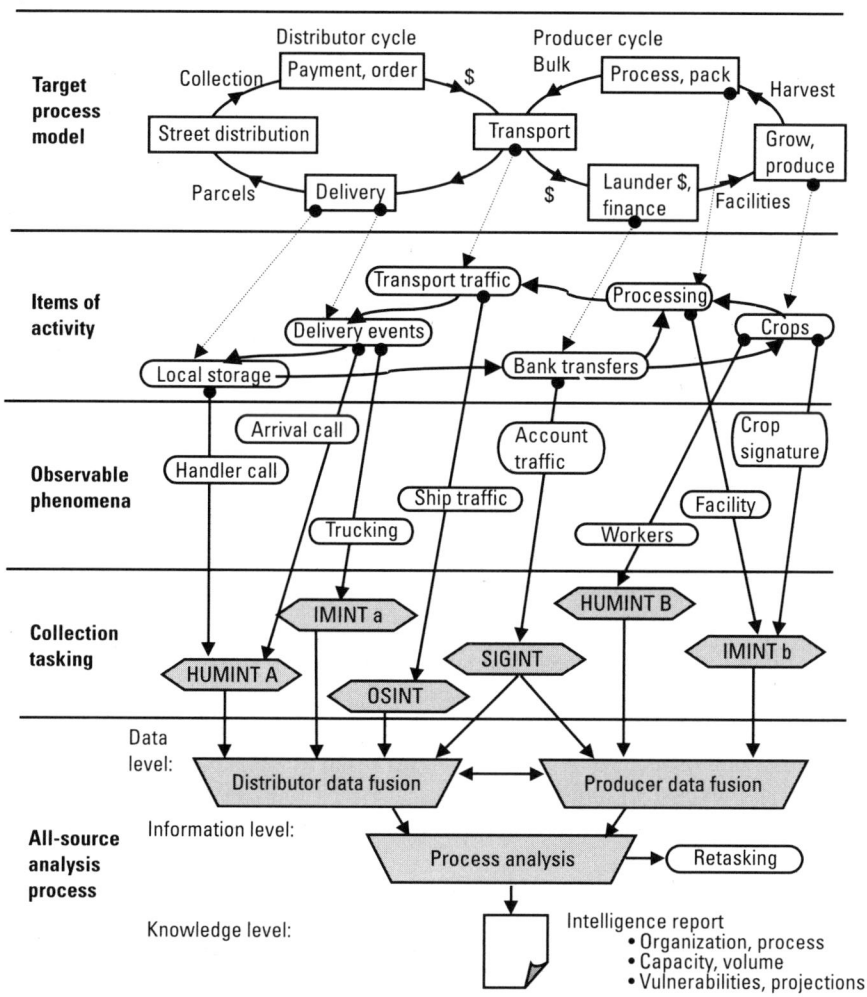

Figure 2.4 Target process modeling, decomposition, collection planning, and composition (all-source analysis) for a hypothetical surveillance and analysis of a drug operation.

The observable phenomena from each of these events are identified and assigned to technical (and, in this case, HUMINT) collectors. The collectors include OSINT (shipping traffic logs), ground-based video surveillance of shipping depots (IMINT a), airborne IMINT—observing crop activities and potential processing facilities (IMINT b), and SIGINT analysis of electronic transfers of funds via court-authorized intercepts.

The example illustrates the complementary nature of HUMINT and technical sources, whereby two HUMINT sources are required to guide the technical intelligence sources. HUMINT A provides insight into trucking routes to be used, allowing video surveillance to be focused on most-likely traffic points. HUMINT B, closely related to crop workers, monitors the movements of harvesting crews, providing valuable cueing for airborne sensors to locate crops and processing facilities. The technical sources also complement the HUMINT sources by providing verification of uncertain cues and hypotheses for the HUMINT sources to focus attention. The collected data is analyzed for the existence of evidence and the synchronization of events to verify process cycles. The analysis process delivers a report that describes the organization, the process flow, capacity, volume and projected output, and the vulnerabilities that may be exploited by law enforcement.

2.5 KM in the Intelligence Process

The intelligence process must deal with large volumes of source data, converting a wide range of text, imagery, video, and other media types into organized information, then performing the analysis-synthesis process to deliver knowledge in the form of intelligence products. IT is providing increased automation of the information indexing, discovery, and retrieval (IIDR) functions for intelligence, especially the exponentially increasing volumes of global open-source data [17]. The functional information flow in an automated or semiautomated facility (depicted in Figure 2.5) requires digital archiving and analysis to ingest continuous streams of data and manage large volumes of analyzed data. The flow can be broken into three phases:

1. Capture and compile;
2. Preanalysis;
3. Exploitation (analysis-synthesis).

Capture and compile includes the acquisition of volumes of multimedia data and the conversion to digital form for storage and analysis. Electronic data (network sources) are directly formatted, while audio, video, and paper documents must be converted to digital form. Foreign sources may be translated by natural-language analysis to convert to a common language base.

The preanalysis phase *indexes* each data item (e.g., article, message, news segment, image, book or chapter) by assigning a reference for storage; generating an abstract that summarizes the content of the item and metadata with a description of the source, time, reliability-confidence, and relationship to other items (*abstracting*); and extracting critical descriptors of content that characterize the contents (e.g., keywords) or meaning (*deep indexing*) of the item for subsequent analysis. Spatial data (e.g., maps, static imagery, or video imagery) must

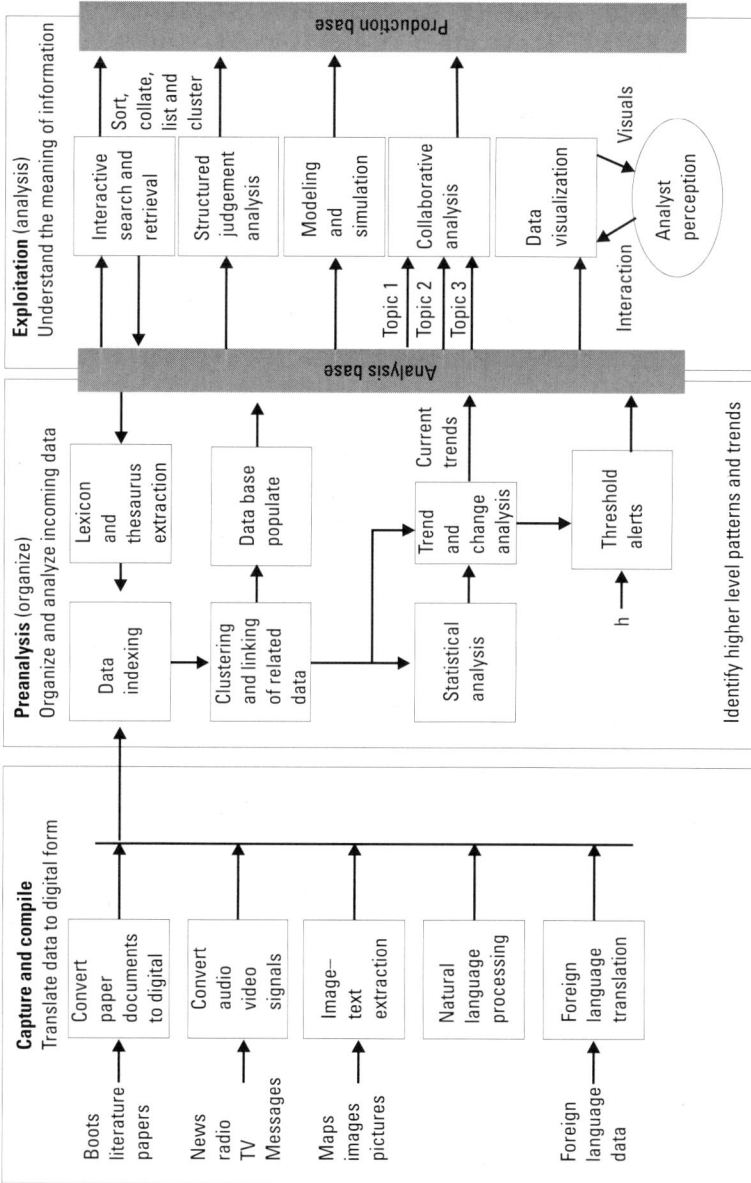

Figure 2.5 Intelligence processing and analysis flow includes three distinct phases to develop the production intelligence base.

be indexed by spatial context (spatial location) and content (imagery content). The indexing process applies standard subjects and relationships, maintained in a lexicon and thesaurus that is extracted from the analysis information base. Following indexing, data items are clustered and linked before entry into the analysis base. As new items are entered, statistical analyses are performed to monitor trends or events against predefined templates that may alert analysts or cue their focus of attention in the next phase of processing. For example, if analysts are interested in relationships between nations A and B, all reports may be scored for a *tension factor* between those nations, and alerts may be generated on the basis of frequency, score intensity, and sources of incoming data items.

The third phase of processing, exploitation, presents data to the HUMINT analyst for examination using visualization tools to bring to focus the most meaningful and relevant data items and their inter-relationships. The categories of automated tools that are applied to the analysis information base include the following tools [18]:

- *Interactive search and retrieval* tools permit analysts to search by content, topic, or related topics using the lexicon and thesaurus subjects.

- *Structured judgment analysis* tools provide visual methods to link data, synthesize deductive logic structures, and visualize complex relationships between data sets. These tools enable the analyst to hypothesize, explore, and discover subtle patterns and relationships in large data volumes—knowledge that can be discerned only when all sources are viewed in a common context.

- *Modeling and simulation* tools model hypothetical activities, allowing modeled (expected) behavior to be compared to evidence for validation or projection of operations under scrutiny.

- *Collaborative analysis* tools permit multiple analysts in related subject areas, for example, to collaborate on the analysis of a common subject.

- *Data visualization* tools present synthetic views of data and information to the analyst to permit patterns to be examined and discovered.

2.6 Intelligence Process Assessments and Reengineering

The U.S. IC has been assessed throughout and since the close of the Cold War to study the changes necessary to adapt to advanced collection capabilities, changing security threats, and the impact of global information connectivity and information availability. Published results of these studies provide insight into the areas of intelligence effectiveness that may be enhanced by organizing the community into a KM enterprise. We focus here on the technical aspects of the changes rather than the organizational aspects recommended in numerous studies.

2.6.1 Balancing Collection and Analysis

Intelligence assessments have evaluated the utility of intelligence products and the balance of investment between collection and analysis. One internal CIA study conducted in 1971 recognized that the early advances in space-based collection must be complemented by analytic methods to gain an overall improvement in the utility of intelligence:

> During the past decade alone, … spectacular increases in collection activities have occurred. Where satellite photography is concerned, the increases have led to greatly improved knowledge about military capabilities of potential enemies. But expanded collection by means other than photography has not brought about a similar reduction in our uncertainly about the intentions, doctrines, and political processes of foreign powers. Instead, the growth in raw intelligence—and here satellite photography must be included—has come to serve as a proxy for improved analysis, inference, and estimation [19].

The U.S. Congress, monitoring the relative contribution of collection and analysis, has commented on the balance of investments in each of these areas. In 1996 the House Intelligence Committee noted its "desire to focus more attention on the 'downstream' activities of processing, exploitation and dissemination of intelligence data and analysis. The Committee strongly registered its conviction that collection costs must be reduced over the long-term and funding increased for numerous processing activities. Moreover, we remain very concerned about the Community's ability to utilize the anticipated volume of information from planned collection increases" [20].

2.6.2 Focusing Analysis-Synthesis

An independent study [21] of U.S. intelligence recommended a need for intelligence to sharpen the focus of analysis-synthesis resources to deal with the increased demands by policymakers for knowledge on a wider ranges of topics, the growing breadth of secret and open sources, and the availability of commercial open-source analysis. The study offered several recommendations for analysis-synthesis [22]:

- Retain the focus of critical national and military intelligence analytic resources on the most crucial national security threats and hard targets whose understanding is only amenable to secret sources, methods, and analyses.

- Exploit the growing availability of university experts, think tanks, and commercial (private-sector open-source) analysis, developing means of collaborating with these resources.

- Apply competitive analysis-synthesis—duplicative analysis to ensure independent perspectives and judgments—for only the most critical or ambiguous targets.

2.6.3   Balancing Analysis-Synthesis Processes

One assessment conducted by the U.S. Congress reviewed the role of analysis-synthesis and the changes necessary for the community to reengineer its processes from a Cold War to a global awareness focus. Emphasizing the crucial role of analysis, the commission noted:

> The *raison d'etre* of the Intelligence Community is to provide accurate and meaningful information and insights to consumers in a form they can use at the time they need them. If intelligence fails to do that, it fails altogether. The expense and effort invested in collecting and processing the information have gone for naught [23].

The commission identified the KM challenges faced by large-scale intelligence analysis that encompasses global issues and serves a broad customer base. The commission's major observations provide insight into the emphasis on people-related (rather than technology-related) issues that must be addressed for intelligence to be valued by the policy and decision makers that consume intelligence:

1. *Build relationships.* A concerted effort is required to build relationships between intelligence producers and the policymakers they serve. Producer-consumer relationships range from assignment of intelligence liaison officers with consumers (the closest relationship and greatest consumer satisfaction) to holding regular briefings, or simple producer-subscriber relationships for general broadcast intelligence. Across this range of relationships, four functions must be accomplished for intelligence to be useful:

   - Analysts must understand the consumer's level of knowledge and the issues they face.

   - Intelligence producers must focus on issues of significance and make information available when needed, in a format appropriate to the unique consumer.

- Consumers must develop an understanding of what intelligence can and—equally important—cannot do.
- Both consumer and producer must be actively engaged in a dialogue with analysts to refine intelligence support to decision making.

2. *Increase and expand the scope of analytic expertise.* The expertise of the individual analysts and the community of analysts must be maintained at the highest level possible. This expertise is in two areas: domain, or region of focus (e.g., nation, group, weapon systems, or economics), and analytic-synthetic tradecraft. Expertise development should include the use of outside experts, travel to countries of study, sponsorship of topical conferences, and other means (e.g., simulations and peer reviews).

3. *Enhance use of open sources.* Open-source data (i.e., publicly available data in electronic and broadcast media, journals, periodicals, and commercial databases) should be used to complement (cue, provide context, and in some cases, validate) special, or closed, sources. The analyst must have command of all available information and the means to access and analyze both categories of data in complementary fashion.

4. *Make analysis available to users.* Intelligence producers must increasingly apply dynamic, electronic distribution means to reach consumers for collaboration and distribution. The DoD Joint Deployable Intelligence Support System (JDISS) and IC *Intelink* were cited as early examples of networked intelligence collaboration and distribution systems.

5. *Enhance strategic estimates.* The United States produces national intelligence estimates (NIEs) that provide authoritative statements and forecast judgments about the likely course of events in foreign countries and their implications for the United States. These estimates must be enhanced to provide timely, objective, and relevant data on a wider range of issues that threaten security.

6. *Broaden the analytic focus.* As the national security threat envelope has broadened (beyond the narrower focus of the Cold War), a more open, collaborative environment is required to enable intelligence analysts to interact with policy departments, think tanks, and academia to analyze, debate, and assess these new world issues.

In the half decade since the commission recommendations were published, the United States has implemented many of the recommendations. Several examples of intelligence reengineering include:

- *Producer-consumer relationships.* The introduction of collaborative networks, tools, and soft-copy products has permitted less formal interaction and more frequent exchange between consumers and producers. This allows intelligence producers to better understand consumer needs and decision criteria. This has enabled the production of more focused, timely intelligence.

- *Analytic expertise.* Enhancements in analytic training and the increased use of computer-based analytic tools and even simulation are providing greater experience—and therefore expertise—to human analysts.

- *Open source.* Increased use of open-source information via commercial providers (e.g., Lexis Nexis™ subscription *clipping* services to tailored topics) and the Internet has provided an effective source for obtaining background information. This enables special sources and methods to focus on validation of critical implications.

- *Analysis availability.* The use of networks continues to expand for both collaboration (between analysts and consumers as well as between analysts) and distribution. This collaboration was enabled by the introduction and expansion of the classified Internet *(Intelink)* that interconnects the IC [24].

- *Broadened focus.* The community has coordinated open panels to discuss, debate, and collaboratively analyze and openly publish strategic perspectives of future security issues. One example is the "Global Trends 2015" report that resulted from a long-term collaboration with academia, the private sector, and topic area experts [25].

2.7 The Future of Intelligence

Since the end of the Cold War, the effects of technology and globalization have caused numerous reassessments of the role and structure of national and military intelligence organizations. The changing world and the technology that is available to nations and individuals has changed the threats to nation-states. Both have empowered nonstate actors to carry out acts of massive physical and psychological destruction capable of adversely impacting political, economic, and even global stability. The U.S. DCI enumerated the new kinds of threats of the twenty-first century that "keep [him] awake at night" [26]:

- International terrorism that combines organized crime and ideological rationale for targeting others with weapons of mass destruction;

- The channels for and proliferation of weapons of mass destruction and their delivery systems;

- Rogue nation-states that pose threats to their neighbors, regions, and the United States;

- Information warfare threats to governments and supporting private infrastructures;

- "Traditional concerns" regarding fragile states in volatile regions, failing nations, and nation-states in transition.

This change in the threats has caused national intelligence organizations worldwide to consider the implications for changes in the intelligence process. At a 2001 conference of national intelligence leaders in Priverno, Italy [27], the shifts in perspective were discussed, and the need for evolutionary or revolutionary approaches was acknowledged by U.S., European, and Russian speakers. Summarizing the conference, U.S. intelligence officer Carole Dumaine noted three critical areas of change [28]:

1. *Global intelligence cooperation.* National intelligence organizations must collaborate and include academic- and business-sector contributors to provide true in-depth global background intelligence.

2. *Open-source intelligence.* Organizations must "move beyond" the primary focus on secret sources and methods and develop means to embrace and integrate open sources into analysis.

3. *New analysis.* Analytic communities must create new cultures of collaboration and "reflection" that will enable them to understand non-state threats.

The two primary dimensions of future threats to national (and global) security include the source (from nation-state actors to no-state actors) and the threat-generating mechanism (continuous results of rational nation-state behaviors to discontinuities in complex world affairs). These threat changes and the contrast in intelligence are summarized in Table 2.4 [28]. Notice that these changes coincide with the transition from sensor-centric to network- and knowledge-centric approaches to intelligence introduced in Chapter 1.

These changes, similar to those faced by the business community, are imposed by a rapidly changing global environment that involves the complex interaction of many actors. The potential for surprise is great in such complexity, and intelligence in the nation-state and in business must be agile, anticipatory, and adaptive to this rapidly changing world. Within the U.S. IC, the alternatives between evolution and revolution in approaches to intelligence remain a critical subject of serious debate [29]. In either approach, intelligence must focus on knowledge creation in an enterprise environment that is prepared

Table 2.4
The Changing Intelligence Environment

	Traditional Focus	New Focus
Threat Dimensions	A few large powerful nation-state threats	Many diverse and empowered nonstate actor threats
	Threats caused by continuity of world affairs	Threats resulting from discontinuities in world affairs
Characteristics of Intelligence	**Centralized Intelligence**	**Distributed Intelligence**
	Focus on collection and secret sources	Focus on analysis, collaboration with others, open and closed sources
	Targets are known, continuous, predictable	Targets are unknown, discontinuous, and unpredictable
	Intelligence management on tactical, operational, measurable objectives	Intelligence management on strategic, anticipatory, adaptive objectives
	Hierarchical analysis organization and control	Networked analysis organization and collaboration
	Focus on intelligence as a product	Focus on intelligence as a service

to rapidly reinvent itself to adapt to emergent threats. The U.S. *National Strategy for Homeland Security* recognizes the need for changes and has recommended significant policy, organizational, and infrastructure changes in U.S. intelligence to respond to terrorist threats. The *Strategy* asserts, "The United States will take every necessary action to avoid being surprised by another terrorist attack. We must have an intelligence and warning system that can detect terrorist activity before it manifests itself in an attack so that proper preemptive, preventive, and protective action can be taken" [30]. The following chapters introduce the key KM practices, systems, and technologies that will enable the kind of intelligence organizational, operational, and infrastructure capacity and agility necessary to achieve such objectives.

Endnotes

[1]   From definition (2) in Joint Pub 1-02.

[2]   Executive Order 12333 provides guidelines for the conduct of intelligence activities. The U.S. Senate Select Committee on Intelligence provides copies of the major laws at its Web site: http://intelligence.senate.gov/statutes.htm.

[3]   The "DCI Strategic Intent" is a 1998 classified statement of mission and vision for the IC to provide direction for transformation to a collaborative enterprise with effective application of people, resources, and IT.

[4]   The intelligence cycle follows the description of the U.S. CIA; note that the U.S. DoD Joint Pub 2-0 defines six steps in the cycle by including: 1) planning and direction, 2) collection, 3) processing and exploitation, 4) analysis and production, 5) dissemination and integration, and 6) evaluation and feedback. See Joint Publication 2-0 *Doctrine for Intelligence Support to Joint Operations*, March 2000, in particular Chapter 2—"The Intelligence Cycle," accessed on-line on October 30, 2002 at http://www.dtic.mil/doctrine/jel/new_pubs/jp2_0.pdf.

[5]   By fragile, we refer to the potential for loss of value if revealed to the subject of surveillance. Even the most sophisticated sources and methods may often be easily defeated by denial or deception-if revealed.

[6]   Shulsky, A. N., *Silent Warfare—Understanding the World of Intelligence*, second edition, Washington D.C.: Brasey's, pp. 63–69.

[7]   Interview with Dr. Joseph Markowitz in *Open Source Quarterly*, Vol. 1, No. 2, pp. 8–15.

[8]   Herman, M., *Intelligence Power in Peace and War*, Cambridge England, Cambridge University Press, 1996, Chapter 4: "Collection Sources."

[9]   Computer network operations are comprised of an offensive component (computer network attack), a defensive component (computer network defense), and the intelligence function (computer network exploitation).

[10]   U.S. Congressional Commission, *Preparing for the 21st Century: An Appraisal of Intelligence*, Washington, D.C.: Government Printing Office, March 1, 1996.

[11]   *Making Intelligence Smarter, The Future of U.S. Intelligence*, Independent Task Force of Council on Foreign Relations, New York, 1996.

[12]   *Strategic Assessment: 1996*, National Defense University, Washington D.C., 1996, Chapter 6: "Intelligence."

[13]   Fuld, L. M., *The New Competitor Intelligence: The Complete Resource for Finding, Analyzing, and Using Information About Your Competitors*, New York: John Wiley and Sons, 1994.

[14]   "Attachment 3: Sources of Intelligence, A.3.1 Human Intelligence (HUMINT)," in *USAF Intelligence Targeting Guide*, AF Pamphlet 14-210, February 1, 1998.

[15]   Ameringer, C. D., *U.S. Foreign Intelligence*, Lexington MA: Lexington Books, 1990, pp. 13–14.

[16] Holden-Rhodes, J. F., *Sharing the Secrets: Open Source Intelligence and the War on Drugs*, Westport, CT: Praeger, 1997.

[17] *Preparing US Intelligence for the Information Age*, Director Central Intelligence, STIC 95-003, June 1995.

[18] "Part II Analytic Tools To Cope with the Open Source Explosion," in *Preparing US Intelligence for the Information Age*, Director Central Intelligence, STIC 93-007, December 1993, and "Part III, Analytic Tools Recommendations for Open Source Information," in *Preparing US Intelligence for the Information Age*, Director Central Intelligence, STIC 95-002, April 1995.

[19] *A Review of the Intelligence Community*, F-1992-02088 CIA, March 19, 1971, sanitized and downgraded from top secret for public release to The Princeton Collection, May 1998, p. 3.

[20] "Committee Findings and Recommendations," U.S. Congress House Intelligence Committee FY-1996 Markup Report, June 1995.

[21] Hedley, Jr., J. H., "Checklist for the Future of Intelligence," Institute for the Study of Diplomacy, Georgetown University, Washington D.C., 1995. See also "IC21—The Intelligence Community in the 21st Century, U.S. House of Representatives, Permanent Select Committee on Intelligence, March 4, 1996.

[22] Hedley, Jr., J. H., "Checklist for the Future of Intelligence," Institute for the Study of Diplomacy, Georgetown University, Washington D.C., 1995. See the section entitled, "Sharpening the Focus," accessed on-line at http://sfswww.Georgetown.edu/sfs/programs/isd/files/intell.htm.

[23] "Improving Intelligence Analysis," in *Preparing for the 21st Century: An Appraisal of U.S. Intelligence*, U.S. Congress Commission on the Roles and Capabilities of the U.S. Intelligence Community, Washington, D.C.: Government Printing Office, March 1, 1996.

[24] Martin, F. T., *Top Secret Intranet: How U.S. Intelligence Built Intelink—The World's Largest, Most Secure Network*, New York: Prentice Hall, 1998.

[25] *Global Trends 2015: A Dialogue about the Future with Non-Government Experts*, Washington D.C.: National Intelligence Council, December 2000.

[26] Tenet, G. J., "The CIA and Security Challenges of the New Century," *International Journal of Intelligence and CounterIntelligence*, Vol. 13, No. 2, Summer 2000, p. 138.

[27] Conference on Intelligence in the 21st Century, Priverno, Italy, February 14–16, 2001, accessed on-line at http://future-intel.it/programma.html.

[28] Dumaine, C., "Intelligence in the New Millennium," CIA Directorate of Intelligence, AFCEA Spring Intelligence Conference, April 18, 2001. Table 2.4 is based on this unclassified paper.

[29] For representative viewpoints, see: Medina, C. A., "What to Do When Traditional Models Fail," and Ward, S. R., "Evolution Beats Revolution," in *Studies in Intelligence*, Vol. 46, No. 3, Washington D.C.: CIA, 2002 Unclassified Edition, accessed on-line on October 3, 2002 at http://www.cia.gov/csi/studies/vol46no3/index.html.

[30] The White House, *The National Strategy for Homeland Security*, U.S. Office of Homeland Security, July 2002, p. viii. See also "Intelligence and Warning, "pp. 15–19, for specific organizational, infrastructure, and policy changes.

3

Knowledge Management Processes

KM is the term adopted by the business community in the mid 1990s to describe a wide range of strategies, processes, and disciplines that formalize and integrate an enterprise's approach to organizing and applying its knowledge assets. Some have wondered what is truly new about the concept of managing knowledge. Indeed, many pure knowledge-based organizations (insurance companies, consultancies, financial management firms, futures brokers, and of course, intelligence organizations) have long "managed" knowledge—and such management processes have been the core competency of the business.

Several factors distinguish the new strategies that we develop in this chapter—and each of these has key implications for both public and private intelligence enterprises. The scope of knowledge required by intelligence organizations has increased in depth and breadth as commerce has networked global markets and world threats have diversified from a monolithic Cold War posture. The global reach of networked information, both open and closed sources, has produced a deluge of data—requiring computing support to help human analysts sort, locate, and combine specific data elements to provide rapid, accurate responses to complex problems. Finally, the formality of the KM field has grown significantly in the past decade—developing theories for valuing, auditing, and managing knowledge as an intellectual asset; strategies for creating, reusing, and leveraging the knowledge asset; processes for conducting collaborative transactions of knowledge among humans and machines; and network information technologies for enabling and accelerating these processes.

3.1   Knowledge and Its Management

In the first chapter, we introduced the growing importance of knowledge as the central resource for competition in both the nation-state and in business. Because of this, the importance of intelligence organizations providing strategic knowledge to public- and private-sector decision makers is paramount. We can summarize this importance of intelligence to the public or private enterprise in three assertions about knowledge.

First, knowledge has become the central asset or resource for competitive advantage. In the Tofflers' third wave, knowledge displaces capital, labor, and natural resources as the principal reserve of the enterprise. This is true in wealth creation by businesses and in national security and the conduct of warfare for nation-states.

Second, it is asserted that the management of the knowledge resource is more complex than other resources. The valuation and auditing of knowledge is unlike physical labor or natural resources; knowledge is not measured by "head counts" or capital valuation of physical inventories, facilities, or raw materials (like stockpiles of iron ore, fields of cotton, or petroleum reserves). New methods of quantifying the abstract entity of knowledge—both in people and in explicit representations—are required. In order to accomplish this complex challenge, knowledge managers must develop means to capture, store, create, and exchange knowledge, while dealing with the sensitive security issues of knowing when to protect and when to share (the trade-off between the restrictive "need to know" and the collaborative "need to share").

The third assertion about knowledge is that its management therefore requires a delicate coordination of people, processes, and supporting technologies to achieve the enterprise objectives of security, stability, and growth in a dynamic world:

- *People.* KM must deal with cultures and organizational structures that enable and reward the growth of knowledge through collaborative learning, reasoning, and problem solving.

- *Processes.* KM must also provide an environment for exchange, discovery, retention, use, and reuse of knowledge across the organization.

- *Technologies.* Finally, IT must be applied to enable the people and processes to leverage the intellectual asset of actionable knowledge.

Definitions of KM as a formal activity are as diverse as its practitioners (Table 3.1), but all have in common the following general characteristics:

Table 3.1
Representative Diversity of KM Definitions

A Sampling of KM Definitions

"A conscious strategy of getting the right knowledge to the right people at the right time and helping people share and put information into action in ways that strive to *improve organizational performance.*" —O'Dell and Grayson [1]

"… an emerging discipline that stresses a formalized, integrated approach to managing an enterprise's tangible and intangible information assets. …KM is a coordinated attempt to tap the unrealized potential for sharing and reuse that lies in an enterprise's collective consciousness." —The Gartner Group [2]

"The leveraging of intellectual capital to increase the organization's capacity for collective action which creates business value."—Motorola University [2]

"The notion of putting the combined knowledge of the firm at an employee's fingertips is the essence of knowledge management. The basic goal: to take key pieces of data from various sources, such as groupware, databases, applications and people's minds, and make them readily available to users in an organized, logical form that represents knowledge." —Sharon Watson [3]

"A systematic process for acquiring, creating, integrating, sharing, and using information, insights, and experiences, to achieve organizational goals." —U.S. DoD Rapid Improvement Team for Acquisition KM [2]

A systematic process for acquiring, creating, integrating, sharing, and using information, insights, and experiences, to make the right business decisions and achieve organizational goals. Objectives to:

- Facilitate natural communities of practice

- Develop an architecture for systematic and integrated knowledge sharing both within and across communities of practice

- Convert knowledge into a usable tool for the acquisition professional,

- Provide a disciplined and organized methodology for constant improvement and development of knowledge domains

"All with the goal of encouraging innovation and producing successful results." —U.S. Marine Corps System Command's Rapid Improvement KM Team [2]

"Create a capability where the acquisition worker can locate acquisition knowledge on demand, from any source, at any time, from any location with a high degree of confidence that information is accurate and relevant." —U.S. Navy's Acquisition Reform Office Vision for Acquisition Knowledge Management Systems [2]

- *KM is based on a strategy* that accepts knowledge as the central resource to achieve business goals and that knowledge—in the minds of its people, embedded in processes, and in explicit representations in knowledge bases—must be regarded as an intellectual form of capital to be leveraged. Organizational values must be coupled with the growth of this capital.

- *KM involves a process* that, like a supply chain, moves from raw materials (data) toward knowledge products. The process is involved in acquiring (data), sorting, filtering, indexing and organizing (information), reasoning (analyzing and synthesizing) to create knowledge, and finally disseminating that knowledge to users. But this supply chain is not a "stovepiped" process (a narrow, vertically integrated and compartmented chain); it horizontally integrates the organization, allowing collaboration across all areas of the enterprise where knowledge sharing provides benefits.

- *KM embraces a discipline and cultural values* that accept the necessity for sharing purpose, values, and knowledge across the enterprise to leverage group diversity and perspectives to promote learning and intellectual problem solving. Collaboration, fully engaged communication and cognition, is required to network the full intellectual power of the enterprise.

The U.S. National Security Agency (NSA) has adopted the following "people-oriented" definition of KM to guide its own intelligence efforts:

> Strategies and processes to create, identify, capture, organize and leverage vital skills, information and knowledge to enable people to best accomplish the organizational mission [4].

The U.S. DoD has recognized the sharp contrast in the industrial and knowledge age models of national security (Table 3.2) and the change in perspective from emphasizing weapons and sensor platforms in hierarchies to an emphasis on a knowledge-based warfighting enterprise operating in networks. The network-centric model recognizes the enterprise comprised of human (knowledge) resources, which requires shared knowledge creation, sharing, and viewing [5]. The DoD has further recognized that KM is *the* critical enabler for information superiority:

> The ability to achieve and sustain information superiority depends, in large measure, upon the creation and maintenance of reusable knowledge bases; the ability to attract, train, and retain a highly skilled work force proficient in utilizing these knowledge bases; and the development of core business processes designed to capitalize upon these assets [6].

The processes by which abstract knowledge results in tangible effects can be examined as a net of influences that effect knowledge creation and decision making (Figure 3.1). Of course, all competing enterprises apply knowledge; here we are seeking to understand how knowledge contributes the deciding marginal

Table 3.2
DoD Contrast in National Security Business Model Perspectives

Industrial Age Model (Platform Centric)	Knowledge Age Model (Network Centric)
Producer valued	Customer valued
People viewed as costs	People viewed as assets
Individual focus	Enterprise focus
Function-based operations	Process-based operations
Isolated activities within functions	Integrated processes facilitate sharing
Local view	Global view
Reducing operations to decrease cost and increase profits	Systems-thinking approach to increasing productivity and profits
Individual responsibility	Shared responsibility
Scarce resources	Infinite resources (human, structural, and intellectual)
Span of control	Span of influence

benefit to an organization. The flow of influences in the figure illustrates the essential contributions of shared knowledge.

1. *Dynamic knowledge.* At the central core is a comprehensive and dynamic understanding of the complex (business or national security) situation that confronts the enterprise. This understanding accumulates over time to provide a breadth and depth of shared experience, or organizational memory.

2. *Critical and systems thinking.* Situational understanding and accumulated experience enables dynamic modeling to provide forecasts from current situations—supporting the selection of adapting organizational goals. Comprehensive understanding (perception) and thorough evaluation of optional courses of actions (judgment) enhance decision making. As experience accumulates and situational knowledge is refined, critical explicit thinking and tacit sensemaking about current situations and the consequences of future actions is enhanced.

3. *Shared operating picture.* Shared pictures of the current situation (common operating picture), past situations and outcomes (experience), and forecasts of future outcomes enable the analytic workforce to collaborate and self-synchronize in problem solving.

Operations	Knowledge influence flow	Impact on actions

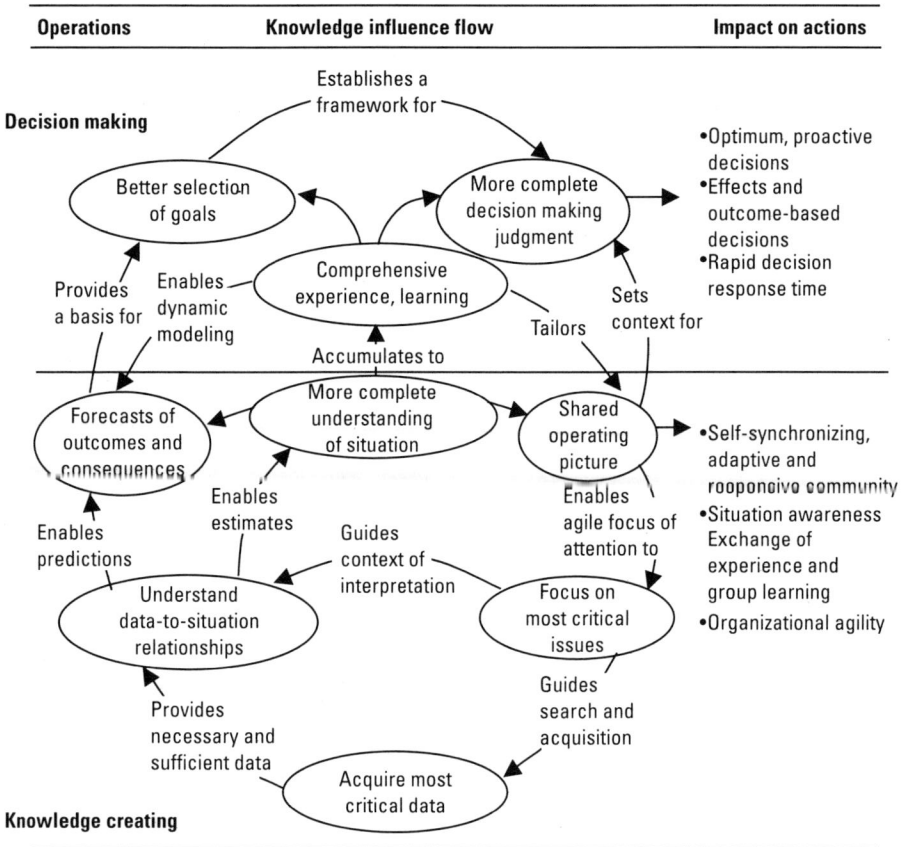

Figure 3.1 The influence flow of knowledge to action.

4. *Focused knowledge creation.* Underlying these functions is a focused data and experience acquisition process that tracks and adapts as the business or security situation changes.

While Figure 3.1 maps the *general* influences of knowledge on goal setting, judgment, and decision making in an enterprise, an understanding of how knowledge influences a *particular* enterprise in a particular *environment* is necessary to develop a KM strategy. Such a strategy seeks to enhance organizational knowledge of these four basic areas as well as information security to protect the intellectual assets. Examples of business and military applications of such a five-part strategy are summarized in Table 3.3. Note that the first four strategy areas correspond to the four contributions of knowledge discussed in the previous paragraph.

Table 3.3
Knowledge-Enhancement Strategy Components

Strategy	Knowledge Enhancements	Business Intelligence Enhancements	Military Intelligence Enhancements
1. Dynamic Knowledge—remain aware of the situation and acquire the right data	Improve the quantity, quality, accuracy, rate of update, and range of data types to achieve full understanding of processes to permit precision control	Statistical sampling Total quality management (TQM)—Taguchi analysis methods Sales (demand) and supply chain (supply) data warehousing Market trend analysis	Intelligence data warehousing Data fusion to detect known targets, threats Data mining to discover relationships, changes, abnormal patterns of behavior
2. Support Critical, Systems Thinking—provide aids to perception and decision	Support exploratory thinking of alternative hypotheses, future courses of action (options assessment), and consequences	Market dynamics modeling Supply and demand forecasting aids Cost-risk analysis	Multiple competing hypothesis analysis Multiple course of action (COA) assessment Commander decision aids
3. Shared Operating Picture—distribute and apply the knowledge effectively	Provide timely and widely distributed information to all process participants—in appropriate formats with appropriate content	Electronic mail Collaborative electronic interaction tools Multiple-access business database	Intelligence distribution (intelligence links) Collaborative decision making aids Real-time common operating picture (COP)
4. Focus Knowledge Creation—optimize the information supply chain	Refine the process of converting data to actionable knowledge: speed, accuracy, uncertainty management, and decision support	Data warehousing Data fusion Data mining Statistical process control	Sensor system refinements in coverage, detection, precision, revisit rate, and dwell Multisensor coverage
5. Protection of Intellectual Capital—ensure the protection of information	Protect the source data, information extraction, warehousing, and distribution from corruption, exploitation (eavesdropping), and deterioration	Industrial information security (INFOSEC) Database backup Commercial encryption Internet security (firewalls, encryption) E-mail security	Military INFOSEC Operational security Key distribution Encryption Intrusion detection

3.2   Tacit and Explicit Knowledge

In the first chapter, we offered a brief introduction to hierarchical taxonomy of data, information, and knowledge, but here we must refine our understanding of knowledge and its construct before we delve into the details of management processes. In particular, the earlier definition was a very general process definition, neither distinguishing different kinds of knowledge nor making distinctions between two views of knowledge—as an object or as a process (action).

In this chapter, we distinguish between the knowledge-creation processes within the knowledge-creating hierarchy (Figure 3.2). The hierarchy illustrates the distinctions we make, in common terminology, between explicit (represented and defined) processes and those that are implicit (or tacit; knowledge processes that are unconscious and not readily articulated).

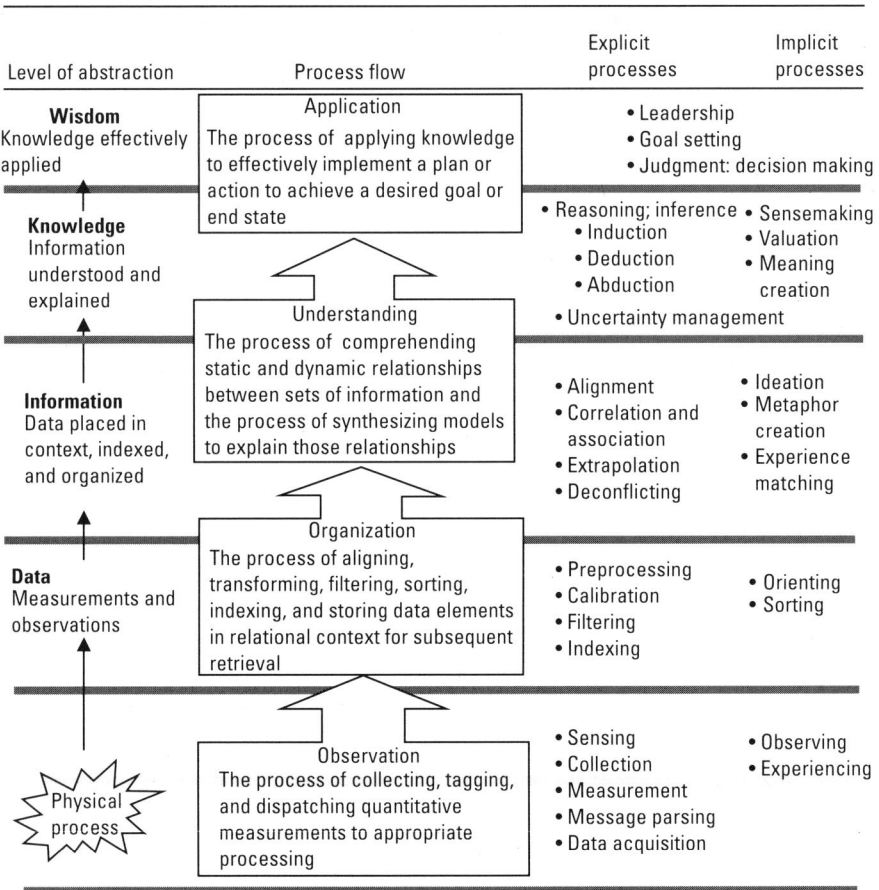

Level of abstraction	Process flow	Explicit processes	Implicit processes
Wisdom Knowledge effectively applied	**Application** The process of applying knowledge to effectively implement a plan or action to achieve a desired goal or end state	• Leadership • Goal setting • Judgment: decision making	
Knowledge Information understood and explained	**Understanding** The process of comprehending static and dynamic relationships between sets of information and the process of synthesizing models to explain those relationships	• Reasoning; inference • Induction • Deduction • Abduction • Uncertainty management	• Sensemaking • Valuation • Meaning creation
Information Data placed in context, indexed, and organized	**Organization** The process of aligning, transforming, filtering, sorting, indexing, and storing data elements in relational context for subsequent retrieval	• Alignment • Correlation and association • Extrapolation • Deconflicting	• Ideation • Metaphor creation • Experience matching
Data Measurements and observations	**Observation** The process of collecting, tagging, and dispatching quantitative measurements to appropriate processing	• Preprocessing • Calibration • Filtering • Indexing	• Orienting • Sorting
Physical process		• Sensing • Collection • Measurement • Message parsing • Data acquisition	• Observing • Experiencing

Figure 3.2   The knowledge-creating hierarchy.

3.2.1 Knowledge As Object

The most common understanding of knowledge is as an object—the accumulation of things perceived, discovered, or learned. From this perspective, data (raw measurements or observations), information (data organized, related, and placed in context), and knowledge (information explained and the underlying processes understood) are also objects. The KM field has adopted two basic distinctions in the categories of knowledge as object [7]:

1. *Explicit knowledge.* This is the better known form of knowledge that has been captured and codified in abstract human symbols (e.g., mathematics, logical propositions, and structured and natural language). It is tangible, external (to the human), and logical. This documented knowledge can be stored, repeated, and taught by books because it is impersonal and universal. It is the basis for logical reasoning and, most important of all, it enables knowledge to be communicated electronically and reasoning processes to be automated. The development of language, logic, and mathematics has enabled scientific data to be captured, human thought to be recorded, and each to be logically analyzed external to mind. Newspapers and novels, HTML content, scientific data, and engineering data all convey explicit knowledge that can be stored, retrieved, and analyzed.

2. *Tacit knowledge.* This is the intangible, internal, experiential, and intuitive knowledge that is undocumented and maintained in the human mind. It is a personal knowledge contained in human experience. Philosopher Michael Polanyi pioneered the description of such knowledge in the 1950s, considering the results of Gestalt psychology and the philosophic conflict between moral conscience and scientific skepticism. In *The Tacit Dimension* [8], he describes a kind of *knowledge that we cannot tell.* This tacit knowledge is characterized by intangible factors such as perception, belief, values, skill, "gut" feel, intuition, "know-how," or instinct; this knowledge is unconsciously internalized and cannot be explicitly described (or captured) without effort. Polanyi described perception as the "most impoverished form of tacit knowing" [9], and he asserted that there exist higher creative forms of tacit knowing. This kind of knowledge forms the bridge between perception and the higher forms of (conscious) reasoning that we can tell about more easily. This is the personal knowledge that is learned by experience, honed as a skill, and often applied subconsciously.

These two forms can be contrasted (Table 3.4) as two means of knowledge representations as well as two modes of human thought. Some have described

Table 3.4
The Bases of Explicit and Tacit Knowledge

	Explict	**Tacit**
Knowledge Constructs and Modes of Human Thought	Explicit knowledge represented as an abstraction, context free:	Tacit knowledge expressed as a narrative, rich in context:
	Mathematical models and logical (context-free) constructs	Narrative interactive exchanges between storyteller and listener
	Objective and independent of listener context	Social constructs and experiential context
		Subjective and dependent upon listener experience
Knowledge Description	Physical science; the behavior and interaction of mass and energy in the material world	Metaphysics; the behavior and interaction of people, ideas, and minds
Historical Basis	Descartes (*Discourse on Method*) (the physical sciences)	Pascal (*Pensées*) (metaphysics/the mind)
Knowledge Exchange	Objective symbology conveys explicit knowledge	Emotional narration conveys tacit knowledge
Knowledge Presenter	IT presents objective knowledge in the form of documents, equations, and numerical and graphical visualizations	Humans (*knowledge artists* or *storytellers*) describe concepts and perceptions from their own perspective to life to an idea or concept
Protection	Protected by information security (INFOSEC) measures	Protected by operational security (OPSEC) measures

explicit knowledge as "know-what" and tacit as "know-how," distinguishing the ability of tacit knowledge to put explicit knowledge into practice.

The science and mathematics of the Enlightenment Age emerged from the rich development of explicit representations of the physical world. René Descartes' *Discourse on Method* is often cited as representative of the basis for this approach to understanding the world. Descartes' reductionist problem-solving method proceeded by stating assumptions, breaking the problem into component parts, working on understanding relationships and functions by moving from simple to more complex, and finally integrating the solution into a whole by a logical chain of reasoning [10]. The Cartesian approach seeks to describe the physical world free of context, objectively and in pure abstraction. But tacit knowledge is not of the physical sciences; it is of the mind and the interaction of minds. For this reason, tacit knowledge is context rich and subjective. In contrast to explicit knowledge of the physical sciences (physics), tacit knowledge, a realm of the mind, is understood in the realm of metaphysics. Blasé Pascal, a contemporary of Descartes, is likewise cited as emphasizing the tacit knowledge of the human "heart." In his *Pensées,* Pascal wrote, "The heart has its reasons, which reason does not know" [11], emphasizing the kind of knowledge that is different than context-free logic.

Explicit knowledge is better understood and represented than knowledge of the tacit kind. Progress in the cognitive sciences and has increased our insight into the capture and representation of this knowledge and the processes underlying its creation, but much has yet to be learned. Logician Keith Devlin has contrasted these perspectives of knowledge, appealing for new analytic techniques to understand the tacit kind of knowledge. He concludes:

> Though the conclusion I eventually draw is that the existing techniques of logic and mathematics—indeed of the traditional scientific method in general—are inadequate for understanding the human mind, I do not see this as a cause for dismay. Rather, I rejoice to be living in an age when a major intellectual challenge is forcing us to develop new analytic techniques...mathematicians and scientists have come to realize that the truly difficult problems of the information age are not technological; rather they concern ourselves—what it is to think, to reason, and to engage in conversation. Meeting these challenges will almost certainly require new kinds of science—or, if you want to reserve the title "science" for the traditions begun by Galileo, Bacon, and Descartes—new analytic techniques, new conceptual tools with which to analyze and understand the workings of the human mind [12].

Devlin contrasts the current and promising new analytic techniques for explicit and tacit representations of both mind and knowledge (Table 3.5). An understanding of the relationship between knowledge and mind is of particular interest to the intelligence discipline, because these analytic techniques will serve two purposes:

1. *Mind as knowledge manager.* Understanding of the processes of exchanging tacit and explicit knowledge will, of course, aid the KM process itself. This understanding will enhance the efficient exchange of knowledge between mind and computer—between internal and external representations.

2. *Mind as intelligence target.* Understanding of the complete human processes of reasoning (explicit logical thought) and sensemaking (tacit, emotional insight) will enable more representative modeling of adversarial thought processes. This is required to understand the human mind as an intelligence target—representing perceptions, beliefs, motives, and intentions [13]. (In Section 5.5, intelligence applications of mental models are described more fully.)

Table 3.5
Representations and Approaches to Understanding Mind

	Cartesian Emphasis (Explicit Representation)	Pascal's Emphasis (Tacit Representation)
Approaches to Understand the Human Mind	Descartes (*Discourse on Method*, 1637)—1. accept only things clear and without doubt by reasoning, 2. reduce problem to component parts, 3. reason from simple to complex, 4. check and verify	Pascal ("Thoughts," *Pensées*, 1600)—"The heart has its reasons that reason does not know"—human understanding includes influences of both the mind (logic, mathematics) and the heart (intuition, judgment)
Basis of Approach	**Reductionism**	**Holism**
	Investigation of the mind as something that is objective, dispassionate, and rational, and study of content that is independent of context (context free)	Investigation of the mind is subjective and includes representation of: Meaning—content in context, cultural knowledge; Structure of thought or conversation
Mind	Consciousness explained as a dualism of mind distinct from body; reasoning distinct from feeling	Consciousness explained as monism; mind and body are one, with integration of reasoning and feeling
Knowledge	Explicit (or external to mind) representation of knowledge	Tacit (or internal to mind) representation of knowledge
Elements of Approach to Analysis	1. Cognition—study of mind as rational rule execution, context-free algebra of thought independent of body (Descartes)	1. New cognitive science—study of mind as tacit subconscious patterns and conscious reasoning; an integrated study of mind
	2. Hard math—study of thought in syllogisms and calculi to represent linguistic patterns of thought (rules). Aristotle (*Organon*), Leibniz (*De Arte Combinatoria*), Venn (symbolic logic), Boole (*Laws of Thought*—propositional logic), Frege, Peirce (predicate logic), *Montague* (intentional logic—to study meaning as signifier (sign) in the mind)	2. Soft mathematics—study of language and thought in which meaning is not intrinsic to language but requires tacit knowledge of greater context (Tarski). *Situation theory* limits all meaning to a part of the world (Barwise). Complexity identifies high-level holistic patterns in dense, highly interactive (nonlinear) situations.
	3. Language—study of language out of context (Chomsky, de Saussure)	3. New language—study as joint communication *using* language; using common (mutual) knowledge or joint knowledge of common ground.
	4. Artificial intelligence (AI)—purely rational logical intellect-based expert systems (Minsky)	4. New intelligence—represents intellect and emotion; emergent behavior of complex systems (Damasio, Dennett)

Davidow and Malone have categorized knowledge, both tacit and explicit, in the *Virtual Corporation* in four general classes based on the way in which the knowledge is applied (Table 3.6) [14]. The categories move from explicit static and dynamic descriptions (models and simulations, respectively [15]) to more tacit representations that are "actionable." These categories are helpful to distinguish the movement from explicit representations (data and information) toward mixed tacit and explicit knowledge, which leads to action. Behavioral knowledge, for example, can be represented in explicit simulations that immerse the analyst in an effort to provide tacit experience with the dynamics of the simulated environment. The basis of the simulation may be both content and

Table 3.6
Categories of Knowledge in Business and National Intelligence

Category	Level of Understanding	Business Application Examples	Intelligence Application Examples
Explicit Content Information	Historical record describing the existence, location, and state of physical items (inventory) and abstract entities (accounts)	Inventory of materials, products Customer and billing records Market research data	Force inventory Orders of battle Orders, personnel records, intelligence reports
Explicit Form Information	Static description of the physical shape and composition of objects and characteristics of events	Product description Real estate property description CAD/CAM and geographic information system (GIS) models	Target models of discriminants for automatic target recognition (ATR) Force model descriptions
Tacit and Explicit Behavioral Knowledge	Experience and dynamic description of the behavior of an object or system of objects—behavioral models and simulations	Skills and expertise of subject matter experts (SMEs) Engineering simulations Market dynamic models	Skills and expertise of experienced analysts Economic models Weapon simulations Battle management simulation tools
Tacit and Explicit Actionable Knowledge	Insight, experience, and reasoning processes that provide decision-making advice and perform control of operations	Expert judgment of executive officers Industrial robotics Machine vision for inspection Automated stock trading	Expert judgment of senior intelligence officers Alternative outcomes decision aids Automated sensor management

form information about the environment. The result is both tacit and explicit actionable knowledge: insight into alternatives and consequences, risks and pay-offs, and areas of uncertainty. All of these form a sound basis for judgment to take action.

Table 3.6 provides representative examples of each category for business and intelligence applications.

Previously, we have used the terms *resource* and *asset* to describe knowledge, but it is not only an object or a commodity to be managed. Knowledge can also be viewed as a dynamic, embedded in processes that lead to action. In the next section, we explore this complementary perspective of knowledge.

3.2.2   Knowledge As Process

Knowledge can also be viewed as the action, or dynamic process of creation, that proceeds from unstructured content to structured understanding. This perspective considers knowledge as action—as *knowing*. Because knowledge explains the basis for information, it relates static information to a dynamic reality. Knowing is uniquely tied to the creation of meaning.

The knowing processes, both explicit and tacit, move from components (data) toward integrated understanding of meaning—relating the abstractions of knowledge to the real world (Figure 3.3). The two paths, though separate columns in the table, are not independent but are interactive. (Polanyi believed that all explicit knowledge, or its interpretation, is rooted in tacit knowledge.)

The explicit knowing process is referred to as *reasoning*; as described earlier, it is attributed to the Western emphasis on logic, reductionism, and dualism. This knowing process emphasizes the *abstraction* of truth in the intellect of

Knowledge form and process	Explicit reasoning	Tacit sense making	Contribution of the abstraction
Knowledge	Intellect Hypotheses Explanations Beliefs Education	Insight, sense Imagination Understandings Perceptions Experience	Meaning: relative to reality action: basis of action in reality
Information	Relationships, links, indexed data	"Images" Metaphors Ideas	Context: relative to each other
Data	Text Symbolic Numeric	Experiences Feelings Emotions	Content: independent abstractions

Figure 3.3 Reasoning and sensemaking—knowledge in action.

the individual. By contrast, the tacit knowing process has been called *sensemaking,* a more holistic form of knowing more closely related to the Eastern emphasis on holistic intuition and oneness. In contrast to dualism of mind-body, this view emphasizes humanity-nature oneness (and therefore mind-body and self-other oneness). This knowing process focuses on the "action" of truth in the character of the individual.

Karl Weick introduced the term *sensemaking* to describe the tacit knowing process of *retrospective rationality*—the method by which individuals and organizations seek to rationally account for things by going back in time to structure events and explanations holistically [16]. We do this, to "make sense" of reality, as we perceive it, and create a base of experience, shared meaning, and understanding.

To model and manage the *knowing* process of an organization requires attention to both of these aspects of knowledge—one perspective emphasizing cognition, the other emphasizing culture and context. The general knowing process includes four basic phases that can be described in process terms that apply to tacit and explicit knowledge, in human and computer terms, respectively (Figure 3.4):

1. *Acquisition.* This process acquires knowledge by accumulating data through human observation and experience or technical sensing and measurement. The capture of e-mail discussion threads, point-of-sales transactions, or other business data, as well as digital imaging or signals analysis are but examples of the wide diversity of acquisition methods.

2. *Maintenance.* Acquired explicit data is represented in a standard form, organized, and stored for subsequent analysis and application in digital databases. Tacit knowledge is stored by humans as experience, skill, or expertise, though it can be elicited and converted to explicit form in terms of accounts, stories (rich explanations), procedures, or explanations.

3. *Transformation.* The conversion of data to knowledge and knowledge from one form to another is the creative stage of KM. This knowledge-creation stage involves more complex processes like internalization, intuition, and conceptualization (for internal tacit knowledge) and correlation and analytic-synthetic reasoning (for explicit knowledge). In the next subsection, this process is described in greater detail.

4. *Transfer.* The distribution of acquired and created knowledge across the enterprise is the fourth phase. Tacit distribution includes the sharing of experiences, collaboration, stories, demonstrations, and hands-on training. Explicit knowledge is distributed by mathematical,

Knowledge process	Acquire knowledge	Maintain knowledge	Transform knowledge	Transfer knowledge
Tacit knowledge functions (human terms)	• Listen • Experience • Observe	• Remember • Recall • Capture • Collect	Socialize • Internalize, feel • Develop intuition • Envision, conceptualize • Self-organize, create • Storytelling, exchange ↑ Internalize ●	• Share • Demonstrate • Share experience • Emulate
Explicit knowledge functions (computer terms)	• Sense • Data entry • Measure • Capture • Collect	• Store • Catalog, index • Update, refresh • Search • Protect	● Externalize ↓ • Appraise, assess, evaluate • Validate, verify • Correlate, link • Exploit, analyze or reason (deduce, abduce, induce) • Synthesize, abstract • Compile	• Disseminate • Exchange • Collaborate • Control workflow • Push (subscription, broadcast) • Pull (requisition)
Davenport and Prusak model (organization terms)	**Generation** • Acquire, rent experts • Dedicate resources, team • Fusion—combine diversity • Adaptation to crisis • Networking of people		**Combine** Codification and coordination • Mapping (structure) • Modeling (dynamics) • Narrating (storytelling)	**Transfer** • Exchange people • Mentor • Transfer and absorb • Reduce friction, enhance trust

Figure 3.4 Practical transaction processes of KM.

graphical, and textual representations, from magazines and textbooks to electronic media.

Figure 3.4 also shows the correlation between these four phases and the three phases of *organizational knowing* (focusing on culture) described by Davenport and Prusak in their text *Working Knowledge* [17]:

1. *Generation.* Organizational networks generate knowledge by social processes of sharing, exploring, and creating tacit knowledge (stories, experiences, and concepts) and explicit knowledge (raw data, organized databases, and reports). But these networks must be properly

organized for diversity of both experience and perspective and placed under appropriate stress (challenge) to perform. Dedicated cross-functional teams, appropriately supplemented by outside experts and provided a suitable challenge, are the incubators for organizational knowledge generation.

2. *Codification and coordination.* Codification explicitly represents generated knowledge and the structure of that knowledge by a *mapping* process. The map (or ontology) of the organization's knowledge allows individuals within the organization to locate experts (tacit knowledge holders), databases (of explicit knowledge), and tacit-explicit networks. The coordination process models the dynamic flow of knowledge within the organization and allows the creation of narratives (stories) to exchange tacit knowledge across the organization.

3. *Transfer.* Knowledge is transferred within the organization as people interact; this occurs as they are mentored, temporarily exchanged, transferred, or placed in cross-functional teams to experience new perspectives, challenges, or problem-solving approaches.

3.2.3 Knowledge Creation Model

A widely adopted and insightful model of the processes of creating and exchanging knowledge, or knowledge conversion, within an organization was developed, by Ikujiro Nonaka and Hirotaka Takeuchi in *The Knowledge-Creating Company* [18]. The model is very helpful in understanding how analysts interact with computer support (automation) to create intelligence within the intelligence organizations. Nonaka and Takeuchi describe four modes of conversion, derived from the possible exchanges between two knowledge types (Figure 3.5):

1. *Tacit to tacit—socialization.* Through social interactions, individuals within the organization exchange experiences and mental models, transferring the know-how of skills and expertise. The primary form of transfer is narrative—storytelling—in which rich context is conveyed and subjective understanding is compared, "reexperienced," and internalized. Classroom training, simulation, observation, mentoring, and on-the-job training (practice) build experience; moreover, these activities also build teams that develop shared experience, vision, and values. The socialization process also allows consumers and producers to share tacit knowledge about needs and capabilities, respectively.

2. *Tacit to explicit—externalization.* The articulation and explicit codification of tacit knowledge moves it from the internal to external. This can be done by capturing narration in writing, and then moving to the

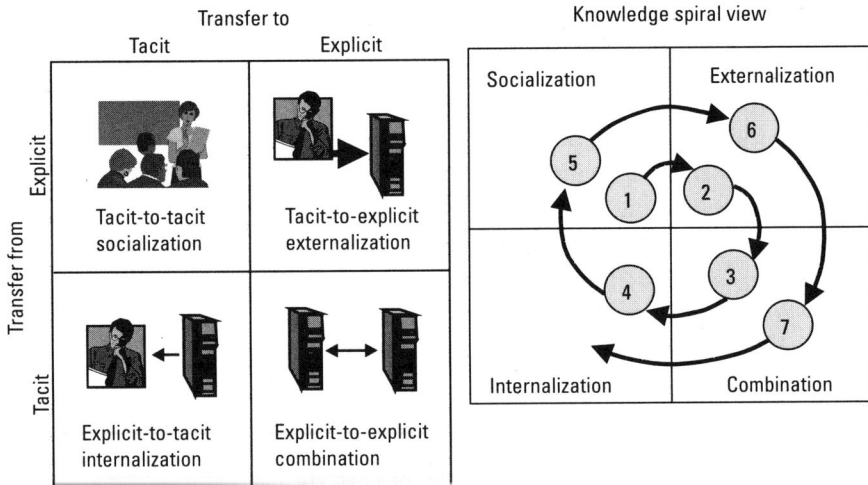

Figure 3.5 Knowledge-conversion process model.

construction of metaphors, analogies, and ultimately models. Externalization is the creative mode where experience and concept are expressed in explicit concepts—and *the effort to express is in itself a creative act.* (This mode is found in the creative phase of writing, invention, scientific discovery, and, for the intelligence analyst, hypothesis creation.)

3. *Explicit to explicit—combination.* Once explicitly represented, different objects of knowledge can be characterized, indexed, correlated, and combined. This process can be performed by humans or computers and can take on many forms. Intelligence analysts compare multiple accounts, cable reports, and intelligence reports regarding a common subject to derive a combined analysis. Military surveillance systems combine (or *fuse*) observations from multiple sensors and HUMINT reports to derive aggregate force estimates. Market analysts search *(mine)* sales databases for patterns of behavior that indicate emerging purchasing trends. Business developers combine market analyses, research and development results, and cost analyses to create strategic plans. These examples illustrate the diversity of the combination processes that combine explicit knowledge.

4. *Explicit to tacit—internalization.* Individuals and organizations internalize knowledge by *hands-on* experience in applying the results of combination. Combined knowledge is tested, evaluated, and results

in new tacit experience. New skills and expertise are developed and integrated into the tacit knowledge of individuals and teams.

Nonaka and Takeuchi further showed how these four modes of conversion operate in an unending spiral sequence to create and transfer knowledge throughout the organization (Figure 3.5). The internalization mode naturally leads to further socialization, and process leads to further tacit sharing, creativity, and knowledge expansion. The spiral model represents the concept of an ever-learning organization, expanding in knowledge and the application of that knowledge to the dynamic business environment.

Based on this model, Nonaka and Takeuchi identified five enabling conditions that promote creation within the spiral (Table 3.7). These conditions promote the cohesion of organizational purpose, freedom of thought, and breadth of perspective necessary to permit the organization to transfer knowledge between tacit and explicit forms and to explore new perspectives without boundaries. These conditions can best be seen in small teams (e.g., intelligence-analysis teams, crisis-analysis teams, and decision-making teams), although they apply across large organizations. Intention (also described as *shared vision and commitment* in numerous management texts) provides organizational cohesion of purpose and reduces the friction from competitions for different objectives. Autonomous teams are given the freedom to explore alternative solutions beyond current mindsets; access to information (i.e., people, databases, and processes) is not restricted. Organizations that have redundancy of information (in people, processes, and databases) and diversity in their makeup (also in people, processes, and databases) will enhance the ability to move along the spiral. The modes of activity benefit from a diversity of people: socialization requires some who are stronger in dialogue to elicit tacit knowledge from the team; externalization requires others who are skilled in representing knowledge in explicit forms; and internalization benefits from those who experiment, test ideas, and learn from experience, with the new concepts or hypotheses arising from combination. These redundancies and diversities also apply to processes and information sources, which provide different perspectives in each stage of the spiral.

Organizations can also benefit from *creative chaos*—changes that punctuate states of organizational equilibrium. These states include static presumptions, entrenched mindsets, and established processes that may have lost validity in a changing environment. Rather than destabilizing the organization, the injection of appropriate chaos can bring new-perspective reflection, reassessment, and renewal of purpose. Such change can restart tacit-explicit knowledge exchange, where the equilibrium has brought it to a halt.

Underlying this model is Nonaka and Takeuchi's important assertion that the basis of this creative process is the tacit knowledge of individuals:

Table 3.7
Conditions that Enable Knowledge Creation

Enabling Condition	Condition Definition	Implementation in the Intelligence Enterprise
Intention	Organization's shared vision and aspiration to meet its goals	Intelligence vision and implementing strategy to achieve the vision articulated to, understood by, and embraced by entire organization
		Organizational commitment to vision (goal) and strategy (plan)
Autonomy	Individual liberty of team members in thought, exploration, and action	Establish loose team charters
		Allow teams to establish their own boundaries
		Provide teams broad access to data; allow independence in choosing areas of focus
Redundancy	Internal overlapping of information about activities, responsibility, and purpose.	Share redundant information from multiple perspectives
		Create and maintain alternative and competing hypotheses
		Conduct internal, competing analyses
		Rotate personnel to different organizational assignments to expand perspectives (e.g., analysis, field operations, field visits, customer liaison)
Requisite Variety	Internal diversity of the organization; diversity is matched to the complexity and variety of the environment	Maintain a flat, highly networked organizational structure
		Assign diverse disciplines to problems matched to the problem scope (e.g., cross-functional analytic teams such as analysis, operations, academics, or field personnel)
Creative Chaos	Introduction of actions to stimulate beneficial interaction between the organization and its environment	Punctuation of habitual states of behavior and noncreative equilibrium conditions
		Reconsideration of existing premises and frames of discernment (mindsets)
		Reflection on purpose

… an organization cannot create knowledge by itself. Tacit knowledge of individuals is the basis of organizational knowledge creation. The organization has to mobilize tacit knowledge created and accumulated at the individual level [19].

3.3 An Intelligence Use Case Spiral

While Nonaka and Takeuchi focused on knowledge creation in the business and product-development areas, we can see how the knowledge-conversion spiral describes the exchanges within a typical intelligence application. To illustrate how the spiral operates in an intelligence environment, we follow a future fictional, yet representative, crisis situation in which U.S. intelligence is confronted with a crisis in a failing nation-state that threatens U.S. national interests. We follow a distributed crisis intelligence cell, using networked collaboration tools, through one complete spiral cycle to illustrate the spiral. This case is deliberately chosen because it stresses the spiral (no face-to-face interaction by the necessarily distributed team, very short time to interact, the temporary nature of the team, and no common "organizational" membership), yet illustrates clearly the phases of tacit-explicit exchange and the practical insight into actual intelligence-analysis activities provided by the model.

3.3.1 The Situation

The crisis in small but strategic Kryptania emerged rapidly. Vital national interests—security of U.S. citizens, U.S. companies and facilities, and the stability of the fledgling democratic state—were at stake. Subtle but cascading effects in the environment, economy, and political domains triggered the small political liberation front (PLF) to initiate overt acts of terrorism against U.S. citizens, facilities, and embassies in the region while seeking to overthrow the fledgling democratic government. The PLF waged information operations—spreading rumors via e-mail, roaming AM radio broadcasts, and publishing black propaganda on the Internet. The PLF also corrupted Kryptanian government information systems to support false claims of political corruption. A crisis intelligence analysis cell is rapidly formed, comprised of the following globally distributed participants:

- Five intelligence officers in Washington, D.C., including a team leader and four analysts with experience in the country and language skills;

- Six political scientists with expertise in Kryptania in four universities (three in Europe, one in the region);

- Two Kryptanian expert consultants at a regional think tank abroad;

- Four field intelligence officers (surged to seven within 3 days) in Kryptania;

- Six Kryptanian government security officials in Kryptania.

The crisis team is formed and all participants are notified and issued public/private keys (at their appropriate access levels) to crisis collaboration portals/collaboration workspaces on computer networks. The first portal is a secure collaborative workspace (a specially secured virtual private network on the Internet) for *sensitive but unclassified* (SBU) information access by the academics and consultants. A separate multilevel security (MLS) portal is formed on classified networks for those with access to classified intelligence data. A secure *data pump* moves the SBU data onto the classified portal; an automatic classification reviewer sanitizes and passes unclassified data down to the SBU portal. The custom portal/collaboration capability provides the distributed team a workspace allowing:

1. Interactive discussion areas (*spaces* or *e-rooms*), organized by issue areas: collection, threats, gaps, and analysis;

2. Teleconference capability (secure audio or video conferencing for use between individuals one to one, for broadcast one to many, or for broadcast to the entire group);

3. General e-mail between members, instant mail, or broadcast postings for the entire group;

4. Group bulletin board organized by issue areas;

5. Structured group repository (database) to allow members to post acquired data, intermediate products of analysis, and annotations on intermediate products;

6. Shared application tools that allow shared analysis (passing control from user to user) and annotation of data.

In addition, the portal provides MLS access to the crisis team's shared knowledge base, including:

1. Alerts and changes in situation status and impact on team priorities and mission;

2. Organized open source *news* and *intelligence* headlines provided at multiple levels of security (open source to classified, respectively) and continuously updated. Headlines are linked to full reports for drill down; related reports are automatically cross-linked. Open-source reports are annotated (i.e., source authority, pedigree, or confidence);

3. Basic country data for Kryptania and other regional countries (e.g., maps, government organizations, points of contact, and political, military, economic, business, and technical data);

4. Links to open (Internet) and closed sources (intelligence net) of information;

5. Access to relevant open- and closed-source intelligence databases;

6. Schedules (e.g., planned team same-time socialization meetings or report due milestones);

7. Analytic tools that can be accessed and applied to the group or individual data.

The team composition includes a diverse mix of intelligence officers, trusted academics, and Kryptanian government officials (requisite variety and redundancy within the limits of security), along with a common vision to understand and mitigate the threat. The team is provided a loose charter to identify specific threat patterns, organizations, and actions (autonomy); the current crisis provides all the creative chaos necessary for the newly formed team.

This first spiral of knowledge creation (Figure 3.6) occurs within the first several days of the team's formation.

3.3.2 Socialization

Within 10 hours of the team formation, all members participate in an on-line SBU kickoff meeting (same-time, different-place teleconference collaboration) that introduces all members, describes the group's intelligence charter and procedures, explains security policy, and details the use of the portal/collaboration

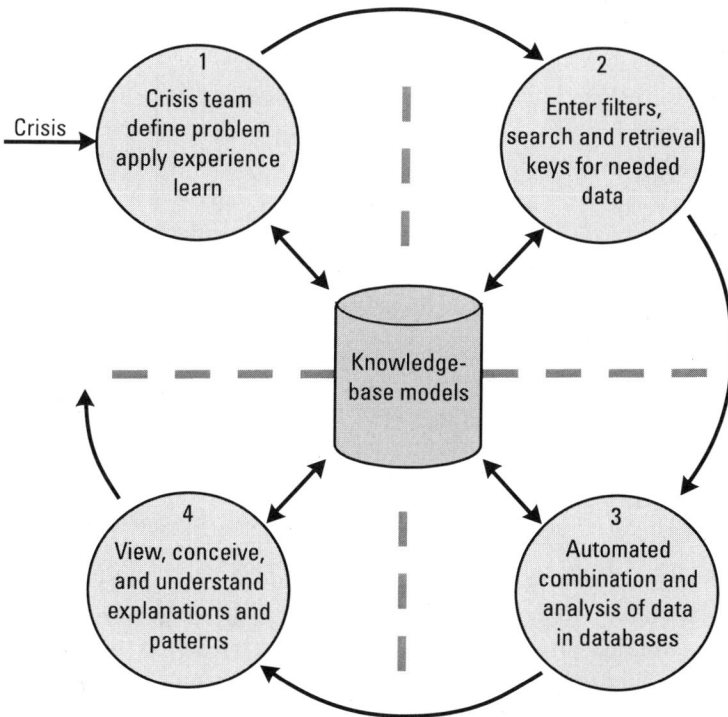

Figure 3.6 Intelligence-use case spiral example.

workspace created for the team. The team leader briefs the current situation and the issues: areas of uncertainly, gaps in knowledge or collection, needs for information, and possible courses of events that must be better understood. The group is allowed time to exchange views and form their own subgroups on areas of contribution that each individual can bring to the problem. Individuals express concepts for new sources for collection and methods of analysis. In this phase, the dialogue of the team, even though not face to face, is invaluable in rapidly establishing trust and a shared vision for the critical task over the ensuing weeks of the crisis. Over the course of the next day, several total-group and subgroup teleconferences sustain the dialogue and begin to allow the members to exchange tacit perspectives of Kryptania, approaches to understanding the threats, and impressions of where the crisis might lead. This process of dialogue exposes the diversity of mental models about the threat and even the different interpretations of the group's charter (the organizational intention). As this happens, the team begins to request additional source or types of information on the portal and starts to record requests, impressions, needed actions, and needs for charter clarifications (questions about the "boundaries" of the problem and restrictions on access) on the bulletin board.

3.3.3   Externalization

The initial discussions lead to the creation of initial explicit *models* of the threat that are developed by various team members and posted on the portal for all to see, including:

1. Structure charts of the PLF and possible linked financial supporters and organized crime operations;
2. Lists of likely sources of the black propaganda;
3. Map of Kryptania showing cities of greatest influence by the PLF and supporters;
4. Time history of PLF propaganda themes and terrorist activities;
5. Causal chains of past actions and hypotheses of possible future course of FLP actions.

The team collaboratively reviews and refines these models by updating new versions (annotated by contributors) and suggesting new submodels (or linking these models into supermodels). This externalization process codifies the team's knowledge (beliefs) and speculations (to be evaluated) about the threat. Once externalized, the team can apply the analytic tools on the portal to search for data, link evidence, and construct hypothesis structures. The process also allows the team to draw on support from resources outside the team to conduct supporting collections and searches of databases for evidence to affirm, refine, or refute the models.

3.3.4 Combination

The codified models become *archetypes* that represent current thinking—current prototype hypotheses formed by the group about the threat (who—their makeup; why—their perceptions, beliefs, intents, and timescales; what—their resources, constraints and limitations, capacity, feasible plans, alternative courses of action, vulnerabilities). This prototype-building process requires the group to structure its arguments about the hypotheses and combine evidence to support its claims. The explicit evidence models are combined into higher level explicit explanations of threat composition, capacity, and behavioral patterns. Initial (tentative) intelligence products are forming in this phase, and the team begins to articulate these prototype products—resulting in alternative hypotheses and even recommended courses of action for the United States and Kryptania.

3.3.5 Internalization

As the evidentiary and explanatory models are developed on the portal, the team members discuss (and argue) over the details, internally struggling with acceptance or rejection of the validity of the various hypotheses. Individual team members search for confirming or refuting evidence in their own areas of expertise and discuss the hypotheses with others on the team or colleagues in their domain of expertise (often expressing them in the form of stories or metaphors) to *experience* support or refutation. This process allows the members to further refine and develop internal belief and confidence in the predictive aspects of the models. As accumulating evidence over the ensuing days strengthens (or refutes) the hypotheses, the process continues to internalize those explanations that the team has developed that are most accurate; they also internalize confidence in the sources and collaborative processes that were most productive for this ramp-up phase of the crisis situation.

3.3.6 Socialization

As the group periodically reconvenes, the subject focuses away from "what we must do" to the evidentiary and explanatory models that have been produced. The dialogue turns from issues of startup processes to model-refinement processes. The group now socializes around a new level of the problem: Gaps in the models, new problems revealed by the models, and changes in the evolving crisis move the spiral toward new challenges to create knowledge about vulnerabilities in the PLF and supporting networks, specific locations of black propaganda creation and distribution, finances of certain funding organizations, and identification of specific operation cells within the Kryptanian government. All of these refined issues challenge the team and begin a new spiral of exploration and creation by the team.

3.3.7 Summary

This example illustrates the emergent processes of knowledge creation over the several day ramp-up period of a distributed crisis intelligence team. The full spiral moved from team members socializing to exchange the tacit knowledge of the situation toward the development of explicit representations of their tacit knowledge. These explicit models allowed other supporting resources to be applied (analysts external to the group and on-line analytic tools) to link further evidence to the models and structure arguments for (or against) the models. As the models developed, team members discussed, challenged, and internalized their understanding of the abstractions, developing confidence and hands-on experience as they tested them against emerging reports and discussed them with team members and colleagues. The confidence and internalized understanding then led to a drive for further dialogue—initializing a second cycle of the spiral.

3.4 Taxonomy of KM

Using the fundamental tacit-explicit distinctions, and the conversion processes of socialization, externalization, internalization, and combination, we can establish a helpful taxonomy of the processes, disciplines, and technologies of the broad KM field applied to the intelligence enterprise. A basic taxonomy that categorizes the breadth of the KM field (Table 3.8) can be developed by distinguishing three areas of distinct (though very related) activities:

1. *People.* The foremost area of KM emphasis is on the development of intellectual capital by people and the application of that knowledge by those people. The principal knowledge-conversion process in this area is socialization, and the focus of improvement is on human operations, training, and human collaborative processes. The basis of collaboration is human networks, known as communities of practice—sharing purpose, values, and knowledge toward a common mission. The barriers that challenge this area of KM are cultural in nature.

2. Processes. The second KM area focuses on human-computer interaction (HCI) and the processes of externalization and internalization. Tacit-explicit knowledge conversions have required the development

of tacit-explicit representation aids in the form of information visualization and analysis tools, thinking aids, and decision support systems. This area of KM focuses on the efficient networking of people and machine processes (such autonomous support processes are referred to as *agents*) to enable the shared reasoning between groups of people and their agents through computer networks. The barrier to achieving robustness in such KM processes is the difficulty of creating a shared *context* of knowledge among humans and machines.

3. *Processors.* The third KM area is the technological development and implementation of computing networks and processes to enable explicit-explicit combination. Network infrastructures, components, and protocols for representing explicit knowledge are the subject of this fast-moving field. The focus of this technology area is networked computation, and the challenges to collaboration lie in the ability to sustain growth and interoperability of systems and protocols.

Table 3.8
Basic KM Taxonomy for the Intelligence Enterprise

	Intelligence Enterprise KM: Acquiring, Creating, Maintaining, and Applying Knowledge to Achieve Organizational Objectives		
Perspective of Knowledge Management	**People Operational View**	**Processes Human-Computer Interaction View**	**Processors Technical View**
Knowledge Conversion	Socialization: tacit-to-tacit transactions	Externalization and internalization: transactions between tacit and explicit	Combination: explicit-to-explicit transactions
Focus of the Enterprise	Operations, business processes, training	Tools, thinking aids, decision support, knowledge representation and visualization	Infrastructure, knowledge, protocols
Basis of Collaboration	Networks of people (communities of practice): shared purpose, values, practice, knowledge	Networks of people and agents: shared reasoning and representation of tacit and explicit knowledge	Networked computation: shared configuration of content in networks and nodes (computers)
Barriers to Collaboration and Interoperation	Culture (trust, values, vision)	Context	Content and its structure

Note that these three areas correspond to three basic descriptive views of the enterprise that will be subsequently introduced in Chapter 9.

The taxonomy can be further extended (Table 3.9) to consider the disciplines and supporting tools and technologies in each of these three areas:

1. *People.* The objective of people-oriented disciplines is to create a knowledge-based organization that learns, shares, and creates knowledge collaboratively. The tools and technologies applied to this

Table 3.9
Taxonomy of Disciplines and Supporting Tools and Technologies

Perspective of Knowledge Management	People Operational View	Processes Human-Computer Interaction View	Processors Technical View
Objective	Collaborative, learning organization	Efficient HCI	Effective human-computer networks
Disciplines and Areas of Research and Development	Collaboration for: • Knowledge sharing • Problem solving eLearning Virtual teaming	HCI Human-agent collaboration Knowledge presentation	Data capturing, representing, and warehousing Cognitive (reasoning) AI and machine learning Networked computing
Automation Support Tools and Technologies	Virtual team establishment and support across time and space Automatic experience capturing and linking (cases) to problems Auto training and eLearning	Data, information, and high-dimensionality knowledge presentation to humans, virtual, and artificial reality High-level abstract interaction between human and machine agents Human-machine problem solving and workflow Data representation, knowledge mapping to index, correlating, and linking (externalizing and internalizing) knowledge Search and retrieval	Data fusion and data mining Decision support aids Analytic (thinking) tools Creativity and problem-solving support tools Multimedia Information retrieval, summarization, and abstraction

discipline range from collaborative services to create virtual (distributed) teams and supporting services to eLearning tools to integrate learning into the work process.

2. *Processes.* HCI and related disciplines have the objective of achieving efficient human-machine interaction, enabling humans-agent teams to smoothly exchange tacit and explicit knowledge. Tools that support this process include virtual- and artificial-reality visualizations (and multisensory presentations), human-machine conversation, and autonomous agent services to search and explore large data volumes.

3. *Processors.* Effective computer networks are the objective of the diverse computing disciplines that support KM: enterprise architecting, networked computing infrastructure, data warehousing, services for information management, collaboration, cognitive (reasoning) support, and knowledge distribution.

Because the KM field can also be described by the many domains of expertise (or disciplines of study and practice), we can also distinguish five distinct areas of focus (Table 3.10) that help describe the field. The first two disciplines view KM as a competence of people and emphasize making people knowledgeable:

1. *Knowledge strategists.* Enterprise leaders, such as the chief knowledge officer (CKO), focus on the enterprise mission and values, defining value propositions that assign contributions of knowledge to value (i.e., financial or operational). These leaders develop business models to grow and sustain intellectual capital and to translate that capital into organizational values (e.g., financial growth or organizational performance). KM strategists develop, measure, and reengineer business processes to adapt to the external (business or world) environment.

2. *Knowledge culture developers.* Knowledge culture development and sustainment is promoted by those who map organizational knowledge and then create training, learning, and sharing programs to enhance the socialization performance of the organization. This includes the cadre of people who make up the core competencies of the organization (e.g., intelligence analysis, intelligence operations, and collection management). In some organizations a chief learning officer (CLO) is designated this role to oversee enterprise human capital, just as the chief financial officer (CFO) manages (tangible) financial capital.

The next three disciplines view KM as an enterprise capability and emphasize building the infrastructure to make knowledge manageable:

3. *KM applications.* Those who apply KM principles and processes to specific business applications create both processes and products (e.g., software application packages) to provide component or end-end services in a wide variety of areas listed in Table 3.10. Some commercial KM applications have been sufficiently modularized to allow them to be outsourced to application service providers (ASPs) [20] that

Table 3.10
The Disciplines of KM

Knowledge Perspective	Discipline	The KM Disciplines: Key Areas of Focus
Making People Knowledgeable (KM as a Competence)	1. Knowledge strategy	Chief information officer (CIO)/CKO mission, values, value propositions
		Intellectual capital, knowledge metrics
		Knowledge capital management: human capital (know-how) and structural capital (business process, know-what)
		eBusiness process engineering and reengineering
		Business modeling business process rules
	2. Knowledge (learning) culture developers	Chief learning officer (CLO)
		Knowledge sharing, exchange, and collaboration
		Virtual teams, communities of practice
		Best practices, training, e-learning
		Problem solving, storytelling
Making Knowledge Manageable (KM as a Capability)	3. KM applications	Program management (PM), intellectual capital management (ICM)
		Supply chain management (SCM)
		Customer relationship management (CRM)
		Content/document management (CM/DM)
		Business and competitive intelligence (BI/CI)
	4. Enterprise architecture	Data storage, warehousing
		KM services, tools (e.g., collaboration, cognition)
		KM architectures
	5. Technology and tools	Knowledge capture, search, mapping
		Knowledge storage and dissemination
		Content management
		Collaboration, personalization
		Problem solving, decision aiding, decision making
		Fusion and mining, analysis

"package" and provide KM services on a per-operation (transaction) basis. This allows some enterprises to focus internal KM resources on organizational tacit knowledge while outsourcing architecture, infrastructure, tools, and technology.

4. *Enterprise architecture.* Architects of the enterprise integrate people, processes, and IT to implement the KM business model. The architecting process defines business use cases and process models to develop requirements for data warehouses, KM services, network infrastructures, and computation.

5. *KM technology and tools.* Technologists and commercial vendors develop the hardware and software components that physically implement the enterprise. Table 3.10 provides only a brief summary of the key categories of technologies that make up this broad area that encompasses virtually all ITs.

Within the community of intelligence disciplines, each of these five areas can be identified in the conventional organizational structure, but all must be coordinated to achieve an enterprisewide focus on knowledge creation and sharing. In subsequent chapters, we detail each of these discipline areas as applied to the intelligence enterprise.

3.5   Intelligence As Capital

We have described knowledge as a resource (or commodity) and as a process in previous sections. Another important perspective of both the resource and the process is that of the *valuation* of knowledge. The value (utility or usefulness) of knowledge is first and foremost quantified *by its impact* on the user in the real world. In business, this impact is financial and so we will examine commercial approaches to valuing knowledge financially. But the value of intelligence goes far beyond financial considerations in national and MI application. In these cases, the value of knowledge must be measured in its impact on national interests: the warning time to avert a crisis, the accuracy necessary to deliver a weapon, the completeness to back up a policy decision, or the evidential depth to support an organized criminal conviction. Knowledge, as an abstraction, has no intrinsic value—its value is measured by its impact in the real world.

In financial terms, the valuation of the intangible aspects of knowledge is referred to as capital—*intellectual capital.* These intangible resources include the personal knowledge, skills, processes, intellectual property, and relationships that can be leveraged to produce assets of equal or greater importance than other organizational resources (land, labor, and capital).

Quantifying and measuring the value of knowledge intangibles can be based on purely financial measures using methods developed by Karl-Erik Sveiby to estimate the intellectual capital value of knowledge-based corporations [21]. Consider the market valuation of a representative consultancy business (Figure 3.7) to identify the components of tangible net book value and intangible intellectual capital using Svelby's method.

In this example, the difference in the market value ($100 million) of the business and the tangible assets of the business is $50 million. With $40 million in short- and long-term debt, the visible equity in the business is $10 million (net book value.) But the intangible, intellectual capital is the $50 million difference between the tangible assets and the market value. What is this capital value in our representative business? It is comprised of four intangible components:

1. *Customer capital.* This is the value of established relationships with customers, such as trust and reputation for quality. This identified image represents the *brand equity* of the business—intangible capital that must be managed, nurtured, and sustained to maintain and grow the customer base. Intelligence tradecraft recognizes this form of capital in the form of credibility with consumers—"the ability to speak to an issue with sufficient authority to be believed and relied upon by the intended audience" [22].

2. *Innovation capital.* Innovation in the form of unique strategies, new concepts, processes, and products based on unique experience form this second category of capital. In intelligence, new and novel sources and methods for unique problems form this component of intellectual capital.

3. *Process capital.* Methodologies and systems or infrastructure (also called structural capital) that are applied by the organization make up its process capital. The processes of collection sources and both collection and analytic methods form a large portion of the intelligence organization's process (and innovation) capital; they are often fragile (once discovered, they may be forever lost) and are therefore carefully protected.

4. *Human capital.* The people, individually and in virtual organizations, comprise the human capital of the organization. Their collective tacit knowledge—expressed as dedication, experience, skill, expertise, and insight—form this critical intangible resource.

It is the role of the CKO in knowledge-based organizations to value, account for (audit), maintain, and grow this capital base, just as the CFO oversees the visible tangible assets. But this intangible capital base requires the

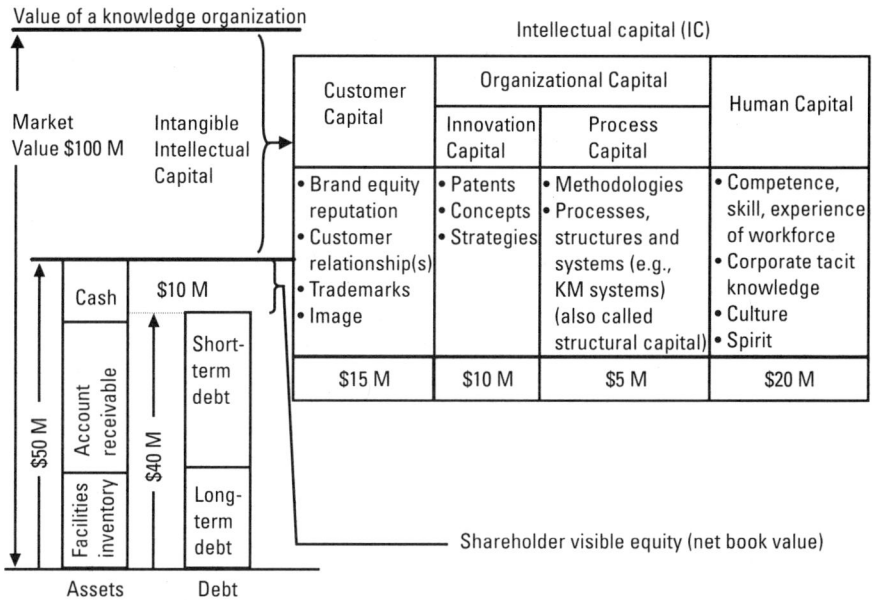

Figure 3.7 The components of intellectual capital.

definition of organizational values—both financial and nonfinancial—in order to audit the total value of the knowledge-based organization.

Organizations must begin by explicitly defining organizational values in a statement called the *value proposition*—the business case or rationale for achieving business goals (e.g., returns, improvements, or benefits) through KM [1]. An organization may have one or more propositions; there may be a primary focus, with multiple secondary foci—but all must explicitly couple value (qualitative benefits of significance to the organization's mission) to quantitative measures.

O'Dell and Grayson have defined three fundamental categories of value propositions in *If Only We Knew What We Know* [23]:

1. *Operational excellence.* These value propositions seek to boost revenue by reducing the cost of operations through increased operating efficiencies and productivity. These propositions are associated with business process reengineering (BPR), and even business transformation using electronic commerce methods to revolutionize the operational process. These efforts contribute operational value by raising performance in the operational value chain.

2. *Product-to-market excellence.* The propositions value the reduction in the time to market from product inception to product launch. Efforts

that achieve these values ensure that new ideas move to development and then to product by accelerating the product development process. This value emphasizes the transformation of the business, *itself* (as explained in Section 1.1).

3. *Customer intimacy.* These values seek to increase customer loyalty, customer retention, and customer base expansion by increasing intimacy (understanding, access, trust, and service anticipation) with customers. Actions that accumulate and analyze customer data to reduce selling cost while increasing customer satisfaction contribute to this proposition.

For each value proposition, specific impact measures must be defined to quantify the degree to which the value is achieved. These measures quantify the benefits, and utility delivered to stakeholders. Using these measures, the value added by KM processes can be observed along the sequential processes in the business operation. This sequence of processes forms a *value chain* that adds value from raw materials to delivered product. Table 3.11 compares the impact measures for a typical business operation to comparable measures in the intelligence enterprise.

It should be noted that these measures are applicable to the steady-state operation of the learning and improving organization. Different kinds of measures are recommended for organizations in transition from legacy business models. During periods of change, three phases are recognized [24]. In the first phase, users (i.e., consumers, collection managers, and analysts) must be convinced of the benefits of the new approach, and the measures include metrics as simple as the number of consumers taking training and beginning to use services. In the crossover phase, when users begin to transition to the systems, measurers change to usage metrics. Once the system approaches steady-state use, financial-benefit measures are applied. Numerous methods have been defined and applied to describe and quantify economic value, including:

1. Economic value added (EVA) subtracts cost of capital invested from net operating profit;

2. Portfolio management approaches treats IT projects as individual investments, computing risks, yields, and benefits for each component of the enterprise *portfolio*;

3. Knowledge capital is an aggregate measure of management value added (by knowledge) divided by the price of capital [25];

4. Intangible asset monitor (IAM) [26] computes value in four categories—tangible capital, intangible human competencies, intangible internal structure, and intangible external structure [27].

Table 3.11
Business and Intelligence Impact Measures

Value Proposition	Business KM Impact Measures	Intelligence KM Impact Measures
Customer Intimacy	Customer retention Number of calls handled per day Cross-selling penetration to increase revenue from existing customers	Intelligence consumer satisfaction, retention, and growth Number of consumer requests received per day; response time Percentage of correct anticipation of consumer needs
Product Leadership	Revenues from commercialization of new product Percentage of revenues of new product Time-to-market cycle Ratio of successful to failed product launches Number product launches per year	Target identification and location accuracies Source breadth and analytic depth of analytic products Time-to-operation of crisis portals Number of reports (by type) posted per year
Operational Excellence	Cost per unit Productivity and yields Number defects or instances of poor quality Production cycle time Inventory carrying costs Safety record	Cost per target located, serviced False alarms, missed targets Warning failures Response to query cycle time Collect/process/analysis yield ratios Security record

The balanced scorecard (BSC) is a widely applied method (similar to IAM) that provides four views, goals, strategies, and measures of the value in the organization [28]. The BSC goes beyond financial measures and can be readily applied to national and competitive intelligence organizations to illustrate the use of these measures to value the intelligence enterprise. Linked by a common vision and strategy, the four views address complementary perspectives of expected outcomes beginning with three *cause* views before concluding with the financial *effect* view (Figure 3.8).

The first scorecard area focuses on the organizational staff and its ability to share and create knowledge. The learning and growth view sets goals and measures organizational ability to continuously learn, improve, and create value.

This requires a measurement of training, the resulting learning and subsequent success in application, sharing of knowledge (collaboration), and satisfaction (morale and retention). The second view is the view of internal operations

Balanced scorecard valuation structure			
Vision and strategy			
• Describe the future state (vision) of the intelligence enterprise: the missions, capabilities, processes, relationships, and resources • Define the means (strategy) to achieve the vision: timeline, milestones, resources, risks and alternative trajectories • Identify specific vision goals and strategy measures			
Customer (external)	Internal	Financial	Growth and learning
• Define the intelligence customer base • Define value delivered to customers • Identify specific goals and measures	• Define major measures of performance (MOP's) and effectiveness (MOE's) • Identify specific MOP and MOE metrics and goal values	• Define the financial contribution of intelligence to stakeholders (cost, risk reduction, etc.) • Identify specific goals and measures	• Identify how intelligence value is created by change and innovation in changing threat environment • Identify specific goals and measures

Figure 3.8 BSC valuation model.

that examines performance in terms of detailed functional measures like cycle times, quality of operations (e.g., return rates and defect rates), production yields, performance, and transaction and per-operations costs. These measures are, of course, related to the optional implementation of the organization and relate to efficiency and productivity. It is in this area that BPR attempts to achieve speed, efficiency, and productivity gains.

By contrast, the third scorecard view measures performance from the external perspective of customers, which is directly influenced by the scores in the prior two views. Customers view the performance of the organization in terms of the value of products and services. Timelines, performance, quality (accuracy), and cost are major factors that influence customer satisfaction and retention. This view measures these factors, recognizing that customer measures directly influence financial performance. (Note that throughout this chapter, we distinguish the commercial customer and intelligence consumer, although they are analogous while comparing commercial and intelligence applications. The United States defines the intelligence consumer as an authorized person who uses intelligence or intelligence information directly in the decision-making process or to produce other intelligence [29].)

The financial view sets goals and measures quantitative financial metrics. For a commercial company, this includes return metrics (e.g., return on equity, return on capital employed, or return on investment) that measure financial returns relative to a capital base. Traditional measures consider only the financial capital base, but newer measures consider return as a function of both tangible (financial, physical) and intangible (human capital) resources to provide a measure of overall performance using all assets.

The four views of the BSC provide a means of "balancing" the measurement of the major causes and effects of organizational performance but also provide a framework for modeling the organization. Consider a representative BSC cause-and-effect model (Figure 3.9) for an intelligence organization that uses the four views to frame the major performance drivers and strategic outcomes expected by a major KM initiative [30]. The strategic goals in each area can be compared to corresponding performance drivers, strategic measures, and the causal relationships that lead to financial effects. The model explicitly exposes the presumed causal basis of the strategy and two categories of measures:

1. *Performance drivers* are "lead indicators" that precede the desired strategic effects. In Figure 3.9, training and introduction of a collaborative network (both directly measurable) are expected to result in the strategic outcome of increased staff analytic productivity.
2. *Strategic outcomes* are the "lag indicators" that should follow if the performance drivers are achieved—and the strategy is correct. Notice that this model presumes many hidden causal factors (e.g., analytic staff morale and workplace efficiency), which should be explicitly defined when establishing the scorecard.

The model provides a compact statement of strategy (on the left column), measures (in the boxes), and goals (specific quantitative values can be annotated in the boxes). In this representative example, the organization is implementing collaborative analytic training, a collaborative network and integrated lessons learned database, expecting overall analytic productivity improvements. Internal operations will be improved by cross-functional collaboration across intelligence sources (INTs) and single-source analysts, the availability of multi-INT data on the collaborative net, and by the increased source breadth and analytic depth provided by increased sharing among analysts. These factors will result in improved accuracy in long-term analyses for policymakers and more timely responses to crisis needs by warfighters. The results of these learning and internal improvements may be measured in customer satisfaction performance drivers, and then in their perception of the value added by intelligence (the accuracy and timeliness effects on their decisions).

Strategic objective	Performance drivers	Strategic outcome measures

Financial perspective

F1—Improve operating performance on budgeted investment
F2—Reduce warning risk

Return on investment

I&W Rate | Threat awareness

Customer perspective

C1—Increase value added to warfighter
C2—Increase value added to long term policymaker consumers

Target/ID and BDA Accuracy

Satisfaction measured by use/decisions

Issue / request anticipation

Warfighter Intel value added

Policy maker Intel value added

Internal perspective

I1—Increase analytic breadth and depth by applying nets and tools
I2—Make multi-INT data libraries accessible with recent data
I3—Improve efficiency of coordinated multi-INT collection and processing

Analytic breadth and depth

Multi-int data availability

Collection/ processing efficiency

Term estimate accuracy

Crisis response time

Learning perspective

L1—Develop and apply analytic lessons learned
L2—Improve cross functional collaboration on intelligence networks
L3—Provide collaborative analysis training

Lessons learned knowledge base

Collaboration net availability use

Collaborative analysis training

Analytic staff productivity

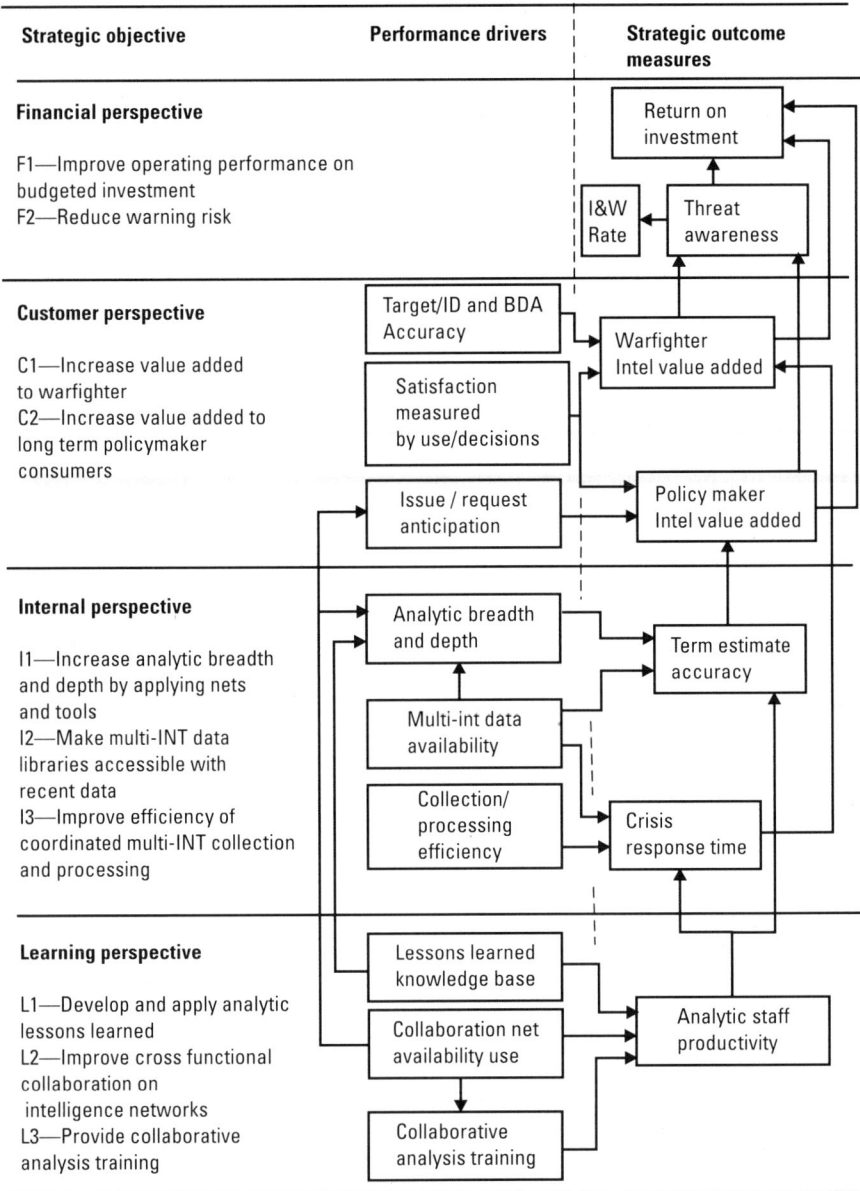

Figure 3.9 Representative intelligence cause-and-effect relationships between BSC measures.

Ultimately, these effects lead to overall improved intelligence organizational financial performance, measured by the capital value of threat awareness compared to the capital invested in the organization. (Capital invested includes existing tangible and intangible assets, not just annual budget.) Notice that the threat awareness is directly related to the all-critical value of reducing I&W risks (intelligence failures to warn that result in catastrophic losses—these are measured by policymakers).

3.6 Intelligence Business Strategy and Models

The commercial community has explored a wide range of business models that apply KM (in the widest sense) to achieve key business objectives. These objectives include enhancing customer service to provide long-term customer satisfaction and retention, expanding access to customers (introducing new products and services, expanding to new markets), increasing efficiency in operations (reduced cost of operations), and introducing new network-based goods and services (eCommerce or eBusiness). All of these objectives can be described by value propositions that couple with business financial performance. (However, as the dot com revolution has demonstrated, models and propositions must be confirmed by real market behaviors to achieve financial success.)

The strategies that leverage KM to achieve these objectives fall into two basic categories. The first emphasizes the use of analysis to understand the value chain from first customer contact to delivery. Understanding the value added to the customer by the transactions (as well as delivered goods and services) allows the producer to increase value to the customer. Values that may be added to intelligence consumers by KM include:

- *Service values.* Greater value in services are provided to policymakers by anticipating their intelligence needs, earning greater user trust in accuracy and focus of estimates and warnings, and providing more timely delivery of intelligence. Service value is also increased as producers personalize (tailor) and adapt services to the consumer's interests (needs) as they change.

- *Intelligence product values.* The value of intelligence products is increased when greater value is "added" by improving accuracy, providing deeper and more robust rationale, focusing conclusions, and building increased consumer confidence (over time).

The second category of strategies (prompted by the eBusiness revolution) seeks to transform the value chain by the introduction of electronic transactions between the customer and retailer. These strategies use network-based advertising, ordering, and even delivery (for information services like banking, investment, and news) to reduce the "friction" of physical-world retailer-customer transactions. These strategies introduce several benefits—all applicable to intelligence:

- *Disintermediation.* This is the elimination of intermediate processes and entities between the customer and producer to reduce transaction friction. This friction adds cost and increases the difficulty for buyers to locate sellers (cost of advertising), for buyers to evaluate products (cost of travel and shopping), for buyers to purchase products (cost of sales) and for sellers to maintain local inventories (cost of delivery). The elimination of "middlemen" (e.g., wholesalers, distributors, and local retailers) in *eRetailers* such as Amazon.com has reduced transaction and intermediate costs and allowed direct transaction and delivery from producer to customer with only the eRetailer in between. The effect of disintermediation in intelligence is to give users greater and more immediate access to intelligence products (via networks such as the U.S. Intelink) and to analysis services via intelligence portals that span all sources of intelligence.

- *Infomediation.* The effect of disintermediation has introduced the role of the information broker (*infomediary*) between customer and seller, providing navigation services (e.g., shopping agents or auctioning and negotiating agents) that act on the behalf of customers [31]. Intelligence communities are moving toward greater cross-functional collection management and analysis, reducing the *stovepiped* organization of intelligence by collection disciplines (i.e., imagery, signals, and human sources). As this happens, the traditional analysis role requires a higher level of infomediation and greater automation because the analyst is expected (by consumers) to become a broker across a wider range of intelligence sources (including closed and open sources).

- *Customer aggregation.* The networking of customers to producers allows rapid analysis of customer actions (e.g., queries for information, browsing through catalogs of products, and purchasing decisions based on information). This analysis enables the producers to better understand customers, aggregate their behavior patterns, and react to (and perhaps anticipate) customer needs. Commercial businesses use these capabilities to measure individual customer patterns and mass market trends to more effectively personalize and target sales and new product developments. Intelligence producers likewise are enabled to analyze warfighter and policymaker needs and uses of intelligence to adapt and tailor products and services to changing security threats.

These value chain transformation strategies have produced a simple taxonomy to distinguish eBusiness models into four categories by the level of transaction between businesses and customers (Figure 3.10) [32]. Each model has direct implications for intelligence product and service delivery:

1. *Business to business (B2B).* The large volume of trade between businesses (e.g., suppliers and manufacturers) has been enhanced by network-based transactions (releases of specifications, requests for quotations, and bid responses) reducing the friction between suppliers and producers. High-volume manufacturing industries such as the automakers are implementing B2B models to increase competition among suppliers and reduce bid-quote-purchase transaction friction. This is equivalent to process models that enable efficient electronic transactions between intelligence source providers (e.g., the IMINT, SIGINT, or other stovepipes), which allow cross-source cueing, coordinated multiple-source collection, data fusion, and mining.

2. *Business to customer (B2C).* Direct networked outreach from producer to consumer has enabled the personal computer (e.g., Dell Computer) and book distribution (e.g., Amazon.com) industries to disintermediate local retailers and reach out on a global scale directly to customers. Similarly, intelligence products are now being delivered *(pushed)* to consumers on secure electronic networks, via *subscription* and *express order* services, analogous to the B2B model.

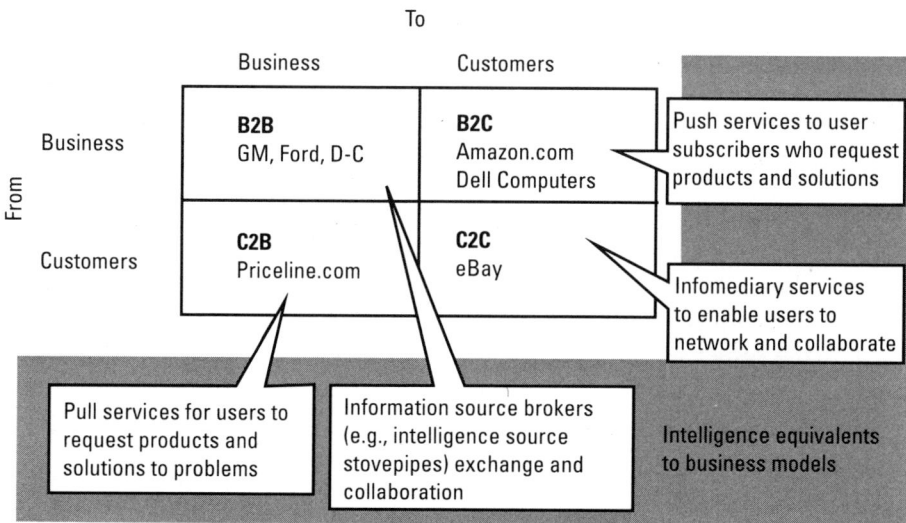

Figure 3.10 Taxonomy of basic eBusiness models applied to intelligence.

3. *Customer to business (C2B).* Networks also allow customers to reach out to a wider range of businesses to gain greater competitive advantage in seeking products and services. Businesses such as Priceline.com (travel services) and Lendingtree.com (financial services) employ the C2B model to enable customers to secure rapid quotations and secure immediate purchases on relatively volatile products (remaining airline seats and changing loan rates, respectively). Similarly, the introduction of secure intelligence networks and on-line intelligence product libraries (e.g., common operating picture and map and imagery libraries) allows consumers to *pull* intelligence from a broader range of sources. (This model enables even greater competition between source providers and provides a means of measuring some aspects of intelligence utility based on actual use of product types.)

4. *Customer to customer (C2C).* The C2C model automates the mediation process between consumers, enabling consumers to locate those with similar purchasing-selling interests. eBay.com is the primary commercial example of C2C, brokering between diverse buyers and sellers worldwide on an limitless variety of items, based only on free-flow supply and demand. Intelligence nets that locate and connect collectors, analysts, operators, and even consumers with common interests introduce the C2C model, for all are consumers of intelligence at varying levels.

3.7   Intelligence Enterprise Architecture and Applications

The intelligence enterprise can be compared to the commercial business models and applications, considering the analogies introduced in Chapter 1. Just like commercial businesses, intelligence enterprises:

- Measure and report to stakeholders the returns on investment. These returns are measured in terms of intelligence performance (i.e., knowledge provided, accuracy and timeliness of delivery, and completeness and sufficiency for decision making) and outcomes (i.e., effects of warnings provided, results of decisions based on knowledge delivered, and utility to set long-term policies).

- Service customers, the intelligence consumers. This is done by providing goods (intelligence products such as reports, warnings, analyses, and target folders) and services (directed collections and analyses or tailored portals on intelligence subjects pertinent to the consumers).

- Require intimate understanding of business operations and must adapt those operations to the changing threat environment, just as businesses must adapt to changing markets.

- Manage a supply chain that involves the anticipation of future needs of customers, the adjustment of the delivery of raw materials (intelligence collections), the production of custom products to a diverse customer base, and the delivery of products to customers *just in time* [33].

Consider the general business enterprise model (Figure 3.11) that can directly represent an intelligence enterprise. The enterprise maintains a business strategy to define, measure, and monitor the value of goods and services delivered to customers. The model includes both a *front office,* which services customers (intelligence consumers) and a *back office,* which includes the supply chain (intelligence chain) and the supporting business intelligence operations that monitor the supply chain and adapt to customer needs.

The enterprise strategy guides the entire enterprise by the value proposition, goals, and measures defined in the last section. The components of the BSC, for example, can be correlated to the functions within the architecture. Internal goals influence the BI and SCM components, while customer goals guide the implementation of CRM functions. Learning goals guide the implementation of organizational development across all functional areas. Financial

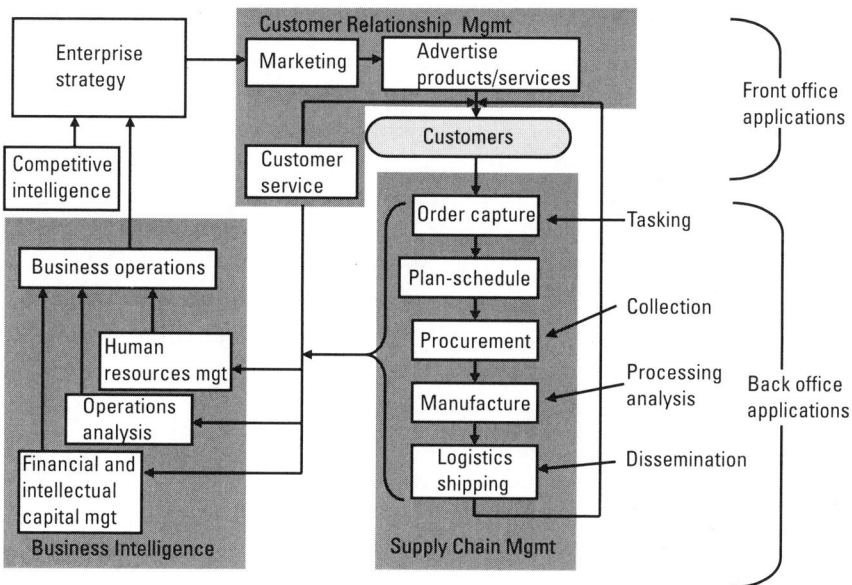

Figure 3.11 Enterprise architecture model related to the intelligence business model.

goals are achieved as an effect of these other activities; for this reason, attention to the architecture is critical to establish the organization's functional base, which minimizes transaction friction in front- and back-office functions.

3.7.1   Customer Relationship Management

CRM processes that build and maintain customer loyalty focus on managing the relationship between provider and consumer. The short-term goal is customer satisfaction; the long-term goal is loyalty. Intelligence CRM seeks to provide intelligence content to consumers that anticipates their needs, focuses on the specific information that supports their decision making, and provides *drill down* to supporting rationale and data behind all conclusions. In order to accomplish this, the consumer-producer relationship must be fully described in models that include:

- Consumer needs and uses of intelligence—applications of intelligence for decision making, key areas of customer uncertainty and lack of knowledge, and specific impact of intelligence on the consumer's decision making;
- Consumer transactions—the specific actions that occur between the enterprise and intelligence consumers, including urgent requests, subscriptions (standing orders) for information, incremental and final report deliveries, requests for clarifications, and issuances of alerts.

Automated CRM capabilities have been deployed in the electronic retailing community with growing success, employing the knowledge of electronic transactions (from customers online browsing habits in online catalogs to their personal purchasing history) to better understand customer interests. Similarly, as intelligence has embraced wider dissemination on electronic networks, there are strong analogies between commercial retail CRM and potential intelligence CRM functions (Table 3.12). CRM offers the potential to personalize intelligence delivery to individual decision makers while tracking their changing interests as they browse subject offerings and issue requests through their own custom portals.

3.7.2   Supply Chain Management

The SCM function monitors and controls the flow of the supply chain, providing internal control of planning, scheduling, inventory control, processing, and delivery. Building on earlier generation enterprise resource planning (ERP) functions, SCM functions can also extend beyond the enterprise to coordinate the supply chain with suppliers (at the front end) and external customers (at the

Table 3.12
Business Customer and Intelligence Consumer Relationship Analogies

Business CRM	Intelligence CRM
The customer: a purchaser of goods and services.	The intelligence consumer: an authorized person who uses intelligence or intelligence information directly in the decision-making process or to produce other intelligence
Business CRM Functions	**Comparable Intelligence CRM Functions**
1. Track customer catalog browsing to understand interests and trends	1. Track consumer intelligence portal browsing to understand interests and trends
2. Track and record customer transaction history: inquiries, shopping (browsing offerings), purchases, returns, satisfaction survey responses	2. Track and record consumer transactions: inquiries for reports, searches for online data, requests for intelligence tasking (e.g. topics or urgencies), uses of intelligence (e.g., decisions made based on intelligence, benefits of intelligence, or feedback)
3. Analyze individual customer buying patterns	3. Analyze individual consumer intelligence request patterns
4. Personalize sales promotions to customer interests (targeted marketing); suggest products based on prior purchase patterns	4. Personalize news, reports, and alerts to consumer interests; anticipate and deliver new products based on previous interests and current trends
5. Provide common access to customer profile to marketing, sales reps, delivery, and customer service to present coordinated delivery of service	5. Provide common access to consumer profile to tasking, analysis, and production to deliver coordinated intelligence to consumers from all elements of the enterprise that interface with the consumer
6. Analyze entire customer base: identify customer groups, purchasing trends, and behaviors to manage sales campaigns; introduce new products and reach new markets	6. Analyze entire consumer base to identify interest trends, concerns, and issues: identify consumer groups with common interests; develop products and services tailored to groups; identify individual consumers with near identical interests and offer collaborative analysis and products

delivery end). SCM is the core of B2B business models, seeking to integrate front-end suppliers into an extended supply chain that optimizes the entire production process to slash inventory levels, improve on-time delivery, and reduce the order-to-delivery (and payment) cycle time. In addition to throughput efficiency, the B2B models seek to aggregate orders to leverage the supply chain to gain greater purchasing power, translating larger orders to reduced prices. The key impact measures sought by SCM implementations include:

- Cash-to-cash cycle time (time from order placement to delivery/ payment);
- Delivery performance (percentage of orders fulfilled on or before request date);
- Initial fill rate (percentage of orders shipped in supplier's first shipment);
- Initial order lead time (supplier response time to fulfill order);
- On-time receipt performance (percentage of supplier orders received on time).

Like the commercial manufacturer, the intelligence enterprise operates a supply chain that "manufactures" all-source intelligence products from raw sources of intelligence data and relies on single-source suppliers (i.e., imagery, signals, or human reports). The analogies between business and intelligence SCM are apparent (Table 3.13) and the principles of automation, monitoring, and adaptive control can benefit the high-volume intelligence supply chain in terms of efficiency, product timeliness, and customer satisfaction.

Note that the supply chain in Figure 3.11 distinguishes the tasking, collection, processing, exploitation, and dissemination (TCPED) stages associated with the high-volume national intelligence supply chain. (The TCPED intelligence model is compared to other models in Section 6.2.)

3.7.3   Business Intelligence

The BI function, introduced in Chapter 1, provides all levels of the organization with relevant information on internal operations and the external business environment (via marketing) to be exploited (analyzed and applied) to gain a competitive advantage. The BI function serves to provide strategic insight into overall enterprise operations based on ready access to operating data. The objective of BI is to enhance business decision making by providing accurate and timely information to decision makers. In many complex businesses, this has not been the case; near-real-time quantitative data and models of operations have been absent and management decisions have been base on intuition and scant measured data. Similarly, in large intelligence organizations, it has been difficult to quantify overall operating performance due to the lack of dedicated operations metrics capture, storage, and analysis. Wal-Mart has become a commercial legend in BI implementation by warehousing all point of sales, inventory, and supplier data to analyze customer trends and adapt the entire retail supply chain to a high level of response and efficiency. As shown in Figure 3.11, BI is integrated to both CRM and SCM functions in measuring and providing intelligence to management to adopt the strategic operating policies and tactical CRM and SCM operations.

Table 3.13
Business and Intelligence SCM Analogies

Business SCM	Intelligence SCM
The Supply Chain:	The Supply Chain:
Suppliers: upstream producers of raw materials (tier 2) and components (tier 1)	Suppliers: intelligence collectors and single-source processing and analysis
Supply chain—order, plan, procure, produce, ship (delivery, order fulfillment)	Supply chain—plan, task collect, acquire data, analyze all-source data, produce intelligence products, and disseminate
External customers	External intelligence consumers
Business SCM Functions	**Comparable Intelligence SCM Functions**
1. Supplier integration—provide electronic interactions with at least tier 1 suppliers, sharing demand models and supplier capacity and projected deliver data	1. Collection and silo integration—integrate collection planning, tasking to respond to current and projected demands; coordinate multiple-INT collections of data
2. Inventory and warehouse management—automated monitoring and management of individual inventory items and movement through warehouse to transport	2. Intelligence holdings management—data warehouse management to monitor use of holdings, and current/projected demands to assure key data availability
3. Process planning and scheduling—monitor supply and demand in real-time; project demand and schedule supply processes based on current data and statistical models; eliminate inventory *stock outs*	3. Intelligence production planning and scheduling—measure current requests, tasking, processing, and analysis workflow to allocate resources to optimize to priority, timeliness, and depth metrics; eliminate *no response* to consumer requests
4. Delivery and order fulfillment—plan order sequencing to consolidate orders to combine deliveries and deliver on time.	4. Digital production—provide electronic delivery of products with *emerging* results as well as final *point-in-time* delivery
5. Extended customer integration—share supply chain data with consumers: status of current orders in the supply chain and tracking past history of performance	5. Extended consumer integration—share supply chain data with consumer, reporting time to delivery for each request.

The emphasis of BI is on explicit data capture, storage, and analysis; through the 1990s, BI was the predominant driver for the implementation of corporate data warehouses, and the development of online analytic processing (OLAP) tools. (BI preceded KM concepts, and the subsequent introduction of broader KM concepts added the complementary need for capture and analysis of tacit and explicit knowledge throughout the enterprise [34].)

BI implementations within an intelligence organization provide "intelligence about intelligence"—insight into the operation flow through the intelligence cycle. The intelligence BI function should collect and analyze real-time workflow data to provide answers to questions such as:

- What are the relative volumes of requests (for intelligence) by type?
- What is the "cost" of each category of intelligence product?
- What are the relative transaction costs of each stage in the supply chain?
- What are the trends in usage (by consumers) of all forms of intelligence over the past 12 months? Over the past 6 months? Over the past week?
- Which single sources of incoming intelligence (e.g., SIGINT, IMINT, and MASINT) have greatest utility in all-source products, by product category?

Like their commercial counterparts, the intelligence BI function should not only track the operational flows, they should also track the history of operational decisions—and their effects. Both operational and decision-making data should be able to be conveniently navigated and analyzed to provide timely operational insight to senior leadership who often ask the question, "What is the cost of a pound of intelligence?"

3.8   Summary

KM provides a strategy and organizational discipline for integrating people, processes, and IT into an effective enterprise. The development of KM as a discipline has moved through phases of emphasis, as noted by Tom Davenport, a leading observer of the discipline:

> The first generation of knowledge management within enterprises emphasized the "supply side" of knowledge: acquisition, storage, and dissemination of business operations and customer data. In this phase knowledge was treated much like physical resources and implementation approaches focused on building "warehouses" and "channels" for supply processing and distribution. This phase paid great attention to systems, technology and infrastructure; the focus was on acquiring, accumulating and distributing explicit knowledge in the enterprise [35].

Second generation KM emphasis has turned attention to the *demand side* of the knowledge economy—seeking to identify value in the collected data to allow the enterprise to add value from the knowledge base, enhance the knowledge spiral, and accelerate innovation. This generation has brought more focus to people (the organization) and the value of tacit knowledge; the issues of sustainable knowledge creation and dissipation throughout the organization are emphasized in this phase. The attention in this generation has moved from understanding knowledge systems to understanding *knowledge workers.* The third generation to come may be that of KM *innovation,* in which the knowledge process is viewed as a complete life cycle within the organization, and the emphasis will turn to revolutionizing the organization and reducing the knowledge cycle time to adapt to an ever-changing world environment [36].

In this chapter, we introduced a taxonomy that distinguished the processes of KM by the modes of transactions between explicit and tacit knowledge; the subsequent five chapters are organized by this distinction in perspective and mode of transacting knowledge (Table 3.14):

- *People and organizations.* Chapter 4 introduces the characteristics and virtues of the knowledge-based organization. This includes networks of people who share vision and values to collaboratively solve problems, learn, and adapt to the changing threat or business environment. The emphasis is on the socialization of tacit knowledge exchange.

- *Processes and systems.* Chapters 5–7 describe the internalization and externalization transaction processes that exchange tacit and explicit knowledge. Chapters 5 and 6 detail the principles and practice of the core KM competency of intelligence—analysis and synthesis—where

Table 3.14
Structure of KM Presentation in This Text

Perspective of KM	People Operational View	Processes HCI View	Technology Technical View
Focus of the Enterprise	Operations, processes, training	Tools, thinking aids, and visualization	Infrastructure, knowledge protocols
Knowledge Transactions	Socialization tacit-to-tacit transactions	Internalization and externalization transactions between tacit and explicit	Combination explicit-to-explicit transactions
Subsequent Chapters in This Book	Chapter 4: The knowledge-based intelligence organization	Chapter 5: Intelligence analysis and synthesis Chapter 6: Implementing analysis-synthesis Chapter 7: Knowledge transfer and transaction	Chapter 8: Knowledge combination Chapter 9: Enterprise architecture

analysts network with other analysts and machines to create intelligence products.

- *Technology.* Chapters 8 and 9 then describe the information technologies (computing processes, processing nodes, and interconnecting network technologies) that constitute the implementation of the architecture of the intelligence enterprise.

Endnotes

[1]   O'Dell, C. and Grayson, C. J., Jr., *If Only We Knew What We Know,* New York: Free Press, 1998.

[2]   KM definitions are quoted from the DoD Web site: http://center.dau.mil/Topical_ Sessions_templates/Knowledge_Management/Definitions_of_Knowledge_Management.htm.

[3]   Watson, S., "Getting to AHA!" *ComputerWorld,* January 26, 1998.

[4]   NSA adopted the definition from the APQC. See Brooks, C. C.,"Knowledge Management and the Intelligence Community," *Defense Intelligence Journal,* Vo. 9, No.1, Winter 2000, p.17. This issue of the journal is devoted to KM in the U.S. IC.

[5]   Table 3.2 is based on *Advancing Knowledge Management in DoD: A Primer for Executives and Practitioners,* Directorate of eBusiness & Knowledge Management, OASD/C3I, September 2000, p. 2.

[6]   Defense Planning Guidance FY02-07, April 2000, p. 102.

[7]   Some texts refer to *embedded* knowledge as a third category; this knowledge integrated in business processes can include either unconscious human process skills (tacit) or explicitly coded computer programs (explicit).

[8]   Polanyi, M., *The Tacit Dimension,* Garden City, NY: Doubleday, 1966.

[9]   Polanyi, M., *The Tacit Dimension,* Garden City, NY: Doubleday, 1966, p. 7.

[10] Descartes published his *Discourse on Method* in 1637 and described his four-step problem-solving method of analysis and synthesis in "Part II—Principal Rules of the Method."

[11] Pascal, B., *Pensees,* "Part IV, The Means of Belief," 1660, para. 277.

[12] Devlin, K., *Goodbye, Descartes: The End of Logic and the Search for a New Cosmology of the Mind,* New York: John Wiley & Sons, 1997, pp. vii and ix.

[13] This is not to imply that the human mind is a new intelligence target; leadership intentions and customer purchasing intentions are the targets of national and business intelligence, respectively. New representations will permit more accurate modeling of these targets.

[14] Categories are adapted from: Davidow, W. H., and M. S. Malone, *The Virtual Corporation,* Chapter 3, New York: Harper-Collins, 1992.

[15] The U.S. Defense Modeling and Simulation Office distinguishes model as "A physical, mathematical or otherwise logical representation of a system, entity, phenomenon or process," and a simulation as, "A method for implementing a model over time." Models are essentially static representations, while simulations add dynamic (temporal) behavior.

[16] Weick, K., *Sensemaking in Organizations,* Thousand Oaks, CA: Sage Publications, 1995.

[17] Davenport, T. H., and Prusak, L., *Working Knowledge: How Organizations Manage What They Know,* Boston: Harvard Business School Press, 1998.

[18] Nonaka, I., and H. Takeuchi, *The Knowledge-Creating Company: How Japanese Companies Create the Dynamics of Innovation,* New York: Oxford University Press, 1995.

[19] Nonaka, I., and H. Takeuchi, *The Knowledge-Creating Company: How Japanese Companies Create the Dynamics of Innovation,* New York: Oxford University Press, 1995, p. 72.

[20] The ASP business model is also called managed service provider, netsourcing, or total service provider models.

[21] Sveiby, K. E., "The 'Invisible' Balance Sheet," September 8, 1997, updated October 2001, accessed on-line January 1, 2003 at http://www.sveiby.com/InvisibleBalance.htm.

[22] "Effective Use of Intelligence," *Notes on Analytic Tradecraft,* Note 11, CIA Directorate of Intelligence, February 1996, p. 2.

[23] O'Dell, C. and Grayson, C. J., Jr. *If Only We Knew What We Know,* New York: Free Press, 1998, p. 133.

[24] Pastore, R., "Noodling Numbers," *CIO,* May 1, 2000, p. 122.

[25] Strassman, P. A., "The Value of Knowledge Capital," *American Programmer,* March 1998. See also Strassman, P. A., *Knowledge Capital,* New Canaan, CT: The Information Economics Press, 1999.

[26] Sveiby, K. E., *The New Organizational Wealth: Managing and Measuring Knowledge-Based Assets,* San Francisco: Berrett-Koehler, 1997. ICM is similar to the introductory concept illustrated in Figure 3.7.

[27] See Skyrme, D., *Measuring the Value of Knowledge,* London: Business Intelligence, 1998.

[28] Kaplan, R. S. and D. P. Norton, *The Balanced Scorecard,* Boston: Harvard Business School Press, 1996; see also Kaplan, R. S., and D. P. Norton, "Using the Balanced Scorecard as a Strategic Management System," *Harvard Business Review,* January–February 1996.

[29] *A Consumer's Guide to Intelligence,* Washington D.C.: CIA, n.d., p.42.

[30] This example model follows the approach introduced by Kaplan and Norton in *The Balanced Scorecard,* Chapter 7.

[31] Hagel, J., and M. Singer, *Net Worth,* Boston: Harvard Business Review Press, 1999.

[32] This figure is adapted from, "E-Commerce Survey," *The Economist,* February 26, 2000, p. 9.

[33] The concept of *just in time* delivery results from efficient supply chain management and results in reduced inventories (and cost of inventory holdings.) The inventory reduction benefits of just in time delivery of physical products can be high; for intelligence, the benefits of inventory (information) reductions are not as great, but the benefits of making

information available at the right time can provide significant benefits in reducing information overload.

[34] Nylund, A. L., "Tracing the BI Family Tree," *Knowledge Management,* July 1999, pp. 70–71.

[35] See Davenport, T., "Knowledge Management, Round 2," *CIO Magazine,* November 1, 1999, p. 30; for a supporting viewpoint, see also Karlenzig, W., "Senge on Knowledge," *Knowledge Management,* July 1999, p. 22.

[36] Firestone, J. M., *Accelerated Innovation and KM Impact,* White Paper 14, Executive Information Systems, Inc., December 1999.

4

The Knowledge-Based Intelligence Organization

National intelligence organizations following World War II were characterized by compartmentalization (insulated specialization for security purposes) that required individual learning, critical analytic thinking, and problem solving by small, specialized teams working in parallel (*stovepipes* or *silos*). These stovepipes were organized under hierarchical organizations that exercised central control. The approach was appropriate for the centralized organizations and bipolar security problems of the relatively static Cold War, but the global breadth and rapid dynamics of twenty-first century intelligence problems require more agile networked organizations that apply organizationwide collaboration to replace the compartmentalization of the past. Founded on the virtues of integrity and trust, the disciplines of organizational collaboration, learning, and problem solving must be developed to support distributed intelligence collection, analysis, and production.

This chapter focuses on the most critical factor in organizational knowledge creation—the people, their values, and organizational disciplines. The chapter is structured to proceed from foundational virtues, structures, and communities of practice (Section 4.1) to the four organizational disciplines that support the knowledge creation process: learning, collaboration, problem solving, and best practices—called intelligence *tradecraft* (Sections 4.2–4.5, respectively). The emphasis in this chapter is in describing *organizational* qualities and their application in intelligence organizations.

Notice that the *people* perspective of KM presented in this chapter can be contrasted with the process and technology perspectives (Table 4.1) five ways:

Table 4.1
Three KM Perspectives

Perspective of KM	People Operational View	Processes HCI View	Technology Technical View
Focus of the Enterprise	Operations, processes, training	Tools, thinking aids and visualization	Infrastructure, collaboration protocols
Primary Modes of Knowledge Transaction	Socialization: tacit-to-tacit transactions	Internalization and externalization: transactions between tacit and explicit	Combination: explicit-to-explicit transactions
Basis of Collaboration	Shared purpose, values, and tacit knowledge	Shared business processes, tacit, and explicit knowledge	Shared explicit content and virtual spaces for on-line collaboration
Collaboration Enablers	Culture of trust and commitment to shared purpose Communities of practice Organizational learning and problem solving Tradecraft best practices	Collaboration rules of engagement Compatible transaction processes	Asynchronous and synchronous groupware Compatible standards for interoperation
Barriers to Collaboration	Culture	Context	Content

1. *Enterprise focus.* The focus is on the values, virtues, and mission shared by the people in the organization.

2. *Knowledge transaction.* Socialization, the sharing of tacit knowledge by methods such as story and dialogue, is the essential mode of transaction between people for collective learning, or collaboration to solve problems.

3. *Collaboration.* The basis for human collaboration lies in shared purpose, values, and a common trust.

4. *Enablers.* A culture of trust develops communities that share their best practices and experiences; collaborative problem solving enables the growth of the trusting culture.

5. *Barriers.* The greatest barrier to collaboration is the inability of an organization's culture to transform and embrace the sharing of values, virtues, and disciplines.

This chapter purposely precedes the subsequent chapters on KM processes, tools, architectures, and implementing technologies. The numerous implementation failures of early-generation KM enterprises have most often occurred because organizations have not embraced the new business models introduced, nor have they used the new systems to collaborate. As a result, these KM implementations have failed to deliver the intellectual capital promised. These cases were generally not failures of process, technology, or infrastructure; rather, they were failures of organizational culture change to embrace the new organizational model. In particular, they failed to address the cultural barriers to *organizational* knowledge sharing, learning, and problem solving. Numerous texts have examined these implementation challenges [1], and all have emphasized that organizational transformation must precede KM system implementations. So, in this chapter, we focus on the development of the virtues and disciplines that form the foundation of the knowledge-based intelligence organization.

4.1 Virtues and Disciplines of the Knowledge-Based Organization

At the core of an agile knowledge-based intelligence organization is the ability to sustain the creation of organizational knowledge through learning and collaboration. Underlying effective collaboration are values and virtues that are shared by all. The U.S. IC, recognizing the need for such agility as its threat environment changes, has adopted knowledge-based organizational goals as the first two of five objectives in its *Strategic Intent* [2]:

- *Unify the community through collaborative processes.* This includes the implementation of training and business processes to develop an interagency collaborative culture and the deployment of supporting technologies.
- *Invest in people and knowledge.* This area includes the assessment of customer needs and the conduct of events (training, exercises, experiments, and conferences/seminars) to develop communities of practice and build expertise in the staff to meet those needs. Supporting infrastructure developments include the integration of collaborative networks and shared knowledge bases.

Speaking like a corporate officer, the chairman of the U.S. National Intelligence Council emphasized the importance of collaboration to achieve speed and accuracy while reducing intelligence production costs:

… two types of collaborative tools are needed: collaboration in the production process—to increase speed and accuracy—and expertise-based collaboration—to enable teams of analysts to work on a project for several weeks or months. …These new collaborative tools will allow analysts to discuss contentious analytical issues, share information like maps, imagery, and database information, and coordinate draft assessments, all on-line, from their own workspaces, resulting in substantial savings of time and effort [3].

Clearly identified organizational propositions of values and virtues (e.g., integrity and trust) shared by all enable knowledge sharing—and form the basis for organizational learning, collaboration, problem solving, and best-practices (intelligence tradecraft) development introduced in this chapter. This is a necessary precedent before KM infrastructure and technology is introduced to the organization. The intensely human values, virtues, and disciplines introduced in the following sections are essential and foundational to building an intelligence organization whose business processes are based on the value of shared knowledge.

4.1.1   Establishing Organizational Values and Virtues

The foundation of all organizational discipline (ordered, self-controlled, and structured behavior) is a common purpose and set of values shared by all. For an organization to pursue a common purpose, the individual members must conform to a common standard and a common set of ideals for group conduct. These standards and ideals of a society have been recognized as the virtues that enable a society to operate in harmony. The knowledge-based intelligence organization is a society that requires virtuous behavior of its members to enable collaboration. Dorothy Leonard-Barton, in *Wellsprings of Knowledge*, distinguishes two categories of values: those that relate to basic human nature and those that relate to performance of the task [4]. In the first category are *big V* values (also called *moral virtues*) that include basic human traits such as personal integrity (consistency, honesty, and reliability), truthfulness, and trustworthiness. For the knowledge worker's task, the second category (of *little v* values) includes those values long sought by philosophers to arrive at knowledge or justify true belief. Some epistemologies define intellectual virtue as the foundation of knowledge: *Knowledge is a state of belief arising out of intellectual virtue* [5]. Intellectual virtues include organizational conformity to a standard of right conduct in the exchange of ideas, in reasoning and in judgment.

The knowledge-sharing organization (Table 4.2) requires the commitment to both categories of values within individuals, shared within teams, and across the entire organization.

Personal integrity is required of individual participants to objectively and critically think, conduct self assessments, correct mindsets, and learn by

Table 4.2
Organizational Activities-to-Virtues Associations

Organization Level	Activities Performed	Moral Virtues	Intellectual Virtues
Individuals	Critical, creative, and objective thinking Personal adaptation Personal knowledge creation, learning, and growth	Integrity	Agility Creativity Imagination
Teams, Work Groups	Cooperation Collaboration Knowledge sharing Collaborative knowledge creation, problem solving, and process improvement	Truth Trustworthiness	Diversity of minds Openness
Entire Organization	Organizational goal setting, strategy, planning, and decision making Organizational adaptation and growth	Wisdom—seeking highest goal by the best means	Unity of understanding organizational purpose

acknowledging the experience of others. Organizational integrity is dependent upon the individual integrity of all contributor—as participants cooperate and collaborate around a central purpose, the virtue of trust (built upon shared trustworthiness of individuals) opens the doors of sharing and exchange. Essential to this process is the development of *networks of conversations* that are built on communication transactions (e.g., assertions, declarations, queries, or offers) that are ultimately based in personal commitments [6]. Ultimately, the virtue of organizational wisdom—seeking the highest goal by the best means—must be embraced by the entire organization recognizing a common purpose [7].

Trust and cooperative knowledge sharing must also be complemented by an objective openness. Groups that place consensus over objectivity become subject to certain dangerous decision-making errors. Janis, for example, coined the popular term *groupthink* to define the subjective and noncritical tendency of some groups to conform, leading to narrow decision-making processes that fail to objectively consider alternatives [8]. Groupthink occurs when a group values unanimity and cohesion over the objective consideration of alternatives.

Thomas Davenport, author of *Working Knowledge: How Organizations Manage What They Know* [9], has noted that task-oriented (small v) virtues emphasized in knowledge-based organizations necessarily differ from those in the industrial age:

Companies inherited an important set of virtues from the industrial era: diligence, efficiency, replication, control, and operational excellence in general. …Virtues that characterized those companies are important but are becoming proportionately less so. In an age where business models decay with surprising speed, a new set of virtues will be required …creativity, imagination, diversity, speed, openness and the capacity for continual right-angle turns [10].

Here, Davenport emphasizes the virtues of agility (ability to change rapidly), creativity (ability to explore intellectually), and openness (ability to trust inherently) as necessary conditions for competitive organizational performance in a rapidly changing world. These virtues are necessities to enable the concept of organizational revolution introduced in Chapter 1. When organizations must continuously change even the business model itself, creativity, agility, and openness are required throughout the organization to maintain stability in the presence of such dynamics. Business models may change but virtues and purpose provide the stability.

4.1.2   Mapping the Structures of Organizational Knowledge

Every organization has a structure and flow of knowledge—a knowledge environment or *ecology* (emphasizing the self-organizing and balancing characteristics of organizational knowledge networks). The overall process of studying and characterizing this environment is referred to as *mapping*—explicitly representing the network of nodes (competencies) and links (relationships, knowledge flow paths) within the organization. The fundamental role of KM organizational analysis is the mapping of knowledge within an existing organization. As a financial audit accounts for tangible assets, the knowledge mapping identifies the intangible tacit assets of the organization. The mapping process is conducted by a variety of means: passive observation (where the analyst works within the community), active interviewing, formal questionnaires, and analysis. As an ethnographic research activity, the mapping analyst seeks to understand the unspoken, informal flows and sources of knowledge in the day-to-day operations of the organization. The five stages of mapping (Figure 4.1) must be conducted in partnership with the owners, users, and KM implementers.

The first phase is the *definition* of the formal organization chart—the formal flows of authority, command, reports, intranet collaboration, and information systems reporting. In this phase, the boundaries, or focus of mapping interest is established. The second phase audits (identifies, enumerates, and quantifies as appropriate) the following characteristics of the organization:

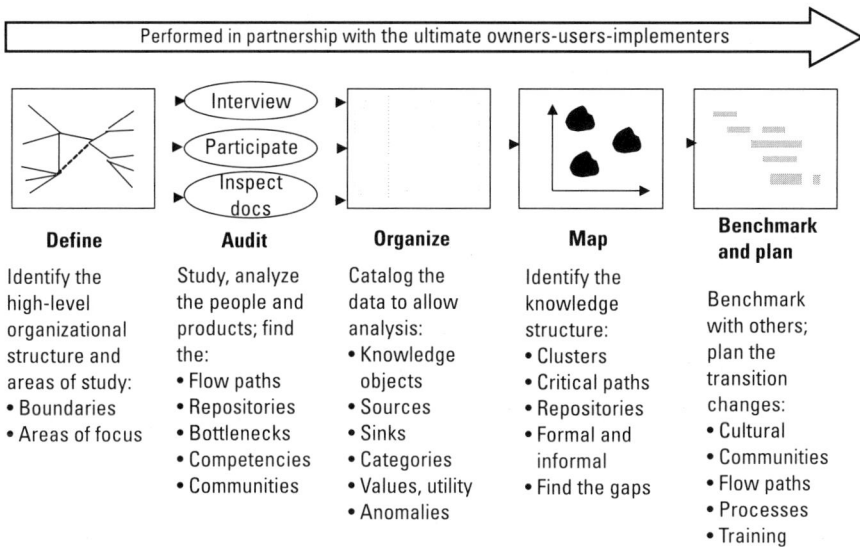

Figure 4.1 Organizational knowledge mapping process.

1. *Knowledge sources*—the people and systems that produce and articulate knowledge in the form of conversation, developed skills, reports, implemented (but perhaps not documented) processes, and databases.

2. *Knowledge flowpaths*—the flows of knowledge, tacit and explicit, formal and informal. These paths can be identified by analyzing the transactions between people and systems; the participants in the transactions provide insight into the organizational network structure by which knowledge is created, stored, and applied. The analysis must distinguish between seekers and providers of knowledge and their relationships (e.g., trust, shared understanding, or cultural compatibility) and mutual benefits in the transaction.

3. *Boundaries and constraints*—the boundaries and barriers that control, guide, or constrict the creation and flow of knowledge. These may include cultural, political (policy), personal, or electronic system characteristics or incompatibilities.

4. *Knowledge repositories*—the means of maintaining organizational knowledge, including tacit repositories (e.g., communities of experts that share experience about a common practice) and explicit storage (e.g., legacy hardcopy reports in library holdings, databases, or data warehouses).

Once audited, the audit data is *organized* in the third phase by clustering the categories of knowledge, nodes (sources and sinks), and links unique to the organization. The structure of this organization, usually a table or a spreadsheet, provides insight into the categories of knowledge, transactions, and flow paths; it provides a format to review with organization members to convey initial results, make corrections, and refine the audit. This phase also provides the foundation for quantifying the intellectual capital of the organization, and the audit categories should follow the categories of the intellectual capital accounting method adopted (e.g., balanced scorecard as described in Chapter 3). The process of identifying the categories of knowledge used by the organization develops a taxonomy (structure) of the knowledge required to operate the organization. The taxonomy, like the Dewey decimal system in a library, forms the structure for indexing the organization's knowledge base, creating directories of explicit knowledge and organizational tacit expertise.

The fourth phase, *mapping*, transforms the organized data into a structure (often, but not necessarily, graphical) that explicitly identifies the current knowledge network. Explicit and tacit knowledge flows and repositories are distinguished, as well as the social networks that support them. This process of visualizing the structure may also identify clusters of expertise, gaps in the flows, *chokepoints*, as well as areas of best (and worst) practices within the network.

Once the organization's current structure is understood, the structure can be compared to similar structures in other organizations by *benchmarking* in the final phase. Benchmarking is the process of identifying, learning, and adapting outstanding practices and processes from any organization, anywhere in the world, to help an organization improve its performance. Benchmarking gathers the tacit knowledge—the know-how, judgments, and enablers—that explicit knowledge often misses [11]. This process allows the exchange of quantitative performance data and qualitative best-practice knowledge to be shared and compared with similar organizations to explore areas for potential improvement and potential risks. This process allows the organization to plan cultural, infrastructure, and technology changes to improve the effectiveness of the knowledge network. The process also allows for comparison of plans with the prior experience of other organizations implementing change.

The U.S. CIA has performed a mapping analysis to support KM enterprise automation initiatives, as well as to satisfy its legal requirements to store records of operations and analyses [12]. The agency implemented a metadata repository that indexes abstracted metadata (e.g., author, subject per an intelligence taxonomy, date, and security level) for all holdings (both hardcopy and softcopy) across a wide variety of databases and library holdings—all at multiple levels of security access. The repository allows intelligence officers to search for holdings on topics of interest, but a multiple-level security feature limits their search reporting to those holdings to which they have access. Because the repository provides only abstracts, access may still be controlled by the originator (i.e., originators of a document maintain security control and may choose to grant access to a complete document on an individual need-to-know basis when requested). The single repository indexes holdings from multiple databases; metadata is automatically generated by tools that read existing and newly created documents. Because the repository provides a pointer to the originating authors, it also provides critical *pointers to people,* or a directory that identifies people within the agency with experience and expertise by subject (e.g., country, intelligence target, or analytic methods). In this sense, the repository indexes tacit knowledge (the people with expertise) as well as explicit knowledge (the documents).

4.1.3   Identifying Communities of Organizational Practice

A critical result of any mapping analysis is the identification of the clusters of individuals who constitute formal and informal groups that create, share, and maintain tacit knowledge on subjects of common interest. A variety of clusters can be identified by the different categories of organizational units that share interests, purpose, and knowledge (Table 4.3).

Functional workgroups are organized by business domain (e.g., an intelligence topic domain: a region of the world or a threat category) or by intelligence

Table 4.3
Fundamental Organizational Clusters of Purpose and Knowledge

Organizational Unit	Description	Intelligence Examples
Entire organization	Formal organization or community of organizations defined by charter, budgets, and authority structure	Intelligence agency or community of agencies, a corporate competitive intelligence unit.
Functional workgroup	Functional unit within the organization; permanent functional responsibility	IMINT analytic team, CI product analysis cell, a national watch unit
Cross-functional project team	Project unit made up of individuals with diverse expertise from within the organization; temporary task (project) responsibility	Crisis response team, IMINT-SIGINT target development team, joint intel-ops team
Community of practice	Formal or ad hoc groups of individuals with common interests who share for the mutual benefit of gaining shared knowledge	Subject matter expert groups (e.g., Asian interest groups), professional organization participants
Informal community	Friends, associates, and colleagues who share knowledge of common interest on an ad hoc basis.	Long-term colleagues who gather to share experience across ops-intel although organizationally dispersed

specialty (e.g., telemetry intelligence, special SIGINT, or radar imagery) to maintain established groups of specialization. Within these groups resides a permanent longer term repository of tacit experience and stored explicit data holdings. The functional workgroup benefits from stability, established responsibilities, processes and storage, and high potential for sharing. Functional workgroups provide the high-volume knowledge production of the organization but lack the agility to respond to projects and crises.

Cross-functional project teams are shorter term *project* groups that can be formed rapidly (and dismissed just as rapidly) to solve special intelligence problems, maintain special surveillance watches, prepare for threats, or respond to crises. These groups include individuals from all appropriate functional disciplines—with the diversity often characteristic of the makeup of the larger organization, but on a small scale—with reach back to expertise in functional departments. Such teams are candidates for virtual collaboration—using network infrastructure and groupware to allow them to form quickly without full-time physical colocation, share their own perspective and data, and communicate while working together toward a common goal.

Communities of organizational practice are the critical organized groups of experts and interested individuals who share knowledge on a regular basis about a particular domain of interest (e.g., Middle Eastern culture, telemetry analysis, foreign languages, social network analysis, ELINT, or digital production with XML). KM researchers have recognized that such organized communities provide a significant contribution to organizational learning by providing a forum for:

- Sharing current problems and issues;
- Capturing tacit experience and building repositories of best practices;
- Linking individuals with similar problems, knowledge, and experience;
- Mentoring new entrants to the community and other interested parties.

Because participation in communities of practice is based on individual interest, not organizational assignment, these communities may extend beyond the duration of temporary assignments and cut across organizational boundaries. In the commercial business world, KM practitioners have implemented message boards, collaboration spaces, and best-practice databases to foster and support such communities. KM researchers have recognized that a key distinction between teams (or workgroups) and communities is this: knowledge is created in teams, but it resides and is diffused and shared through communities [13]. The activities of working, learning, and innovating have traditionally been treated as independent (and conflicting) activities performed in the office, in the classroom, and in the lab. However, studies by John Seely Brown, chief scientist of Xerox PARC, have indicated that once these activities are unified in communities of practice, they have the potential to significantly enhance knowledge transfer and creation [14].

Finally, organizational knowledge mappings generally recognize the existence of informal (or underground) communities of knowledge sharing among friends, career colleagues, and business acquaintances. These informal paths extend outside the organization and can be transient, yet provide valuable sources of insight (because they often provide radically different perspectives) and threatening sources of risk (due to knowledge *leakage*). Security plans and knowledge mappings must consider these informal paths.

4.1.4   Initiating KM Projects

The knowledge mapping and benchmarking process must precede implementation of KM initiatives, forming the understanding of current competencies and processes and the baseline for measuring any benefits of change. KM implementation plans within intelligence organizations generally consider four components, framed by the kind of knowledge being addressed and the areas of investment in KM initiatives (Figure 4.2) [15]:

1. *Organizational competencies.* The first area includes assessment of workforce competencies and forms the basis of an intellectual capital audit of human capital. This area also includes the capture of best

KM initiative

	Organization (people, processes)	Infrastructure (technology, processes)
Explicit	**Competencies** • Education, training • Best practice, business process capture • Human capital analysis	**KM network** • KM Intranets, collaborative groupware, analytic tools • Best practice databases • Knowledge directories
Tactic	**Social collaboration** • Communities of practice enhancement and formation • Knowledge creation socialization events • Human capital analysis	**Virtual collaboration** • Virtual teams and communities of practice • Videoconference and collaborative groupware

(Knowledge type)

Figure 4.2   Basic categories of KM initiatives.

practices (the intelligence business processes, or tradecraft) and the development of core competencies through training and education. This assessment forms the basis of intellectual capital audit.

2. *Social collaboration.* Initiatives in this area enforce established face-to-face communities of practice and develop new communities. These activities enhance the socialization process through meetings and media (e.g., newsletters, reports, and directories).

3. *KM networks.* Infrastructure initiatives implement networks (e.g., corporate intranets) and processes (e.g., databases, groupware, applications, and analytic tools) to provide for the capture and exchange of explicit knowledge.

4. *Virtual collaboration.* The emphasis in this area is applying technology to create *connectivity* among and between communities of practice. Intranets and collaboration groupware (discussed in Section 4.3.2) enable collaboration at different times and places for virtual teams—and provide the ability to identify and introduce communities with similar interests that may be unaware of each other.

4.1.5 Communicating Tacit Knowledge by Storytelling

The KM community has recognized the strength of narrative communication—*dialogue* and *storytelling*—to communicate the values, emotion (feelings, passion), and sense of immersed experience that makeup personalized, tacit knowledge. Such tacit communication is essential in the cultural transformation necessary for collaborative knowledge sharing (the socialization process of the knowledge spiral introduced in Chapter 3) and organizational process innovation. The introduction of KM initiatives can bring significant organizational change because it may require cultural transitions in several areas:

- Changes in purpose, values, and collaborative virtues;

- Construction of new social networks of trust and communication;

- Organizational structure changes (networks replace hierarchies);

- Business process agility, resulting a new culture of continual change (training to adopt new procedures and to create new products).

All of these changes require participation by the workforce and the communication of tacit knowledge across the organization. In the same sense, the collaborative socialization process involves the exchange of tacit experience and insight among collaborators, often expressed in the form of dialogue and stories. Dialogue is the normal mode of exchange between individuals or small groups, while storytelling refers to the more formal method of creating a story to convey tacit knowledge to a large group. Both methods employ a narrative mode of thought and highly interactive knowledge constructs to communicate to active recipients, in contrast with the abstract, analytical mode used by logicians and mathematics to communicate explicit knowledge (Table 4.4).

Storytelling provides a complement to abstract, analytical thinking and communication, allowing humans to share experience, insight, and issues (e.g., unarticulated concerns about evidence expressed as "negative feelings," or general "impressions" about repeated events not yet explicitly defined as threat patterns). KM consultant Stephen Denning, who teaches the methodology for creation of story narratives, describes the complementary nature of these modes:

> Storytelling doesn't replace analytical thinking. It supplements it by ena-
> bling us to imagine new perspectives and new worlds, and is ideally suited

Table 4.4
Analytical and Storytelling Modes of Thought and Communication

Mode of Thought and Communication	Abstract Analytical Mode Communicate Explicit Knowledge	Narrative Storytelling Mode Communicate Tacit Knowledge
Knowledge Constructs	Analytic process (static); little interaction required	Dynamic interactive exchange between teller and listener
	Mathematical models, quantitative and logical (context-free) constructs	Qualitative, social constructs and context
	Physical science context	Mind, emotional context
	Interaction of physical objects	Interaction of people, ideas, minds
Communication Form	$Y = mx + b$	"Once in a deep, dark forest…"
	Conveys explicit knowledge— formal channels	Conveys tacit knowledge— informal channels
	Linearity, precision, predictable structure, context free	Nonlinearity, imprecision, surprise, context rich
Role of the Receiver	Passive spectator	Active participant
	View through a window	Immerse into virtual world
	No deviation from strict rules of interpretation, no gaps in presentation, no creativity	Imagination must collaborate with storyteller to fill in gaps based on personal experience
Knowledge Presenter	IT documents, numerical and graphical visualizations on electronic displays	Humans (knowledge artists, storytellers) who can describe or present what they see differently, bringing life to an idea or concept

to communicating change and stimulating innovation. Abstract analysis is easier to understand when seen through the lens of a well-chosen story and can of course be used to make explicit the implications of a story. …I propose marrying the communicative and imaginative strengths of storytelling with the advantages of abstract and scientific analysis [16].

The *organic school* of KM that applies storytelling to cultural transformation emphasizes a human behavioral approach to organizational socialization, accepting the organization as a complex ecology that may be changed in a large way by small effects. These effects include the use of a powerful, effective story that communicates in a way that spreads credible tacit knowledge across the entire organization [17]. This school classifies tacit knowledge into artifacts, skills, heuristics, experience, and natural talents (the so-called ASHEN classification of tacit knowledge) and categorizes an organizations' tacit knowledge in these classes to understand the flow within informal communities.

The organic approach to organizational culture change (Figure 4.3) first understands the organizational tacit knowledge by capturing anecdotal stories that describe the operation, experiences, and implicit values of the organization. From the oral repository of tacit knowledge in these stories, the KM researcher maps the culture, describing high-level patterns of behavior in archetypes and stories (e.g., exemplar people and roles in the organization and representative knowledge transaction processes, both good and bad). From the analysis of these archetypal stories, the KM researcher can synthesize or select real stories that communicate desired organizational values and create cross-cultural understanding.

This organic research approach has been developed and applied by KM consultant and researcher Dave Snowden, who has immersed himself in organizations to observe the flow of knowledge among communities of practice. Snowdon has noted:

One of the paradoxes is that informal communities are the real dynamos of knowledge. If you build strong boundaries between formal and informal communities, you get increased knowledge flows. But if you try to break

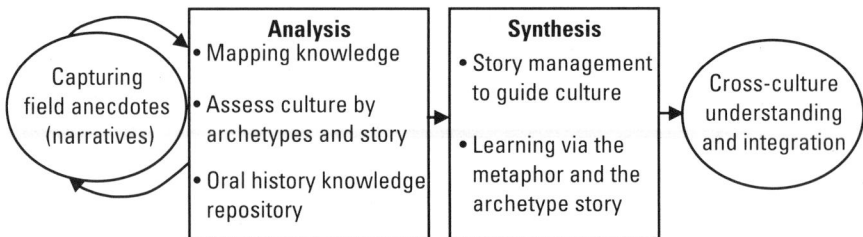

Figure 4.3 Organizational analysis and change-story synthesis.

the boundaries down, the informal knowledge goes offsite because people don't feel secure [18].

Snowdon's studies revealed this need for a delicate balance between sharing and protection of informal and formal information flows. Attempts to "put everything on-line" may bring organizational insecurity (and reluctance to share). Nurturing informal sharing within secure communities of practice and distinguishing such sharing from formal sharing (e.g., shared data, best practices, or eLearning) enables the rich exchange of tacit knowledge when creative ideas are fragile and emergent.

4.2 Organizational Learning

Educators frequently cite futurist Alvin Toffler's remarks on the preeminence of learning in his 1970 bestseller, *Future Shock:* "The illiterate of the 21st Century will not be those who cannot read and write, but those who cannot learn, unlearn, and relearn." Toffler foresaw that rapid technological change and globalization would demand greater agility and flexibility in learning—the rapid creation and application of relevant knowledge that creates value. Such agility and flexibility requires learning to be a continuous and lifelong process—a fundamental discipline of the knowledge-based organization.

Peter Senge's 1990 classic, *The Fifth Discipline*, articulated and popularized the concept of the learning organization "where people continually expand their capacity to create the results they truly desire, where new and expansive patterns of thinking are nurtured, where collective aspiration is set free, and where people are continually learning how to learn together" [19]. Senge asserted that the fundamental distinction between traditional controlling organizations and adaptive self-learning organizations are five key disciplines including both virtues (commitment to personal and team learning, vision sharing, and organizational trust) and skills (developing holistic thinking, team learning, and tacit mental model sharing). Senge's core disciplines, moving from the individual to organizational disciplines, included:

- *Personal mastery.* Individuals must be committed to lifelong learning toward the end of personal and organization growth. The desire to learn must be to seek a clarification of one's personal vision and role within the organization.

- *Systems thinking.* Senge emphasized holistic thinking, the approach for high-level study of life situations as complex systems. An element of learning is the ability to study interrelationships within complex dynamic systems and explore and learn to recognize high-level patterns of emergent behavior. (We discuss this in greater detail in Section 4.4.)

- *Mental models.* Senge recognized the importance of tacit knowledge (mental, rather than explicit, models) and its communication through the process of socialization. The learning organization builds shared mental models by sharing tacit knowledge in the storytelling process and the planning process. Senge emphasized planning as a tacit-knowledge sharing process that causes individuals to envision, articulate, and share solutions—creating a common understanding of goals, issues, alternatives, and solutions.

- *Shared vision.* The organization that shares a collective aspiration must learn to link together personal visions without conflicts or competition, creating a shared commitment to a common organizational goal set.

- *Team learning.* Finally, a learning organization acknowledges and understands the diversity of its makeup—and adapts its behaviors, patterns of interaction, and dialogue to enable growth in personal and organizational knowledge.

It is important, here, to distinguish the kind of *transformational learning* that Senge was referring to (which brings cultural change across an entire organization), from the smaller scale group learning that takes place when an intelligence team or cell conducts a long-term study or must rapidly "get up to speed" on a new subject or crisis. Large-scale organizationwide transformational learning addresses the long-term culture changing efforts to move whole organizations toward collaborative, sharing cultures. Group learning and Senge's personal mastery, on the other hand, includes the profound and rapid growth of subject matter knowledge (intelligence) that can occur when a diverse intelligence team collaborates to study an intelligence target. In the next subsections, we address the primary learning methods that contribute to both.

4.2.1 Defining and Measuring Learning

The process of group learning and personal mastery requires the development of both reasoning and emotional skills. The level of learning achievement can be assessed by the degree to which those skills have been acquired. Researcher Benjamin Bloom and a team of educators have defined a widely used taxonomy of the domains of human learning: knowledge, attitude, and skills (KAS) [20]. These three areas represent the cognitive (or mental skills), affective (attitude or emotional skills), and psychomotor (manual or physical movement) domains of human learning.

The taxonomy of cognitive and affective skills can be related to explicit and tacit knowledge categories, respectively, to provide a helpful scale for measuring the level of knowledge achieved by an individual or group on a particular subject. The levels of learning can be applied to the states of knowledge developed by an intelligence team on a particular problem. Table 4.5 compares the cognitive learning levels, ordered from simple to complex following the Bloom model, for a typical intelligence problem to illustrate the gradation of cognitive intelligence skills.

4.2.2 Organizational Knowledge Maturity Measurement

The goal of organizational learning is the development of maturity at the organizational level—a measure of the state of an organization's knowledge about its domain of operations and its ability to continuously apply that knowledge to increase corporate value to achieve business goals.

Carnegie-Mellon University Software Engineering Institute has defined a five-level People Capability Maturity Model® (P-CMM ®) that distinguishes five levels of *organizational* maturity, which can be measured to assess and

Table 4.5
Cognitive, Explicit Learning Domain Skills Applied to Intelligence

Cognitive Explicit Knowing Mental Reasoning Skills	Intelligence Example: Foreign Threat Analysis
1. Knowing—retaining and recalling data, information, and knowledge	1. Knows the foreign nation-state authorities, government, organization, and military command structure.
2. Comprehending—interpreting problems, translating and relating data to information, assigning meaning	2. Comprehends the relationships and influences between government organizations and actors; comprehends the relative roles of all players
3. Applying—applying concepts from one situation to another, reasoning about cases and analogies	3. Applies experiences of similar and previous governments to reason about formation of policy and intentions.
4. Analyzing—decomposing concepts into components and relationships	4. Analyzes government policy statements and raw intelligence data; links all data to organizations and actions
5. Synthesizing—constructing concepts from components and assigning new meanings	5. Synthesizes models of national leadership intention formation, planning, and decision making
6. Evaluating—making judgments about the values of concepts for decision making	6. Evaluates models and hypotheses, comparing and adapting models as time progresses to asses the utility of models and competing hypotheses

quantify the maturity of the workforce and its organizational KM performance. The P-CMM® framework can be applied, for example, to an intelligence organization's analytic unit (Table 4.6) to measure current maturity and develop strategy to increase to higher levels of performance [21]. The levels are successive *plateaus* of practice, each building on the preceding foundation. The P-CMM® provides a quantitative tool to measure and improve individual competencies, develop effective collaborative teams, and motivate improved organizational performance.

An organization may estimate its maturity, unit by unit, to contribute to intellectual capital estimation and to focus its learning investments (formal and informal). The highest level of optimized performance requires continual measurement of the effectiveness of intelligence processes. One of the benefits of

Table 4.6
Capability Maturity Levels of Intelligence Analysis

Maturity Level	Key Practices Characterizing the Maturity Level	Representative Practices Applied to the Discipline of Intelligence Analysis
1. Initial	Inconsistent management of the workforce Ad hoc approach to problem solving across the organization	Ad hoc mentoring; lack of standard approaches, processes, or training across the analytic workforce No collaboration in learning or analytic problem solving; different analytic standard applied across different units
2. Managed	Focus on management of people	Workforce performance management; evaluation of labor per unit of intelligence product delivered
3. Defined	Focus on management of competency of the workforce	Introduction of analytic processes, training, and evaluation of personnel competency, growth Evaluation of analytic performance; accuracy of intelligence
4. Predictable	Focus on management of capabilities Workforce is empowered, and practices are measured	Standard analytic processes in place, with training and capability measurement Evaluation of effectiveness: metrics used to evaluate analysis utility to customers
5. Optimized	Focus on management of continuous change and improvement Practices are measured and improved to deliver higher value	Continuous characterization of intelligence problem environment and adaptation of mission Continuous measurement and closed-loop adaptation of analytic processes against changing mission and customer values

formal e-learning systems to be discussed in the next section is the ability to measure, capture, and track the achieved skill levels of individuals within the organization to contribute to the measurement of organizational maturity. Similarly, the CRM systems introduced in the last chapter provide a tool to measure the intelligence consumer satisfaction with delivered intelligence.

4.2.3 Learning Modes

The organizational learning process can be formal (e.g., classroom education or training) or informal (e.g., hands-on, day-to-day experience). In the following paragraphs, we describe each of these processes and their roles in organizational learning in the intelligence organization.

4.2.3.1 Informal Learning

We gain experience by informal modes of learning *on the job* alone, with mentors, team members, or while mentoring others. The methods of informal learning are as broad as the methods of exchanging knowledge introduced in the last chapter. But the essence of the learning organization is the ability to translate what has been learned into changed organizational behavior. David Garvin has identified five fundamental organizational methodologies that are essential to implementing the feedback from learning to change; all have direct application in an intelligence organization [22].

1. *Systematic problem solving.* Organizations require a clearly defined methodology for describing and solving problems, and then for implementing the solutions across the organization. Methods for acquiring and analyzing data, synthesizing hypothesis, and testing new ideas must be understood by all to permit collaborative problem solving. (These methods are described in Section 4.4 of this chapter.) The process must also allow for the communication of lessons learned and best practices developed (the intelligence tradecraft) across the organization.

2. *Experimentation.* As the external environment changes, the organization must be enabled to explore changes in the intelligence process. This is done by conducting experiments that take excursions from the normal processes to attack new problems and evaluate alternative tools and methods, data sources, or technologies. A formal policy to encourage experimentation, with the acknowledgment that some experiments will fail, allows new ideas to be tested, adapted, and adopted in the normal course of business, not as special exceptions. Experimentation can be performed within ongoing programs (e.g., use of new analytic tools by an intelligence cell) or in demonstration programs dedicated

to exploring entirely new ways of conducting analysis (e.g., the creation of a dedicated Web-based pilot project independent of normal operations and dedicated to a particular intelligence subject domain).

3. *Internal experience.* As collaborating teams solve a diversity of intelligence problems, experimenting with new sources and methods, the lessons that are learned must be exchanged and applied across the organization. This process of explicitly codifying *lessons learned* and making them widely available for others to adopt seems trivial, but in practice requires significant organizational discipline. One of the great values of communities of common practice is their informal exchange of lessons learned; organizations need such communities and must support formal methods that reach beyond these communities. Learning organizations take the time to elicit the lessons from project teams and explicitly record (index and store) them for access and application across the organization. Such databases allow users to locate teams with similar problems and lessons learned from experimentation, such as approaches that succeeded and failed, expected performance levels, and best data sources and methods.

4. *External sources of comparison.* While the lessons learned just described applied to self learning, intelligence organizations must look to external sources (in the commercial world, academia, and other cooperating intelligence organizations) to gain different perspectives and experiences not possible within their own organizations. A wide variety of methods can be employed to secure the knowledge from external perspectives, such as making acquisitions (in the business world), establishing strategic relationships, the use of consultants, establishing consortia. The process of sharing, then critically comparing qualitative and quantitative data about processes and performance across organizations (or units within a large organization), enables leaders and process owners to objectively review the relative effectiveness of alternative approaches. *Benchmarking* is the process of improving performance by continuously identifying, understanding, and adapting outstanding practices and processes found inside and outside the organization [23]. The benchmarking process is an analytic process that requires compared processes to be modeled, quantitatively measured, deeply understood, and objectively evaluated. The insight gained is an understanding of how best performance is achieved; the knowledge is then leveraged to predict the impact of improvements on overall organizational performance.

5. *Transferring knowledge.* Finally, an intelligence organization must develop the means to transfer people (tacit transfer of skills,

experience, and passion by rotation, mentoring, and integrating process teams) and processes (explicit transfer of data, information, business processes on networks) within the organization. In *Working Knowledge* [24], Davenport and Prusak point out that spontaneous, unstructured knowledge exchange (e.g., discussions at the water cooler, exchanges among informal communities of interest, and discussions at periodic knowledge fairs) is vital to an organization's success, and the organization must adopt strategies to encourage such sharing.

Notice that each of these activities contribute to moving individuals and teams around the learning spiral of Noinaka and Tageuchi (introduced in the last chapter) by encouraging discussion (socialization), explicit description (externalization), analysis and evaluation (combination), and dissemination of results (internalization).

4.2.3.2 Formal Learning

In addition to informal learning, formal modes provide the classical introduction to subject-matter knowledge. For centuries, formal learning has focused on a traditional classroom model that formalizes the roles of instructor and student and formalizes the learning process in terms of courses of study defined by a syllabus and learning completion defined by testing criteria. The advent of electronic storage and communication has introduced additional formal learning processes that allow the process to transcend space-time limitations of the traditional classroom. Throughout the 1980s and 1990s, video, communication, and networking technologies have enabled the capture, enhancement, and distribution of *canned* and *interactive* instructional material. These permit wider distribution of instructional material while enriching the instruction with student interaction (rather than passive listening to lectures). Information technologies have enabled four distinct learning modes that are defined by distinguishing both the time and space of interaction between the learner and the instructor (Figure 4.4) [25]:

1. *Residential learning (RL).* Traditional residential learning places the students and instructor in the physical classroom at the same time and place. This proximity allows direct interaction between the student and instructor and allows the instructor to tailor the material to the students.

2. *Distance learning remote (DL-remote).* Remote distance learning provides live transmission of the instruction to multiple, distributed locations. The mode effectively extends the classroom across space to reach

Time of instruction-learning

		Same	Different
Instructor-student place — **Same**		1. Residential learning	
Instructor-student place — **Different**		2. Distance learning (DL)—remote classroom	3. DL—canned 4. DL—collaboration

Synchronous learning   Asynchronous learning

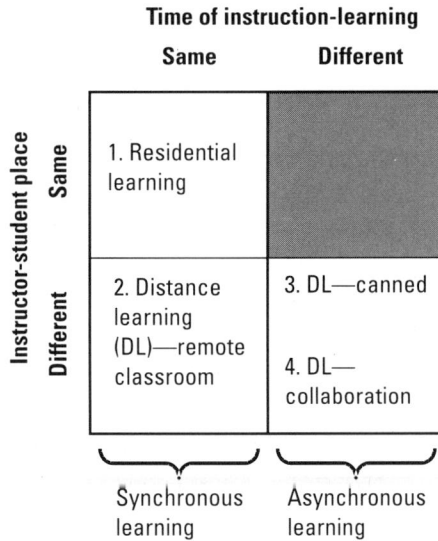

Figure 4.4   The major formal learning modes.

a wider student audience. Two-way audio and video can permit limited interaction between extended classrooms and the instructor.

While RL and DL-remote synchronize instruction and learning at the same time, the next two modes are asynchronous, allowing learning to occur at a time and place separate from the instructor's presentation.

3. *Distance learning canned (DL-canned).* This mode simply packages (or *cans*) the instruction in some media for later presentation at the student's convenience (e.g., traditional hardcopy texts, recorded audio or video, or softcopy materials on compact discs) DL-canned materials include computer-based training courseware that has built-in features to interact with the student to test comprehension, adaptively present material to meet a student's learning style, and link to supplementary materials to the Internet.

4. *Distance learning collaborative (DL-collaborative).* The collaborative mode of learning (often described as *e-learning*) integrates canned material while allowing on-line asynchronous interaction between the student and the instructor (e.g., via e-mail, chat, or videoconference). Collaboration may also occur between the student and software agents (personal *coaches*) that monitor progress, offer feedback, and recommend effective paths to on-line knowledge.

Of course, the DL modes may be combined in a course package to allow periodic synchronous instruction or *live lab* events interspersed between periods of asynchronous learning. The asynchronous mode may also include interactive simulations (e.g., analytic problem games) to develop and evaluate student skills (and measure performance). The advantages of traditional RL include the direct socialization between student and instructor to exchange tacit knowledge, as the instructor adapts to the learning style of the student. The advantage of integrated DL modes, of course, is the ability to deliver cost-effective training to a widely distributed student body that gives students the flexibility to learn at their own time, place, and pace. DL collaborative learning systems can perform preassessments of students, then personalize the lesson plan to a student's skills and styles, then perform postassessments to verify the skills mastered. This data may also be automatically registered in the corporate knowledge map of employee skills.

Intelligence organizations, as premier knowledge institutions, must apply each of these modes to provide the analytic workforce with the tools to rapidly gain the skills necessary to maintain competency in the changing world environment.

4.3 Organizational Collaboration

The knowledge-creation process of socialization occurs as communities (or teams) of people collaborate (commit to communicate, share, and diffuse knowledge) to achieve a common purpose. Collaboration is a stronger term than cooperation because participants are formed around and committed to a common purpose, and all participate in shared activity to achieve the end. If a problem is parsed into independent pieces (e.g., financial analysis, technology analysis, and political analysis), cooperation may be necessary—but not collaboration. At the heart of collaboration is intimate participation by all in the creation of the whole—not in cooperating to merely contribute individual parts to the whole. Cognitive scientists disagree over the ultimate potential of collaboration; in particular, there are divergent views on the concept of *collective intelligence*—whereby a group operates as a coherent, intelligence organism working with one mind [26]. Collaboration is widely believed to have the potential to perform a wide range of functions together:

• Coordinate tasking and workflow to meet shared goals;

• Share information, beliefs, and concepts;

• Perform cooperative problem-solving analysis and synthesis;

• Perform cooperative decision making;

- Author team reports of decisions and rationale.

This process of collaboration requires a team (two or more) of individuals that shares a common purpose, enjoys mutual respect and trust, and has an established process to allow the collaboration process to take place. Four levels (or degrees) of intelligence collaboration can be distinguished, moving toward increasing degrees of interaction and dependence among team members (Table 4.7) [27].

The process of collaboration can occur within a small team of colocated individuals assigned to a crisis team or across a broad community of intelligence planners, collection managers, analysts, and operations personnel distributed across the globe. Collaborative teams can be short lived (e.g., project and crisis teams) or long term (e.g., communities of common intellectual practice). The *means* of achieving collaboration across this wide range of teams all require the creation of a collaborative culture and the establishment of an appropriate environment and workflow (or collaborative business processes). Sociologists have studied the sequence of collaborative groups as they move from inception to decision commitment. Decision emergence theory (DET) defines four stages of collaborative decision making within an individual group: *orientation* of all members to a common perspective; *conflict,* during which alternatives are compared and competed; *emergence* of collaborative alternatives; and finally

Table 4.7
Levels of Cooperative Behavior

Level of Collaboration	Intelligence Analysis Collaboration Examples
1. Awareness	Publication (to the full organization) of knowledge inventory and map of staff competencies (expertise) in all subgroups
	Publication of sources, activities, products, and schedules in all subgroups
2. Coordination	Coordination of scheduled analysis activities, products
	Coordination of tasking against common targets
	Coordination of similar target analyses
3. Active sharing	Sharing of lessons learned
	Sharing of tasking data, analysis in-process status, and intermediate products in shared databases
	Linking of intermediate and final products data across databases
4. Joint activities	Formation of joint tasking and collection teams
	Formation of joint, virtual analytic teams that collaborate analytically and group-author intelligence products

reinforcement, when members develop consensus and commitment to the group decisions [28].

4.3.1 Collaborative Culture

First among the *means* to achieve collaboration is the creation of a collaborating culture—a culture that shares the belief that collaboration (as opposed to competition or other models) is the best approach to achieve a shared goal and that shares a commitment to collaborate to achieve organizational goals. This belief is required to place high value (as described in Section 4.1.1.) on collaboration as a necessary business process. Collaboration also requires mutual trust (versus suspicion) among team members—trust that others will handle shared data appropriately, trust that the rewards for team (rather than individual) accomplishment with be shared, and trust that differences in individuals' roles, contributions, and expertise will be respected. This trust is necessary to enable open-minded interaction among team members, rather than resistive interactions that defend the parties' mindsets [29].

The collaborative culture must also recognize that teams are heterogeneous in nature. Team members have different tacit (experience, personality style) and cognitive (reasoning style) preferences that influence their unique approach to participating in the collaborative process. The Myers-Brigg personality type indicator MBTI®, for example, is one of a number of classification schemes to distinguish different human personalities (tendencies or preferred behaviors) [30]. The Myers-Brigg approach is helpful in characterizing individuals in collaborative groups because it distinguishes four preferences that relate to approaches to problem structuring, problem solving, analysis, and decision making. The four categories (Table 4.8) each include a linear scale that rates an individual's preference between two preference extremes. The table highlights the considerations that must be included in collaborative analytic processes (and supporting groupware tools) to enable collaboration across the range of preferences of participants on collaborative teams. While the MBTI® provides general insight to distinguish style differences, cognitive psychologists point out that it is not a static description; it is important to recognize that an individual's preferences may change and adapt under varying circumstances.

The role of facilitation in collaboration (by both human team leaders and supporting groupware agents) includes:

- Recognition of team member preferences and styles;

- Understanding of the analytic, synthesis (product creation), and communication styles of all team members;

Table 4.8
Myers-Brigg Preference Areas and Influence on Intelligence Analysis

Myers-Brigg Preference Category	Alternative preference Extremes	Influence on Collaborative Intelligence Analysis
Focus of directing time and energy	*Extrovert:* external world of spoken word and interaction with others *Introvert:* inner world of thoughts and emotions	Collaboration must allow interaction between those who "think aloud" and those who "think alone" Tools must provide for the capture of introverted thinking and extroverted speaking
Approach to perceiving a situation and processing information	*Sensing:* prefer to consider facts and experience in the present time in order to be realistic *Intuition:* prefers to explore future uncertain possibilities and ideas	Collaboration between realists and idealists requires the acceptance and sharing of perspectives Shared perspectives should distinguish hard data, hypotheses, and exploratory concepts Collaborative analysis requires facilitation to coordinate perspective switching
Basis for decision making	*Thinking:* preference of logic and reasoning *Feeling:* prefer holistic consideration of personal values and feelings	Collaborative tools must be able to capture and articulate logical as well as value-based rationale for decision making and judgment Collaboration processes must facilitate hypothesis-creation discussions between value-based and context-free perspectives of evidence
Approach to life organization and interface to the outer world	*Judging:* structured, logical planning and control *Perceiving:* flexible and exploratory approach to proceed through life	Collaborative tools must provide structure (in time scheduling, resource allocation, and knowledge organization) yet allow flexibility for perceivers to contribute in unstructured ways Collaborative process structure must remain flexible to allow restructuring to adapt to out-of-sequence analysis

- Coordination of the collaboration process to balance the contributions and participation of introverts and extroverts, idealists and realists, structured planners and explorers, and between logical and holistic decision makers.

The mix of personalities within a team must be acknowledged and rules of collaborative engagement (and even groupware) must be adapted to allow each member to contribute within the constraints and strengths of their individual styles.

Collaboration facilitators may use Myers-Brigg or other categorization schemes to analyze a particular team's structure to assess the team's strengths, weaknesses and overall balance [31].

4.3.2 Collaborative Environments

Collaborative environments describe the physical, temporal, and functional setting within which organizations interact. Through the 1990s, traditional physical environments were augmented by (and in some cases replaced by) *virtual* environments, which were enabled by computation and communication to transcend the time and space limitations of physical meeting places. The term *collaborative virtual environment* (CVE) represents this new class of "spaces," broadly defined as:

> A Collaborative Virtual Environment is a computer-based, distributed, virtual space or set of places. In such places people can meet and interact with others, with [computer] agents or with virtual objects. CVE's might vary in their representational richness from 3D graphical spaces, 2.5D and 2D environments, to text-based environments. Access to CVE is by no means limited to desktop devices but might well include mobile or wearable devices, public kiosks, etc. [32].

The two time-space dimensions of human interaction (Figure 4.5) provide the fundamental framework used to distinguish four separate collaboration modes within real and virtual environments. The time of interaction (same or different times) provides the most basic distinction between modes:

- *Synchronous collaboration* occurs when participants interact at the same time (e.g., video and teleconferences, face-to-face meetings);
- *Asynchronous collaboration* occurs when participants interact with time delay, at different times (e.g., e-mail, bulletin boards).

Figure 4.5 also notes the terminology used to distinguish the traditional physical team (interacting at the same time in the same place) and the virtual

Time of collaboration

		Same	Different	
Physical team →	●	Electronic support systems for physical meetings	● E-mail ● Group workspaces sharing data (files), calendar and applications (tools) ● Bulletin boards	●— Virtual teams

Place of collaboration

Same	Different

- In-person face-to-face
- Collocated; immediate physical access
- Physical meetings

Same (place of collaboration):
- Electronic support systems for physical meetings
- ● E-mail
- ● Group workspaces sharing data (files), calendar and applications (tools)
- ● Bulletin boards

Different (place of collaboration):
- ● Teleconference
- ● Videoconference
- ● Instant messaging; on-line chat spaces
- ● On-line application sharing
- ● E-mail
- ● Group workspaces sharing data (files), calendar and applications (tools)
- ● Bulletin boards

Virtual teams:
- Considerable on-line team communication, interaction, information sharing
- Intermittent on-line access to team members
- On-line meetings

⎰ Synchronous collaboration ⎱   ⎰ Asynchronous collaboration ⎱

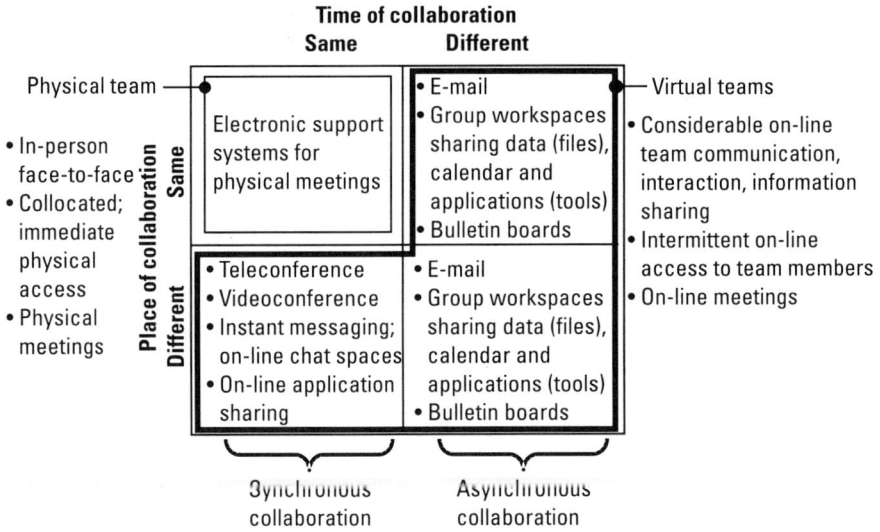

Figure 4.5 Collaboration modes.

team, which employs a virtual environment to transcend either time or space constraints.

Computer-augmented meeting support tools aid colocated physical teams by recording interactions, providing spaces to share explicit knowledge (files and processes), and supporting group analysis and the creation of products. Studies have shown that physical teams benefit from close physical interaction, which allows greater communication of subtle tacit cues (e.g., interest, concern, or surprise) and the deeper development of trust and shared experience than do virtual teams. For the colocated team, these tools may support meetings (formal gatherings) as well as capture the results of day-to-day interactions in the work area.

The virtual collaboration environment, which enables virtual teaming, seeks to provide communication and cognitive support resources to enable the group to effectively communicate across time-space, share knowledge, and collectively reason, decide, and produce (group-author) products. The broad set of on-line tools developed to perform this function, both synchronously (in real time) and asynchronously, is called *groupware*. Groupware includes the kinds of functions enumerated in Table 4.9. (The use of several of these functions was illustrated in the intelligence example in Section 3.3.) The functions and related protocol standards for data collaboration and videoconferencing are maintained by the International Telecommunications Union—Telecom Standardization Sector (ITU-T). The data collaboration functions in the table are defined under the ITU-T T.120 family of standards, while H.3xx standards apply to videoconferencing.

Table 4.9
Basic Groupware Functions

Collaboration Time Mode	Collaboration Function	Function Description
Synchronous	Text conference	One-to-one or one-to-many interactive text among multiple participants (*chat room*)
	Audio conference	One-to-one or one-to-many audio (teleconference) among selected participants (H.3.2.3 or similar protocol)
	Video conference	One-to-one or one-to-many video (videoconference) among selected participants (H.3.2.3 or similar protocol)
	Workspaces (virtual rooms)	General workspace to allow synchronous entry during discussion to show files, notes, discussion items, polls, queries, and general displays of information
	Whiteboard—graphics tool	Synchronous and concurrent creation of graphics and marking of graphics (e.g., maps or photos)
	File sharing	Synchronous sharing of files (search, retrieve, edit, save, manage configuration) and transfer using standard file transfer protocol (FTP) application protocol
	Application sharing	Synchronous and concurrent sharing of applications (e.g., simulation, spreadsheet) allowing control to be passed from user to user or with multiple user control (e.g., simulations) using T. 120 or similar application protocol
Synchronous and Asynchronous	Participation administration	Manage the invitation, access approval and removal of participants to individual meetings; audit and log sessions (content and participants); maintain a database of all collaborators
	Security	Maintain single- or multiple-level security for both access and awareness (i.e., authentication, encryption, and the ability to restrict and protect *access to* and *existence of* participants, knowledge, and virtual events
	Scheduling	Coordinate, plan, and display timing of events (synchronous meetings, milestones, asynchronous file deliveries)
	Workflow management	Display plan and status of activities toward team intermediate milestones and goals
Asynchronous	E-mail	Secure electronic storage and forwarding of mail with attachments [typically over Simple Mail Transfer Protocol (SMTP) application protocol]
	Asynchronous workspaces (virtual rooms)	General workspace to allow asynchronous entry and retrieval of files, notes, discussion items, polls, queries, mail, and team process flow, custom tailored to each user

Table 4.9 (continued)

Asynchronous	Group authoring	Tools to permit group authoring of documents, graphics, and integrated products. This includes application-based products (e.g., Microsoft Word) and web products (e.g., HTML, XML).
	Presence awareness	Display current participants on-line; allow participants to identify available collaborators to send instant messages
	Instant message	Issue pop-up messages to on-line participants to initiate ad hoc chat or conference, transitioning from asynchronous to synchronous mode [requires an instant messaging and presence protocol (IMPP)]
	Bulletin board	Asynchronous workspace to post (broadcast) notices, requests, and general information

Because any large network of diverse contributors may include users with different commercial collaboration products, interoperability between groupware applications is a paramount consideration. A U.S. interoperability study of synchronous and asynchronous collaboration between commercial products evaluated the principal combinations of two basic implementation options [27]:

- Web-based interoperation, in which all users collaborate via a standard Internet browser client. (The client may include unique Java applets, which run on the browser's Java virtual machine, or plug-in applications.)

- Application-based interoperation, in which all users collaborate via T.120/H.323 standards–based applications (i.e., application-unique software that may be required both at the browser or the server, or both).

Of course, the Web-based modes offer greater interoperability with more limited performance (due to more restrictive and limited standards) than the dedicated-standards-based products. The study highlighted the difficult trades between connectivity, flexibility, and platform diversity on one hand and security, technical performance, and collaboration policy control on the other hand. The study noted that intelligence users expect different physical settings when moving from agency to agency for physical meetings, and it is reasonable to expect (and tolerate) a degree of differences in virtual settings (e.g., tool look and feel while using different collaboration applications) when joining different collaboration environments across the community of users.

Collaborative groupware implementations offer two alternative approaches to structuring groups:

- *Centralized collaboration.* Groupware based on a central collaboration server, with distributed software clients, provides the basis for large-volume, long-term, organizationwide collaboration. Shared data is maintained on the central server, which mediates the exchange of data and services between the server and client computers.

- *Distributed collaboration.* Distributed, peer-to-peer (P2P) network collaboration maintains no central server; P2P collaboration allows the direct exchange of data and services between peer computers. P2P collaboration tools synchronize the update of shared data across all peers. Such P2P networks permit rapid creation of workspaces across different organizations and the immediate exchange of data and services with a minimum of setup [33].

The distinction between these modes is often referred to as *center-edge,* distinguishing centralized collaboration at the center of the network from P2P collaboration across enterprises at the "edges" of the network. Rapidly formed virtual intelligence teams may be formed in P2P mode and later transitioned to a centralized mode as the mission stabilizes or the team becomes permanent.

The functions in Table 4.9 allow virtual teams to form, establish collective goals and commitments around a common intelligence problem, share individual contributions of knowledge, and collaboratively work toward a solution. The collaboration functions include administrative, messaging and conferencing, and sharing function categories. Collaborative tools should allow both synchronous and asynchronous modes; allowing groups to work asynchronously in virtual spaces, then conduct synchronous virtual meetings in which objects (files, graphics, schedules, and applications such as simulations) can be opened, displayed, and modified by the group synchronously. Because the collaboration process occurs over the groupware, threaded discussions can be captured to record the flow of conversations for recall, as well as to support group authoring and production of products (e.g., alerts, updates, and analytic reports that may be electronically published to a portal).

4.3.3 Collaborative Intelligence Workflow

The U.S. IC has identified the specific kinds of collaboration that it seeks to achieve, especially between consumers and the collection, operations, and analysis disciplines:

> We aim to narrow the gap between intelligence collectors and intelligence consumers. How do we do this? One way is to assign more DI [Directorate of Intelligence] experts to policy agencies and to negotiating teams. We now have dozens of DI officers dispersed throughout the policy community, offering policymakers "one stop shopping" for intelligence analysis. We have also taken steps, but have a long way to go, to link CIA with

intelligence consumers via a direct electronic connection so that the answer to any intelligence question is just a keystroke away. On the collection side, we are forming close partnerships with our counterparts in the clandestine service—the Directorate of Operations—and with many other collectors of intelligence. Today, analysts and operations officers are increasingly working together—often side-by-side—to ensure CIA meets customer needs. We have also brought together in our 24-hour Operations Center representatives from other agencies responsible for collection. We don't want our Watch Officers to wait passively for vital information during a crisis situation; we want to enable them to get the answers they need quickly and effectively [34].

The emphasis here includes colocation of people (face-to-face teams) and the use of collaborative virtual environments to enable communication and information sharing between those who are not colocated (virtual teams). The implementation of such *links* and *partnerships* can best be illustrated by a representative example of a collaborative intelligence collection and analysis team. Consider a collaborative process within a simple but representative virtual team that includes an intelligence consumer, an all-source analytic cell, single-source analysts, and collection managers to understand the complexity of collaborative interactions in even a small intelligence operation. The flow of collaboration in such a social network (Figure 4.6) includes both synchronous and asynchronous modes of collaboration, performing the intelligence cycle (introduced in Chapter 2) as a concurrent, rather than sequential, supply chain.

4.3.3.1   The Situation

We consider how a study of illegal trafficking in women to the United States might be conducted using a virtual team. (Such an intelligence analysis has been produced by the U.S. State Department to determine the scope of these activities and their effects on U.S. interests. The resulting openly published intelligence report [35] may be compared to the following description to provide the reader additional context for the example.)

4.3.3.2   The Team

This is a *project team,* assigned to solve the particular problem of "understanding the scope of and participants in global trafficking in women to the United States" posed by the State Department consumer. The representative team includes:

- *Intelligence consumer.* The State Department personnel requesting the analysis define high-level requirements and are the ultimate customers for the intelligence product. They specify what information is needed:

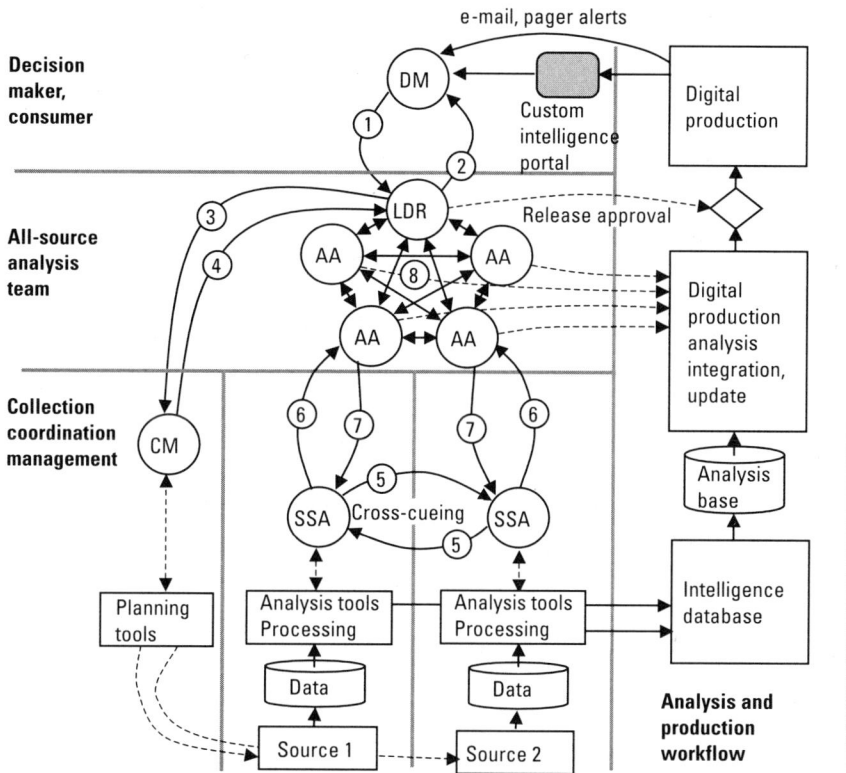

Figure 4.6 Collaborative virtual intelligence team.

the scope or breadth of coverage, the level of depth, the accuracy required, and the timeframe necessary for policy making.

- *All-source analytic cell.* The all-source analysis cell, which may be a distributed virtual team across several different organizations, has the responsibility to produce the intelligence product and certify its accuracy.

- *Single-source analysts.* Open-source and technical-source analysts (e.g., imagery, signals, or MASINT) are specialists that analyze the raw data collected as a result of special tasking; they deliver reports to the all-source team and certify the conclusions of special analysis.

- *Collection managers.* The collection managers translate all-source requests for essential information (e.g., surveillance of shipping lines, identification of organizations, or financial data) into specific collection tasks (e.g., schedules, collection parameters, and coordination between different sources). They provide the all-source team with a status of their ability to satisfy the team's requests.

4.3.3.3 The Collaboration Paths

The numbered paths in Figure 4.6 represent only the major knowledge-sharing transactions, described in the following numbered list, which illustrate the collaboration process in this example. At the right side of the figure is the supporting analysis and production system workflow, which maintains the collected base of raw data, analyzed and annotated information, and analysis results.

1. *Problem statement.* The State Department decision maker (DM), the ultimate intelligence consumer, defines the problem. Interacting with the all-source analytic leader (LDR)—and all-source analysts on the analytic team—the problem is articulated in terms of scope (e.g., area of world, focus nations, and expected depth and accuracy of estimates), needs (e.g., specific questions that must be answered and policy issues) urgency (e.g., time to first results and final products), and expected format of results (e.g., product as emergent results portal or softcopy *document*).

2. *Problem refinement.* The analytic leader (LDR) frames the problem with an explicit description of the consumer requirements and intelligence reporting needs. This description, once approved by the consumer, forms the terms of reference for the activity. The problem statement-refinement loop may be iterated as the situation changes or as intelligence reveals new issues to be studied.

3. *Information requests to collection tasking.* Based on the requirements, the analytic team decomposes the problem to deduce specific elements of information needed to model and understand the level of trafficking. (The decomposition process was described earlier in Section 2.4.) The LDR provides these intelligence data requirements to the collection manger (CM) to prepare a collection plan. This planning requires the translation of information needs to a coordinated set of data-collection tasks for humans and technical collection systems. The CM prepares a collection plan that traces planned collection data and means to the analytic team's information requirements.

4. *Collection refinement.* The collection plan is fed back to the LDR to allow the analytic team to verify the completeness and sufficiency of the plan—and to allow a review of any constraints (e.g., limits to coverage, depth, or specificity) or the availability of previously collected relevant data. The information request–collection planning and

refinement loop iterates as the situation changes and as the intelligence analysis proceeds. The value of different sources, the benefits of coordinated collection, and other factors are learned by the analytic team as the analysis proceeds, causing adjustments to the collection plan to satisfy information needs.

5. *Cross cueing.* The single-source analysts acquire data by searching existing archived data and open sources and by receiving data produced by special collections tasked by the CM. Single-source analysts perform source-unique analysis (e.g., imagery analysis; open-source foreign news report, broadcast translation, and analysis; and human report analysis) As the single-source analysts gain an understanding of the timing of event data, and the relationships between data observed across the two domains, the single-source analysts share these temporal and functional relationships. The cross-cueing collaboration includes one analyst cueing the other to search for corroborating evidence in another domain; one analyst cueing the other to a possible correlated event; or both analysts recommending tasking for the CM to coordinate a special collection to obtain time or functionally correlated data on a specific target. It is important to note that this cross-cueing collaboration, shown here at the single-source analysis level function is also performed within the all-source analysis unit (8), where more subtle cross-source relations may be identified.

6. *Single-source analysis reporting.* Single-source analysts report the interim results of analysis to the all-source team, describing the emerging picture of the trafficking networks as well as gaps in information. This path provides the all-source team with an awareness of the progress and contribution of collections, and the added value of the analysis that is delivering an emerging trafficking picture.

7. *Single-source analysis refinement.* The all-source team can provide direction for the single-source analysts to focus ("Look into that organization in greater depth"), broaden ("Check out the neighboring countries for similar patterns"), or change ("Drop the study of those shipping lines and focus on rail transport") the emphasis of analysis and collection as the team gains a greater understanding of the subject. This reporting-refinement collaboration (paths 6 and 7, respectively) precedes publication of analyzed data (e.g., annotated images, annotated foreign reports on trafficking, maps of known and suspect trafficking routes, and lists of known and suspect trafficking organizations) into the analysis base.

8. *All-source analysis collaboration.* The all-source team may allocate components of the trafficking-analysis task to individuals with areas

of subject matter specialties (e.g., topical components might include organized crime, trafficking routes, finances, and methods), but all contribute to the construction of a single picture of illegal trafficking. The team shares raw and analyzed data in the analysis base, as well as the intelligence products in progress in a collaborative workspace. The LDR approves all product components for release onto the digital production system, which places them onto the intelligence portal for the consumer.

The intelligence customer monitors the emerging results of the study on the custom portal. In the initial days, the portal is populated with an initial library of related subject matter data (e.g., open source and intelligence reports and data on illegal trafficking in general). As the analysis proceeds, analytic results are posted to the portal, leading up to a stable picture of trafficking. If urgent and significant events occur and reports are posted, the consumer may be alerted (via instant message, cell phone, or pager).

This simple example illustrates only two single-source disciplines, although many problems may require several different source specialists (e.g., IMINT, SIGINT, or MASINT). In addition, the analytic team may include a diversity of subject matter experts (e.g., the trafficking analysis team may call upon country-specific analysts, organized crime analysts, financial analysts, and experts in passport fraud).

4.4 Organizational Problem Solving

Intelligence organizations face a wide range of problems that require planning, searching, and explanation to provide solutions. These problems require reactive solution strategies to respond to emergent situations as well as opportunistic (proactive) strategies to identify potential future problems to be solved (e.g., threat assessments, indications, and warnings). The process of solving these problems collaboratively requires a defined strategy for groups to articulate a problem and then proceed to collectively develop a solution. In the context of intelligence analysis, organizational problem solving focuses on the following kinds of specific problems:

- *Planning.* Decomposing intelligence needs for data requirements, developing analysis-synthesis procedures to apply to the collected data to draw conclusions, and scheduling the coordinated collection of data to meet those requirements (as described in Chapter 2).

- *Discovery.* Searching and identifying previously unknown patterns (of objects, events, behaviors, or relationships) that reveal new understanding about intelligence targets. (The discovery reasoning approach is *inductive* in nature, creating new, previously unrevealed hypotheses.)

- *Detection.* Searching and matching evidence against previously known target hypotheses (templates). (The detection reasoning approach is *deductive* in nature, testing evidence against known hypotheses.)

- *Explanation.* Estimating (providing mathematical proof in uncertainty) and arguing (providing logical proof in uncertainty) are required to provide an explanation of evidence. Inferential strategies require the description of multiple hypotheses (explanations), the confidence in each one, and the rationale for justifying a decision. Problem-solving descriptions may include the explanation of explicit knowledge via technical portrayals (e.g., graphical representations) and tacit knowledge via narrative (e.g., dialogue and story). (The approach to reason to the best, or most likely, explanation is called *abduction;* this approach to reasoning is explained in Chapter 5.)

To perform organizational (or collaborative) problem solving in each of these areas, the individuals in the organization must share an awareness of the reasoning and solution strategies embraced by the organization. In each of these areas, organizational training, formal methodologies, and procedural templates provide a framework to guide the thinking process across a group. These methodologies also form the basis for structuring collaboration tools to guide the way teams organize shared knowledge, structure problems, and proceed from problem to solution.

Collaborative intelligence analysis is a difficult form of collaborative problem solving, where the solution often requires the analyst to overcome the efforts of a subject of study (the intelligence target) to both deny the analyst information and provide deliberately deceptive information. In the following paragraphs, we introduce three fundamental yet distinctively different approaches to framing and solving problems and making decisions: each provides a different perspective, yet all are applicable to collaborative analysis [36].

4.4.1 Critical, Structured Thinking

Critical, or structured, thinking is rooted in the development of methods of careful, structured thinking, following the legacy of the philosophers and theologians that diligently articulated their basis for reasoning from premises to conclusions. The Greek philosophers applied dialogue and logical expression to describe their critical reasoning processes. Subsequent philosophers, like Rene DeCartes, in his works *Discourse on Method* and *Rules for the Direction of the*

Mind, established foundational principles for critical reasoning [37]. Critical thinking is based on the application of a systematic method to guide the collection of evidence, reason from evidence to argument, and apply objective decision-making judgment (Table 4.10). The systematic methodology assures completeness (breadth of consideration), objectivity (freedom from bias in sources, evidence, reasoning, or judgment), consistency (repeatability over a wide range of problems), and rationality (consistency with logic). In addition, critical thinking methodology requires the explicit articulation of the reasoning process to allow review and critique by others. These common methodologies form the basis for academic research, peer review, and reporting—as well as for intelligence analysis and synthesis.

These critical thinking strategies are based in Cartesian reductionism because they presume the problem may be decomposed into a finite set of relationships and component parts, which, when solved independently, may be recomposed into a solution of the whole.

Such structured methods that move from problem to solution provide a helpful common framework for groups to communicate knowledge and coordinate a process from problem to solution. The TQM initiatives of the 1980s expanded the practice of teaching entire organizations common strategies for articulating problems and moving toward solutions. A number of general problem-solving strategies have been developed and applied to intelligence applications, for example (moving from general to specific):

- *Kepner-Tregoe™.* This general problem-solving methodology, introduced in the classic text *The Rational Manager* [38] and taught to generations of managers in seminars, has been applied to management, engineering, and intelligence-problem domains. This method carefully distinguishes problem analysis (specifying deviations from expectations, hypothesizing causes, and testing for probable causes) and decision analysis (establishing and classifying decision objectives, generating alternative decisions, and comparing consequences).

- *Multiattribute utility analysis (MAUA).* This structured approach to decision analysis quantifies a utility function, or value of all decision factors, as a weighted sum of contributing factors for each alternative decision. Relative weights of each factor sum to unity so the overall utility scale (for each decision option) ranges from 0 to 1 [39].

- *Alternative competing hypotheses (ACH).* This methodology develops and organizes alternative hypotheses to explain evidence, evaluates the evidence across multiple criteria, and provides rationale for reasoning to the best explanation [40]. (This method is described further in Chapter 6.)

Table 4.10
Basic Elements of Critical Thinking Methodology

Problem-Solving Stages	Basic Critical Thinking Methodology
1. Framing the problem	Explicitly define assumptions, methods, and objectives of inquiry
	Identify the frame of reference for thought (perspective, problem boundaries, constraints, unknowns)
	Identify all relevant bodies of knowledge; alternative views and conclusions
2. Qualifying sources and evidence	Identify the pedigree of all sources and evidence; qualify their reliability
	Verify the methods of collection of evidence
	Assess the accuracy and uncertainty in evidence
3. Reasoning from evidence to argument	Verify the soundness of logical processes (assure the absence of logical fallacies)
	Create alternative arguments with rationale
	Define the propagation of uncertainty from evidence through argumentation to conclusions
4. Judgment to evaluate arguments and make decisions	Define decision rationale and criteria
	Predict decision consequences
	Weigh implications of alternative judgments
	Decide and support with objective rationale

- *Lockwood analytic method for prediction (LAMP).* This methodology exhaustively structures and scores alternative futures hypotheses for complicated intelligence problems with many factors [41]. The process enumerates, then compares the relative likelihood of COAs for all actors (e.g., military or national leaders) and their possible outcomes. The method provides a structure to consider all COAs while attempting to minimize the exponential growth of hypotheses.

A basic problem-solving process flow (Figure 4.7), which encompasses the essence of each of these three approaches, includes five fundamental component stages:

1. *Problem assessment.* The problem must be clearly defined, and criteria for decision making must be established at the beginning. The problem, as well as boundary conditions, constraints, and the format of the desired solution, is articulated.

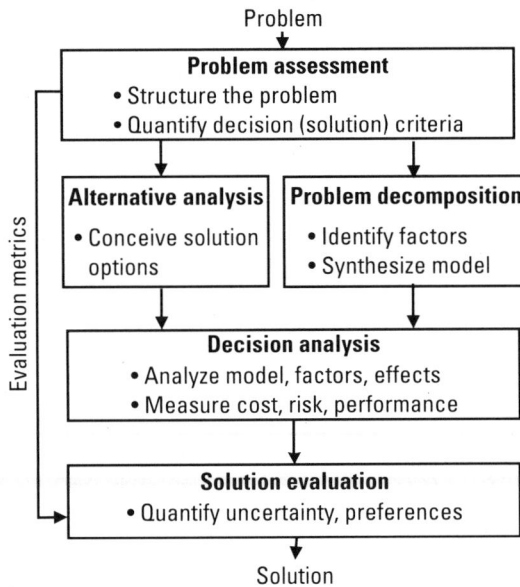

Figure 4.7 A basic problem-solving strategy flow.

2. *Problem decomposition.* The problem is broken into components by modeling the "situation" or context of the problem. If the problem is a corporate need to understand and respond to the research and development initiatives of a particular foreign company, for example, a model of that organization's financial operations, facilities, organizational structure (and research and development staffing), and products is constructed. The decomposition (or analysis) of the problem into the need for different kinds of information necessarily requires the composition (or synthesis) of the model. This models the situation of the problem and provides the basis for gathering more data to refine the problem (refine the need for data) and better understand the context.

3. *Alternative analysis.* In concert with problem decomposition, alternative solutions (hypotheses) are conceived and synthesized. Conjecture and creativity are necessary in this stage; the set of solutions are categorized to describe the range of the solution space. In the example of the problem of understanding a foreign company's research and development, these solutions must include alternative explanations of what the competitor might be doing and what business responses should be taken to respond if there is a competitive threat. The competitor analyst must explore the wide range of feasible solutions and associated constraints and variables; alternatives may range from no research and

development investment to significant but hidden investment in a new, breakthrough product development. Each solution (or explanation, in this case) must be compared to the model, and this process may cause the scope of the model to be expanded in scope, refined, and further decomposed to smaller components.

4. *Decision analysis.* In this stage the alternative solutions are applied to the model of the situation to determine the consequences of each solution. In the foreign firm example, consequences are related to both the likelihood of the hypothesis being true and the consequences of actions taken. The decision factors, defined in the first stage, are applied to evaluate the performance, effectiveness, cost, and risk associated with each solution. This stage also reveals the sensitivity of the decision factors to the situation model (and its uncertainties) and may send the analyst back to gather more information about the situation to refine the model [42].

5. *Solution evaluation.* The final stage, judgment, compares the outcome of decision analysis with the decision criteria established at the onset. Here, the uncertainties (about the problem, the model of the situation, and the effects of the alternative solutions) are considered and other subjective (tacit) factors are weighed to arrive at a solution decision.

This approach underlies the basis for traditional analytic intelligence methods, because it provides structure, rationale, and formality. But most recognize that the solid tacit knowledge of an experienced analyst provides a complementary basis—or an unspoken confidence that underlies final decisions—that is recognized but not articulated as explicitly as the quantified decision data. In the next section, we explore this more holistic approach to problem solving.

4.4.2 Systems Thinking

In contrast with the reductionism of a purely analytic approach (Table 4.11), a more holistic approach to understanding complex processes (noted in Section 4.2) acknowledges the inability to fully decompose many complex problems into a finite and complete set of linear processes and relationships. This approach, referred to as *holism,* seeks to understand high-level patterns of behavior in dynamic or complex adaptive systems that transcend complete decomposition (e.g., weather, social organizations, or large-scale economies and ecologies). Rather than being analytic, systems approaches tend to synthetic—that is, these approaches construct explanations at the aggregate or large scale and compare them to real-world systems under study [43].

Table 4.11
Complementary Problem-Solution Perspectives

	Analytic Thinking	Systems Thinking
Approach	Reductionism—decompose system into component parts and relationships; study the behaviors of components then integrate to explain large-scale system behavior	Holism—study entire system and high-level patterns of behavior without integrating component behaviors; study the basic patterns of large-scale behavior
Basis of the Approach	**Presume completeness,** certainty, and linearity; the behavioral whole is the sum of all parts (large-scale cause is sum of all small-scale causes)	**Presume complexity,** uncertainty and nonlinearity; depth of complexity of behavior exceeds known parts and relationships
	Analyze (decompose) large scale problem into small-scale components; develop independent component solutions and then recompose to an aggregate solution to explain the whole	**Synthesize** (construct) large scale patterns of behavior, assuming the inability to explicitly describe completely all causality and relationships; compare to real-world systems
	Context-free and objective study of explicitly described individual parts	**Context-rich** and subjective use of tacit experience (large-scale patterns)
Application to Intelligence Analysis	Assessment and modeling of physical systems (weapons, communications, manufacturing processes)	Assessment of sociological systems (government and military leadership, economies, populations)

The central distinctions in these approaches are in the degree of complexity of the system being studied, defined by two characteristics:

1. Number of independent causes;
2. Number of relationships between those causes.

Complexity refers the property of real-world systems that prohibits any formalism to represent or completely describe its behavior. In contrast with simple systems that may be fully described by some formalism (i.e., mathematical equations that fully describe a real-world process *to some level of satisfaction for the problem at hand*), complex systems lack a fully descriptive formalism that captures all of their properties, especially global behavior. Dynamic systems are described variously; the most common range of terms include:

- *Simple.* These are dynamic systems whose local and global behavioral properties may be fully described by formalisms and whose behavior may be predicted because the systems may be decomposed into a finite number of causes and interactions between causes.

- *Complex.* This is the general description of *irreducible* (cannot be decomposed) systems, which lack a fully descriptive formalism that captures all of its properties, especially global behavior. These systems may be linear and deterministic.

- *Chaotic.* Chaotic systems are nonlinear deterministic systems that only appear to be random and are characterized by sensitivity to initial conditions and instability.

- *Random.* Random systems are maximally complex.

Reductionist approaches have faired well in explicitly modeling simple systems down to the molecular level in classical physics (e.g., Newton's laws), in predicting planetary motion, in describing fundamental processes in chemistry, and in the engineering description and simulation of highly complicated electronic and mechanical systems. But systems of subatomic scale, human organizational systems, and large-scale economies, where very large numbers of independent causes interact in large numbers of interactive ways, are characterized by inability to model global behavior—and a frustrating inability to predict future behavior. These systems are best understood by a systems-thinking approach that focuses on the study of these high-level emergent behavior patterns, rather than attempting to decompose to infinite detail. We often speak of a subject matter expert's "broad experience," "deep insight," or "wisdom" to refer to the unspoken or unarticulated (tacit) knowledge that is applied—and often not fully justified—in the analysis of a complex problem. The expert's judgment is based not on an external and explicit decomposition of the problem, but on an internal matching of high-level patterns of prior experience with the current situation. The experienced detective as well as the experienced analyst applies such high-level comparisons of current behaviors with previous tacit (unarticulated, even unconscious) patterns gained through experience.

Of interest to the intelligence analyst is the value of holistic study of such complexity to gain insight into the evaluation of complex situations and in the study of strategic surprise. John Casti, in *Complexification: Explaining a Paradoxical World through the Science of Surprise*, has suggested that complexity science offers insight into the general causes of surprise—"when our pictures of reality depart from reality itself" [44]. The use of agent-based simulation tools to create and study emergent behaviors in support of intelligence analysis are described in Chapter 8. Such tools effectively enhance the experience of the analyst by simulating the interaction of many agents (independent but highly interactive causes) over a wide range of conditions to explore the complex space of outcomes (ranging from simple to complex to chaotic behavior). These approaches are inherently synthetic rather than analytic—instead of decomposing observed data, simulations create synthetic data (emergent patterns). These simulations help the analyst explore, discover, and recognize the patterns of complex system behavior at the high level, so they will be recognized holistically in real-world data.

It is important to recognize that analytic and systems-thinking approaches, though in contrast, are usually applied in a complementary fashion by individuals and team alike. The analytic approach provides the structure, record keeping, and method for articulating decision rationale, while the systems approach guides the framing of the problem, provides the synoptic perspective for exploring alternatives, and provides confidence in judgments.

4.4.3   Naturalistic Decision Making

It is important to recognize that humans often make decisions much less formally than by the methods just described. Indeed, in times of crisis, when time does not permit the careful methodologies, humans apply more *naturalistic* methods that, like the systems-thinking mode, rely entirely on the only basis available—prior experience.

The U.S. Commander of Operation Noble Anvil in the 1999 conflict in Kosovo commented candidly on the effects of IT applications in his stressful command and control environment, "Great technology…but needs *controls*…" Admiral James O. Ellis (CINC USN Europe) went on to note that information saturation adds to the "the fog of war" and that the demand for information will always exceed the capability to provide it, but asked the question, "… how much is *enough*?" The admiral clearly recognized the potential for critical information, but acknowledged that that his KM systems were flooding him with more information than his people could apply: "Uncontrolled, [information] will control you and your staffs … and lengthen your decision-cycle times." (Insightfully, the Admiral also noted, "You can only *manage* from your Desktop Computer … you cannot *lead* from it" [45].)

Studies of such *naturalistic* decision-making environments have sought to develop aids to cognition for both rapid, comprehensive situation awareness and decision support. Kline et al. [46] have characterized the general approaches that humans apply to such tasks:

- Decision makers *experience* the situation (holistically) and match the current situation to a repertoire of prior experience patterns—seeking *typicality* or *archetypes* previously experienced. From these matches, solutions (and past outcomes and consequences) are immediately created.

- Decision makers seek *satisfying* solutions, rather than optimizing ones; the focus of energy is placed on elaborating and refining the best immediate approach (often the first imagined option) rather than on creating a diverse set of options.

- *Mental simulation* is applied, based on the decision maker's prior experience, where the decision maker imagines or envisions the likely outcomes of actions.

- The focus of attention is placed more on assessing the situation and acting, rather than on analysis and decision making.

Of course, central to this approach is the decision maker's tacit, prior experience, which is used to pattern-match to the current situation and to mentally *simulate* possible outcomes. While long-term intelligence analysis applies the systematic, critical analytic approaches described earlier, crisis intelligence analysis may be forced to the more naturalistic methods, where tacit experience (via informal on-the-job learning, simulation, or formal learning) and confidence are critical.

4.5 Tradecraft: The Best Practices of Intelligence

The capture and sharing of best practices was developed and matured throughout the 1980s when the total quality movement institutionalized the processes of benchmarking and recording lessons learned. Two forms of best practices and lessons capture and recording are often cited:

1. *Explicit process descriptions.* The most direct approach is to model and describe the best collection, analytic, and distribution processes, their performance properties, and applications. These may be indexed, linked, and organized for subsequent reuse by a team posed with similar problems and instructors preparing formal curricula.

2. *Tacit learning histories.* The methods of storytelling, described earlier in this chapter, are also applied to develop a "jointly told" story by the team developing the best practice. Once formulated, such learning histories provide powerful tools for oral, interactive exchanges within the organization; the written form of the exchanges may be linked to the best-practice description to provide context.

While explicit best-practices databases explain the *how,* learning histories provide the context to explain the *why* of particular processes. The best practices of the U.S. IC have been termed *tradecraft.* In the early 1990s, key lessons of the analytic tradecraft were collected in technical notes used for training. *A Compendium of Analytic Tradecraft Notes* was published in 1996 as a text of analytic best

Table 4.12
An Analytic Tradecraft Taxonomy

Analytic Tradecraft Area	Representative Types of Best Practices Catalogued
Addressing national interests	Methods to view the analytic problem from the policymaker's (consumer's) perspective
	Identifying analysis *value-added* contribution
	Criteria to know policy issues and provide options—without making policy recommendations
	Test to be applied to draft reports: *so-what* and *action-support* tests
Access and credibility	Approaches to gain access to consumers for guidance, tasking, feedback
	Essential characteristics of reports that maintain credibility by identifying the source of facts, stating assumptions, and stating the basis of conclusions and outlook
Articulation of assumptions	Critical methods of articulating the basis for uncertain judgments
	Approaches to identify, study, and apply *drivers* (key variables) and *linchpins* (assumptions) upon which arguments are based
	Analytic methodology to search, identify, establish, and test drivers and linchpins
Outlook	Methodology to apply analytic judgment in developing predictive outlooks
	Approaches to construct and clearly communicate judgments
Facts and sourcing	Guidelines for identifying, distinguishing, and articulating facts, data, and direct/indirect information
	Methods to deal with complexity and deception
Analytic expertise	Measures to specify and identify analytic expertise in reports
Effective summary	Content and format guidelines unique to analytic report summaries
Implementation analysis	Cautions in addressing requested policy implementation alternatives
	Methodologies to consider alternative implementations and consequences
Conclusions and findings	Forms for reporting authoritative analytic findings
Counter intelligence	Framework for analytic awareness of deception (denial and disinformation)
	Basic analytic considerations (the target's means, opportunity, motive for deception) and warning signs in the data (gaps, contradictions, and suspicious confirmations)

practices [47]. These best practices summarize the characteristics of best analyses, providing supporting examples as appropriate. The *Tradecraft Notes* provide a taxonomy of basic analytic practices; Table 4.12 enumerates the initial 10 declassified categories, illustrating key best-practice areas within the Directorate of Intelligence. The CIA maintains a product evaluation staff to evaluate intelligence products, learn from the large range of products produced (estimates, forecasts, technical assessments, threat assessments, and warnings) and maintains the database of best practices for training and distribution to the analytic staff.

4.6　Summary

In this chapter, we have introduced the fundamental cultural qualities, in terms of virtues and disciplines that characterize the knowledge-based intelligence organization. The emphasis has necessarily been on organizational disciplines—learning, collaborating, problem solving—that provide the agility to deliver accurate and timely intelligence products in a changing environment. The virtues and disciplines require support—technology to support collaboration over time and space, to support the capture and retrieval of explicit knowledge, to enable the exchange of tacit knowledge, and to support the cognitive processes in analytic and holistic problem solving.

In subsequent chapters, we will detail these supporting technologies and their application in the KM environment, but first we examine the core of the learning organization's knowledge-creating process: the analysis of intelligence data and the synthesis of intelligence products. In the next two chapters, we describe these core knowledge-creating processes and their implications for implementation in the KM environment. In subsequent chapters, we will introduce the tools, technology, and enterprise infrastructure necessary to support the intensely human analysis-synthesis processes.

Endnotes

[1]　See, for example, Stewart, T. A., *The Wealth of Knowledge: Intellectual Capital and the 21st Century Organization*, New York: Currency-Doubleday 2002; Dixon, N. M., *Common Knowledge: How Companies Thrive by Sharing What They Know*, Boston: Harvard Business School Press, 2000; O'Dell, C., et al., *If Only We Knew What We Know: The Transfer of Internal Knowledge and Best Practice*, New York: Free Press, 1998; Garvin, D. A., *Learning in Action: A Guide to Putting the Learning Organization to Work*, Boston: Harvard Business School Press, 2000.

[2]　The DCI Strategic Intent was published in March 1999. The five major objectives of the Intent and activities in each area were published in the unclassified *Annual Report for the*

United States Intelligence Community, Washington D.C.: CIA, pp. 16–24, accessed on-line on February 28, 2002 at http://www.cia.gov/cia/publications/fy99intellrpt/dci_annual_report_99_1.thml.

[3]   Gannon, J. C., "Strategic Use of Open Source Intelligence" addressed to the Washington College of Law at American University, Washington, D.C., October 6, 2000.

[4]   Leonard-Barton, D., *Wellsprings of Knowledge,* Boston: Harvard Business School Press, 1995, pp. 24–26, 51–53.

[5]   See "Virtue Epistemology" in *Stanford Encyclopedia of Philosophy,* accessed on-line on February 18, 2001 at http://plato.stanford.edu/entries/epistemology-virtue.

[6]   Winograd, T., and F. Flores, *Understanding Computers and Cognition,* New York: Addison Wesley, 1995.

[7]   Wisdom refers to the ability to apply knowledge to select the best ends and to choose the best means to accomplish those ends.

[8]   Janis, I., *Victims of Groupthink: Psychological Study of Foreign-Policy Decisions and Fiascoes,* (2nd ed.), Boston: Houghton Mifflin, 1972.

[9]   Davenport, T. and Prusak, L., *Working Knowledge: How Organizations Manage What They Know,* Boston: Harvard Business School Press, 1998.

[10]   Quoted in *CIO Magazine,* August 15, 2001, p. 118.

[11]   Definition by the APQC, www.apqc.org.

[12]   Varon, E., "The Langley Files," *CIO,* August 1, 2000, pp. 126–130.

[13]   Karlenzig, W., "Senge on Knowledge," *Knowledge Management,* July 1999, pp. 22–23.

[14]   Brown, J. S., and P. Duguid, *The Social Life of Information,* Cambridge, MA: Harvard Business Review, 2000.

[15]   This figure is adapted from Earl, M., and I. Scott, "What Is a Chief Knowledge Officer?" *Sloan Management Review,* Winter 1999, pp. 29–38.

[16]   Denning, S., *The Springboard: How Storytelling Ignites Action in Knowledge-Era Organizations,* Boston: Butterworth Heinemann, 2001, pp. *xvii–xviii.*

[17]   The organic school complements the mechanistic school, which approaches organizations as rational models subject to quantitative analysis and scientific management.

[18]   Barth, S., "The Organic Approach to the Organization: A Conversation with KM Practitioner David Snowdon," *Knowledge Management,* October 2000, p. 24.

[19]   Senge, P., *The Fifth Discipline,* New York: Doubleday, 1990, p. 3. See also Senge's more recent book, *The Dance of Change: The Challenges to Sustaining Change in Learning Organizations,* New York: Currency-Doubleday, 1999.

[20]   Bloom, B. S., B. B. Mesia, and D. R. Krathwohl, *Taxonomy of Educational Objectives,* Volumes 1 and 2, New York: David McKay, 1964.

[21]   Curtis, B., W. Hefley, and S. Miller, *People Capability Maturity Model ® (P-CMM®), Version 2.0,* CMU/SEI –2001-MM-01, Carnegie-Mellon Software Engineering Institute, July 2001.

[22]   Garvin, D., "Building a Learning Organization," in *HBR on Knowledge Management,* Boston: HBR Press, 1998, pp. 47–80.

[23]   Definition adopted by the APQC. See "Benchmarking: Leveraging Best-Practice Strategies," APQC, 1995. Accessed on-line on March 8, 2002, at http://apqc.org/free/whitepapers/dispWhitePaper.cfm?Product ID=663.

[24]   Davenport, T. and Prusak, L., *Working Knowledge: How Organizations Manage What They Know,* Chapter 5, "Knowledge Transfer," Boston: Harvard Business School Press, 1998, pp. 88–106.

[25]   The terminology here is adopted from *Distance Learning,* Army Science Board, 1997 Summer Study Final Report, December 1997, pp. 6–7, 16.

[26]   This is a question of the degree to which groups of individuals can think *collectively* and *coherently.* The fact and efficiency of group communication, interaction, influence, and coordination is not in question; the existence or meaning of group cognition (*groupthink*) is the issue in question. See the divergent views in Smith, J. B., *Collective Intelligence in Computer-Based Collaboration,* Mahwah, NJ: Lawrence Erlbaum Assoc., 1994; and Newell, A., *Unified Theories of Cognition,* Cambridge, MA: Harvard University Press, 1990.

[27]   *Intelligence Community Collaboration, Baseline Study Final Report,* MITRE, December 1999, Section 3.2.2, accessed on-line on April 20, 2002, at http://collaboration.mitre.org/prail.

[28]   Fisher, B. A., and D. Ellis, *Small Group Communication* (3rd ed.), New York: McGraw-Hill, 1990.

[29]   By mindset, we refer to "the distillation of the intelligence analyst's cumulative factual and conceptual knowledge into a framework for making estimative judgments on a complex subject." Mindset includes a commitment to a reference viewpoint on a subject; creating a mindset is a vital and indispensable element of human reasoning but introduces a bias against contradictory evidence or competing mindsets. See Davis, J., "Combating Mind-Set," *Studies in Intelligence,* Vol. 36, No. 5, 1992.)

[30]   The Myers-Briggs Type Indicator® is a psychological instrument to characterize an individual's personality type based on a generalization of Carl Jung's (1875–1961) psychology of personality types. The MBTI® is a questionnaire trademarked and copyrighted by Consulting Psychological Press (1962) that may be used to define a Myers-Briggs type of an individual. The Keirsey temperament model (1978) is a similar classification scheme that distinguishes more subtle features of temperament (interests, orientation, values, self-image, and social roles).

[31]   See, for example, the analysis of a collaborating group in Leonard, D., and S., Straus, "Putting Your Company's Whole Brain to Work," in *HBS on Knowledge Management,* Boston: HBR Press, 1998, pp. 135–136.

[32]   Churchill, E., D. Snowdon, and A. Munro (Eds.), *Collaborative Virtual Environments,* Heidelberg, Germany: Springer Verlag, 2002.

[33]   Groove Product Backgrounder White Paper, Groove Networks, Inc., 2001, accessed on-line on February 15, 2002 at http://www.groove.net/pdf/backgrounder-product.pdf.

[34]   "Intelligence Analysis for the 21st Century," speech by John C. Gannon, (Former) Deputy Director for Intelligence, CIA, at The Fletcher School of Law and Diplomacy, Tufts University, November 18, 1997, accessed on-line on January 31, 2002, at http://www .odci.gov/cia/di/speeches/42826397.html.

[35]   Richard, A. O., *International Trafficking in Women to the United States: A Contemporary Manifestation of Slavery and Organized Crime*, U.S. Dept of State Bureau of Intelligence and Research, DCI Center for the Study of Intelligence Monograph, Washington, D.C.: CIA, November 1999.

[36]   We purposely include predominantly convergent (thinking to narrow choices to arrive at a solution), rather than divergent (exploratory or creative thinking) methodologies of thought here. Divergent creativity is required to synthesize hypotheses, and *lateral* creative thinking methods (e.g., see Edward DeBono's *Serious Creativity*, 1992) are certainly required within the processes described in this chapter.

[37]   Descartes, R., *Discourse on Method*, 1637; his four-step problem-solving method of analysis and synthesis is described in "Part II—Principal Rules of the Method."

[38]   Kepner, C. H., and B. B. Tregoe, *The Rational Manager*, Princeton, NJ: Kepner-Tregoe, 1965.

[39]   Waltz, E., and J. Llinas, *Multisensor Data Fusion*, Boston: Artech House, 1990, pp. 419-423.

[40]   Sawka, K., "Competing Hypothesis Analysis: Building Support for Analytic Findings," *Competitive Intelligence Magazine*, Vol.3, No. 3, July September 1999.

[41]   Lockwood, J., and K. Lockwood, "The Lockwood Analytic Method for Prediction (LAMP)," *Defense Intelligence Journal*, Vol. 3, No.2, Fall 1994, pp. 47–74.

[42]   Decision analysis is an entire discipline of its own, encompassing a broad set of quantitative analysis approaches to arrive at decisions. For an overview of this field, see Watson, S. R., and D., M. Buede, *Decision Synthesis*, Cambridge, England: Cambridge University Press, 1987.

[43]   Philosophers characterize a number of approaches to thinking about the world and attribute these classics to their most well-known teachers: Isaac Newton (1642–1727) introduced the concept of *mechanism* that explained causality in the physical world, René Descartes (1596–1650) introduced *reductionism* to decompose systems into parts for independent analysis, and Francis Bacon (1561–1626) introduced *empiricism,* whereby the systems of nature are observed and explanations (hypotheses) synthesized and then tested (the scientific method).

[44]   Casti, J. L., *Complexification: Explaining a Paradoxical World Through the Science of Surprise*, New York: Harper Collins, 1994, p. 3.

[45]   Ellis, J. O., CINC U.S. Naval Forces Europe, Commander Joint Task Force Noble Anvil, "A View From the Top," After Action Briefing, July 4, 1999, accessed on-line on May 25, 2002, at http://www.d-n-i.net/fcs/ppt/ellis_kosovo_aar.ppt.

[46]   Kline, G. A., Judith Orasanu, and Roberta Calderwood (eds.), *Decision Making in Action*, Norwood, NJ: Ablex, 1993.

[47]   "CIA Opens Door on the Craft of Analysis," Center For Study of Intelligence Newsletter, No. 7, Winter-Spring 1997. The CIA Directorate of Intelligence made available to the public a reprinted and revised edition of *A Compendium of Analytic Tradecraft Notes, Volume I (Notes 1–10),* which have become a standard reference within CIA for practitioners and teachers of intelligence analysis. The revised compendium contains 10 *Tradecraft Notes* issued to analysts during March–December 1995.

5

Principles of Intelligence Analysis and Synthesis

A timeless Sidney Harris cartoon depicts a long-haired professor posed in front a blackboard filled with mathematical equations by his grad student. Neatly scrawled between the board-length equation and the solution is the simple statement, "…then a miracle occurs." The skeptical professor admonishes his student, "I think you should be more explicit here in step two." In this chapter, we will explain the "step two" of the KM process: the jump from accumulated information to the creation of knowledge. At the core of all knowledge creation are the seemingly mysterious reasoning processes that proceed from the known to the assertion of entirely new knowledge about the previously unknown. For the intelligence analyst, this is the process by which *evidence* [1], that data determined to be relevant to a problem, is used to infer knowledge about a subject of investigation—the intelligence target. The process must deal with evidence that is often inadequate, undersampled in time, ambiguous, and carries questionable pedigree.

We refer to this knowledge-creating discipline as *intelligence analysis* and the practitioner as *analyst*. But analysis properly includes both the processes of analysis (breaking things down) and synthesis (building things up). In this chapter, the analytic-synthetic process of reasoning about data to produce knowledge (intelligence) is introduced from a theoretical and functional point of view. Following this chapter on the *principles* of analysis-synthesis, we will move on to discuss the *practice* with practical implementations and applications in the subsequent chapter.

5.1 The Basis of Analysis and Synthesis

The process known as *intelligence analysis* employs both the functions of analysis and synthesis to produce intelligence products. Before describing the functions of this process, we must first distinguish the fundamental problem-solving reasoning processes that underlie intelligence analysis.

The Greek geometer of the third century A.D., Pappus of Alexandria, first distinguished the two fundamental geometric problem-solving approaches by the *direction* of the reasoning processes moving between known facts and desired solutions. In his *Mathematical Collection,* Pappus described these two methods as complementary approaches to connecting known evidence (causes) to solutions (effects):

1. *Analysis* proceeds from a presumed effect (solution) backward, searching for the sequence of antecedent causes that would bring about that effect. Proceeding backward through iterations of antecedent causes and consequent effects, one continues until reaching causes that are known. An effect-to-cause sequence that leads backward to a complete set of known causes (axioms or assumptions) is a proven hypothesis.

2. *Synthesis,* on the other hand, proceeds from known antecedent causes forward toward a solution by linking them, in a construction process, to assemble a cause-effect chain that leads to the solution.

Pappus showed how both analysis and synthesis are used to solve problems by working in both directions to find a solution path (a tree or sequence of causes leading to the desired effect) that fully links a solution to known causes. In geometry, this proceeds from geometric first principles, through theorems to higher order proofs. In a criminal investigation, this leads from a body of evidence, through feasible explanations, to an assembled case. In intelligence, the process leads from intelligence data, through alternative hypotheses, to an intelligence product. Along this trajectory, the problem solver moves forward and backward iteratively seeking a path that connects the known to the solution (that which was previously unknown). Of course, Pappus focused on the decomposition and construction of causal relationships, but the process can be applied to noncausal evidential relationships. Intelligence analysis-synthesis is very interested in financial, political, economic, military, and many other evidential relationships that may not be causal, but provide understanding of the structure and behavior of human, organizational, physical, and financial entities.

Consider how this analysis-synthesis process iterates in a practical intelligence problem addressed by the U.S. government. The U.S. State Department requested an analysis of the illegal global trafficking in women and children to determine the scope of these activities and the effects on U.S. interests. The resulting State Department intelligence report [2] provides insight into how the analysis-synthesis process might proceed within the intelligence cycle introduced in Chapter 2:

- *Collection planning (analysis).* The analyst examines the preexisting evidence in intelligence and open-source reports that caused policymakers to request the study. From the evidence, the analyst identifies the categories of *targets* of the analysis: victims, trafficking organizations, cites and nations involved, and transport vessels and routes of trade to the United States.

- *Collection planning (synthesis).* Primary countries, sources and purchasers (of women and children), and hypothesized (likely) trafficking routes are identified, forming an initial flow model with scant quantitative data. The synthesized *model* includes the following initial components:

  1. A spreadsheet of source and user countries, major trafficking cities and organizations, and estimated quantities of victims;
  2. A map of likely trade flows between major countries and cities;
  3. A list of organizations (known and suspected traffickers).

- *Collection tasking.* The analyst assesses (analyzes) the gaps in knowledge and forms (synthesizes) specific request of intelligence collection against the primary target countries and organization.

- *Processing.* Received intelligence data (HUMINT and technical) are indexed and placed in a database organized for the trafficking case study. Then text reports are indexed by country, names, locations, and organizations for subsequent cross-referencing.

- *Analysis-synthesis.* As collection begins to provide reports on the primary countries of interest, the traffic routes suggest that additional countries should be investigated. The initial models of traffic flow provide coarse estimates of flow rates. As the model is refined and as it is examined, inconsistencies in the estimated levels and flow rates of trafficking also imply that additional data must be collected to refine the accuracy of the model.

- *Retasking.* New tasking is required to refine and validate the model and to explore the role of additional countries and cities suggested by the model.

- *Further analysis-synthesis.* The model is refined to include a financial model of estimated profits, and the modes of trafficking are refined to include additional methods employed: illicit adoption, domestic servants, and maid schemes. The quantitative models are completed, uncertainty and unknowns are noted, and a supporting qualitative assessment is written (synthesized) to articulate the answer to the original policy-maker query.

- *Production.* The report is produced and distributed to those who requested the study and to other interested agencies.

Descriptions of the analysis-synthesis processes can be traced from its roots in philosophy and problem solving to applications in intelligence assessments (Figure 5.1).

Philosophers distinguish between propositions as analytic or synthetic based on the direction in which they are developed. Propositions in which the predicate (conclusion) is contained within the subject are called *analytic* because the predicate can be derived directly by logical reasoning forward from the subject; the subject is said to *contain* the solution. *Synthetic* propositions on the other hand have predicates and subjects that are independent. The synthetic proposition affirms a connection between otherwise independent concepts.

Philosopher Imanuel Kant's (1724–1804) rationalist epistemology distinguished three categories of truth based on the use of an analytic–synthetic distinction (Table 5.1). The most pure knowledge, according to Kant, is contained in those analytic propositions that are based in pure logic. These propositions are built on the most fundamental logical principles necessary to reason (e.g., the principle of identity: "A is A"; the principle of noncontradiction: "x cannot be A and not A at the same time and in the same respect"; the principle of excluded middle: "x either is or is not") and their application to create propositions of truth by deductive logic. Such propositions are analytic and a priori based on the prior fundamental principles. Geometry is also based on the priors, but is synthetic because we create the spatial abstractions to represent things we see in the world. Most empirical knowledge (or claims to truth) is synthetic and *a posteriori* from propositions derived after observing a sample of observations and synthesizing a proposition that extends beyond the sample.

While the Kantian distinctions were defined to establish philosophical positions on the meaning of truth in epistemology (not discussed here), they are practically useful to illustrate the relative *strength* between analytic and synthetic propositions. The empirical scientific method applies analysis and synthesis to develop and then to test hypotheses:

- *Observation.* A phenomenon is observed and recorded as data.
- *Hypothesis creation.* Based upon a thorough study of the data, a working hypothesis is created (by the inductive analysis process or by pure inspiration) to explain the observed phenomena.

Area of study	Analysis	Synthesis
Philosophy The classification analytic and synthetic make distinguish between propositions or judgments	Analytic propositions are those in which the predicate (conclusion) is already contained within the subject. Example: "Terrorists are threats to peace and security"	Synthetic propositions have predicates and subjects that are independent. The synthetic proposition affirms a connection between otherwise independent concepts. Example: "Terrorists use random bombings to accomplish their means."
Problem solving Analysis and synthesis distinguish between processes for solving problems	Analysis is the process of breaking down (decomposing) complex concepts (problems) into constituent components in order to reveal and explain relationships between those component parts. Example: To determine the time of travel, you must determine the distance and velocity: $T=T(d,v)$	Synthesis is the process of assembling component concepts (solution components) into more complex concepts (solutions). Example: Estimate the time of travel by dividing estimated distance by estimated velocity: T=d/v
Intelligence Analysis and synthesis distinguish between processes for examining evidence to explain subjects (targets) of study	Analysis is the process of decomposing intelligence data (evidence) into constituent parts to examine relationships, and discover missing data. The focus is on the problem. Example: To determine a nation's gross national product, decompose national production into the primary components of production, then into primary industries to be studied	Synthesis is the process of assembling feasible solutions (hypotheses) from components of evidence. Example: Estimate gross national product by adding estimates of primary production sources, validated by comparison with import-export data.
The Reasoning Processes	Break down evidence into simple component causes from complex effects; move from concrete events to abstract explanations	Build up a working hypothesis by a logical assembly of evidence; create complex from simple, move from cause to effect.

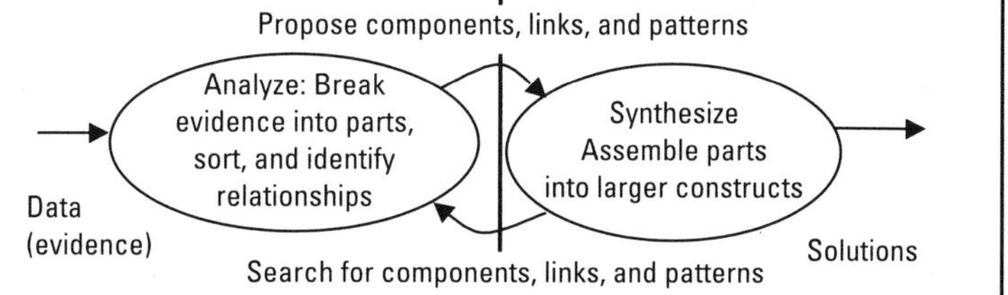

Propose components, links, and patterns

Analyze: Break evidence into parts, sort, and identify relationships

Synthesize Assemble parts into larger constructs

Data (evidence)

Search for components, links, and patterns

Solutions

Figure 5.1 Analysis-synthesis distinctions in three areas of study.

- *Experiment development.* Based on the assumed hypothesis, the expected results (the consequences) of a test of the hypothesis are synthesized (by deduction).

- *Hypothesis testing.* The experiment is performed to test the hypothesis against the data.

Table 5.1
Analytic–Synthetic Distinctions According to Kant

Category	Synthetic		Analytic
Basis of the proposition	A posteriori: knowledge obtained by experience: externally through the senses or internally by the emotion	A priori: knowledge obtained independent of experience	
Proposition category examples	Synthetic a posteriori: empirical laws—derived by induction from a sample of empirical observations	Synthetic a priori: geometry, mathematics	Analytic a priori: fundamental logic— derived by deduction from a priori propositions

- *Verification.* When the consequences of the test are confirmed, the hypothesis is verified (as a theory or law depending upon the degree of certainty).

Of course the intelligence analyst, like the scientist, applies both analysis and synthesis in a complementary fashion to study the components of evidence, assemble candidate hypotheses, test them against the evidence available, and seek additional evidence to confirm, reinforce, or eliminate hypotheses. The analyst iteratively applies analysis and synthesis to move forward from evidence and backward from hypothesis to explain the available data (evidence). In the process, the analyst identifies more data to be collected, critical missing data, and new hypotheses to be explored. This iterative analysis-synthesis process provides the necessary traceability from evidence to conclusion that will allow the results (and the rationale) to be explained with clarity and depth when completed.

But the intelligence analyst is only one contributor in a larger chain of analysis-synthesis operations, which leads to national decisions and subsequent actions. Consider the practical sequence of analysis-synthesis processes that are partitioned between intelligence, operations, and policy. The typical reasoning sequence (Figure 5.2) includes three distinct functions (often performed by three distinct organizations), each requiring an analysis-synthesis loop:

- *Intelligence analysis.* Intelligence collects and breaks down data, guided by the context of the problem , decomposing all elements of data and organizing them into temporal, spatial, and functional frames of

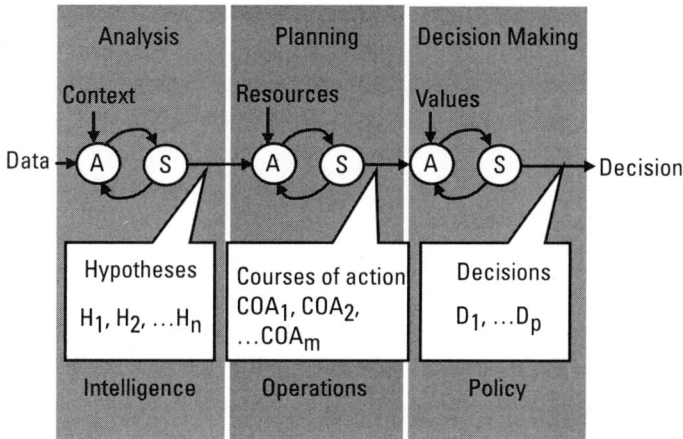

Figure 5.2 Analysis-synthesis in the national decision-making chain.

reference. From this data, hypotheses (explanations or models) are synthesized, ranked, and reported in the intelligence report.

- *Planning.* Operations accepts the intelligence report and analyzes the implications of the hypothesized situation before synthesizing (planning) feasible COAs or responses. These responses depend on the resources available.

- *Decision making.* Policy makers consider the possible COAs in the context of values (cost, risk) to determine the utility of each alternative and make decisions based on a rational selection of highest utility.

Figure 5.2 does not include the feedback loops, which naturally occur between these functions, that represent the collaborative interactions between analysts, operations personnel, and decision makers. These additional interactions between participants provide the necessary context that allows *upstream* analysts to focus their efforts to satisfy *downstream* users. The distinctions in each of these three areas of analysis are summarized in Table 5.2, but the basis of analytic-synthetic reasoning within each is the same. While depicted as a sequential chain, the three functions must collaborate to provide correct upstream insight to focus collection and analysis toward downstream decision making.

The careful distinctions between intelligence and operations or policy-making have long been an area of sensitive discussion in the U.S. government, from the inception of the IC up to today [3]. Sherman Kent, pioneer of American intelligence and author of *Strategic Intelligence for American World Policy*

Table 5.2
Fundamental Analytic Applications

Discipline	Intelligence Analysis	Operational Analysis	Decision Analysis
Analytic Focus	Understand and explain the situation and implications	Understand the range of alternative actions (operations) and their anticipated consequences	Understand the mission purpose and the aggregate implications of policy and operational actions
Analysis Process	Examine and assess evidence	Examine and assess situation hypotheses	Judge large-scale implications of COAs on mission objectives
Synthesis Process	Create hypotheses (models, explanations, situations)	Create COAs	Create policies and decisions to acive mission
Fields of Study	Intelligence analysis	Operations research	Decision and policy analysis

(1949), has been noted for his firm position that "Intelligence must be close enough to policy, plans and operations to have the greatest amount of guidance, and must not be so close that it loses its objectivity and integrity of judgement" [4]. The potential power of intelligence to influence policy was noted in a CIA report discussing intelligence provided to former Secretary of State Henry Kissinger:

> Kissinger has written perceptively of the challenge a DCI faces in walking the fine line between offering intelligence support and making policy recommendations. Probably more than any other National Security Adviser, he was sensitive to the reality that an assessment of the probable implications of any U.S. action can come across implicitly or explicitly, intended or not, as a policy recommendation. He wrote in *White House Years*, "It is to the Director [of Central Intelligence] that the assistant first turns to learn the facts in a crisis and for analysis of events, and since decisions turn on the perception of the consequences of actions the CIA assessment can almost amount to a policy recommendation [5]."

U.S. President George H.W. Bush summed up the general executive office perspective of the distinction: "And when it comes to the mission of CIA and the Intelligence Community, [Director of Central Intelligence] George Tenet has it exactly right. Give the President and the policymakers the best possible intelligence product and stay out of the policymaking or policy implementing except as specifically decreed in the law" [6].

5.2 The Reasoning Processes

Reasoning processes that analyze evidence and synthesize explanations perform *inference* (i.e., they create, manipulate, evaluate, modify, and assert belief). We can characterize the most fundamental inference processes by their *process* and *products:*

- *Process.* The *direction of the inference* process refers to the way in which beliefs are asserted. The process may move from specific (or particular) beliefs toward more general beliefs, or from general beliefs to assert more specific beliefs.
- *Products.* The certainty associated with an inference distinguishes two categories of results of inference. The asserted beliefs that result from inference may be *infallible* (e.g., an analytic conclusion is derived from infallible beliefs and infallible logic is certain) or *fallible judgments* (e.g., a synthesized judgment is asserted with a measure of uncertainty; "probably true," "true with 0.95 probability," or "more likely true than false").

The most basic taxonomy of inferential reasoning processes distinguishes three basic categories of reasoning—induction, abduction, and deduction. Table 5.3, similar to Table 5.1 with Kant's distinctions, provides the structure of this taxonomy and distinguishes the characteristics of each. The form of each method is represented as a common logical syllogism to allow each form to be seen in light of common "everyday" reasoning from premises to conclusion.

It is worth noting that while induction and deduction are the classical formal reasoning forms found in most philosophy and logic texts, abduction is the more recent pragmatic form of reasoning introduced by mathematician and logician C. S. Peirce (1839–1914). Abduction is the less formal but more common approach of inference to achieve the best explanation with uncertain evidence.

5.2.1 Deductive Reasoning

Deduction is the method of inference by which a conclusion is inferred by applying the rules of a logical system to manipulate statements of belief to form new logically consistent statements of belief. This form of inference is infallible, in that the conclusion (belief) must be as certain as the premise (belief). It is belief preserving in that conclusions reveal no more than that expressed in the original premises. Deduction can be expressed in a variety of syllogisms, including the more common forms of propositional logic (Table 5.4).

Table 5.3
The Structure of the Fundamental Inference Processes

Inference: Reasoning Processes That Create and Modify Belief				
Inferential Process	**Induction**		**Abduction:** reasoning to create the best explanation of evidence	**Deduction:** reasoning about premises to derive conclusions
	Inductive generalization: reasoning to apply belief about an observed sample to an entire population	Inductive projection: reasoning to apply a belief about an observed population to a future sample		
Syllogistic Representation	All observed A's are B's ∴ Therefore all A's are B's	All observed A's are B's ∴ Therefore the next observed A will be a B	D is a collection of data H1, H2, ... Hn explains D "best" ∴ Therefore, accepts H k	A or B or C or... but not B or C or... ∴ Therefore, A
Product: Fallibility of Asserted Beliefs	Discovery: conclusions can be stronger than premises; produces fallible knowledge			Detection: conclusions are no stronger than the premises; preserves infallible knowledge
Process: Motion of Reasoning	Moving from specific beliefs to more general beliefs		Moving from general beliefs to more specific beliefs	

From: [7]. With kind permission of Kluwer Academic Publishers.

Texts on formal logic present the variety of logical systems that may be defined to provide foundations for deductive inference [8]. The classical propositional logic system (or calculus) described in Table 5.4 is the basic deductive tool of formal logic; predicate calculus is the system of mathematics that extends deductive principles to the quantitative realm.

5.2.2   Inductive Reasoning

Induction is the method of inference by which a more general or more abstract belief is developed by observing a limited set of observations or instances.

Table 5.4
Several Basic Deductive Propositional Argument Forms

Argument	Form	Simple Example
Modus ponens	Infer by direct deduction: P→Q premise P premise ∴ Q conclusion	If an aircraft has a type 55 radar, it is a fighter Aircraft A has a type 55 radar ∴ Aircraft A is a fighter
Modus tollens	Infer by denying the consequent: P→Q premise -Q premise ∴ -P conclusion	If an aircraft has a type 55 radar, it is a fighter Aircraft A is not a fighter ∴ Aircraft A does not have a type 55 radar
Hypothetical syllogism (chain argument)	Infer by string of IF-THEN statements: P→Q premise Q→R premise R→S premise ∴ PS conclusion	If an aircraft has a type 55 radar, it is a fighter If an aircraft is a fighter it has weapons If an aircraft has weapons it is a threat ∴ If an aircraft has a type 55 radar, it is a threat
Disjunctive syllogism	Infer by denying terms of a disjunctive statement: P ∨ Q premise -Q premise ∴ P conclusion	Either aircraft A or B is a fighter Aircraft A is not a fighter ∴ Aircraft B is a fighter

Symbols used:
P→Q means if P (antecedent) is true, then Q (consequent) is true.
P ∨ Q means either P or Q are true.
-P means negation of the premise that P is not true.
∴ designates *therefore*, and is followed by the conclusion.

Induction moves from specific beliefs about instances to general beliefs about larger and future populations of instances. It is a fallible means of inference.

The form of induction most commonly applied to extend belief from a sample of instances to a larger population, is *inductive generalization:*

All Observed As are Bs
∴ all As are Bs.

By this method, analysts extend the observations about a limited number of targets (e.g., observations of the money laundering tactics of several narcotics rings within a drug cartel) to a larger target population (e.g., the entire drug cartel).

Inductive prediction extends belief from a population to a specific future sample:

All Observed As are Bs
∴ the next observed A will be a B.

By this method, an analyst may use several observations of behavior (e.g., the repeated surveillance behavior of a foreign intelligence unit) to create a general detection template to be used to detect future surveillance activities by that or other such units. The induction presumes future behavior will follow past patterns.

In addition to these forms, induction can provide a means of analogical reasoning (induction on the basis of analogy or similarity) and inference to relate cause and effect. The basic scientific method applies the principles of induction to develop hypotheses and theories that can subsequently be tested by experimentation over a larger population or over future periods of time. The subject of induction is central to the challenge of developing automated systems that generalize and learn by inducing patterns and processes (rules) [9].

The essence of induction is the recognition of a more abstract or general pattern of relationships or behaviors that explains a set of data or observations. In his study of human creativity, Arthur Koestler points out how the essence of human inductive discovery can be observed in three common forms [10]:

1. *Aha!* This is the exclamation at the point of scientific study in which the scientist rapidly discerns a new insight or principle—a discovery. The *Aha!* experience is also called "Eureka!" after the Greek engineer Archimedes's famous exclamation as he realized, while taking a bath, that he could measure the volume of the king's crown and assay its true makeup by immersing it in water—rather than melting it down. It is also the well-known exclamation, "My dear Watson!" as the veteran Sherlock Holmes discovers the clue that reveals the solution to a crime.

2. *Ha ha!* This is the response of laughter to the sudden recognition of irony as the punch line of a joke is told; the hearer realizes the alternative and parallel (but hidden) explanation for the story leading up to the line.

3. *Ahhh...* This expression is the appreciation of the higher, aesthetic beauty of a work of art by realizing the more abstract pattern of meaning not found merely in the details of the artwork, but in the holistic meaning conveyed by the external visual image.

In each of these three cases, Koestler points out that the underlying inductive act is the sudden discovery of a new or novel pattern, previously hidden in the details, to the discoverer. Koesler graphically illustrated the process in a geometric analogy. To understand his analogy, consider the *ha ha* discovery process by representing the story line of a humorous story (Figure 5.3). The sequence of evidence is the series of facts (D1, D2,...) sequentially presented in the story. These facts are projected onto an immediate or common explanation—a frame of discernment to interpret the facts—represented as the plane, A. But, as the punch line (the last piece of evidence, D5) is introduced, it suddenly reveals an entirely different plane in which all of the evidence perfectly fits—revealing a hidden but parallel and ironic explanation for the story. In the geometric analogy, a sinusoid is revealed to fit the data. The cognitive-emotive reaction to the sudden realization is to exclaim "ha ha!"

Koestler uses the term *bisociation* to describe the process of viewing multiple explanations (or multiple *associations*) of the same data simultaneously. In the example in the figure, the data can be projected onto a common plane of discernment in which the data represents a simple curved line; projected onto an

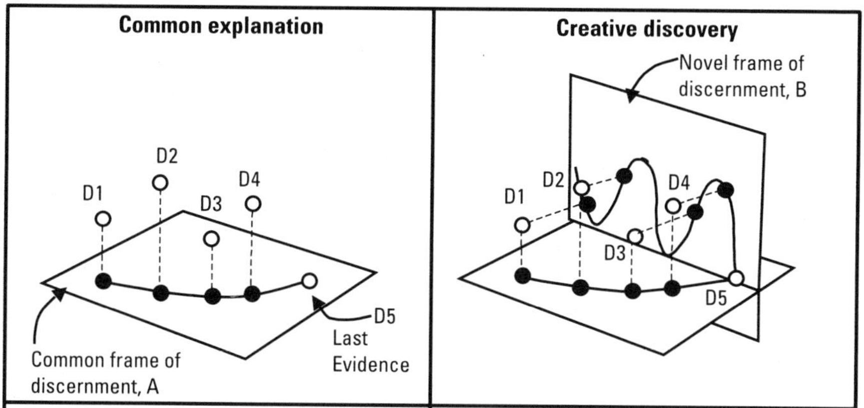

Common explanation	Creative discovery
Disciplined thinking—association of data projected onto a single frame of discernment (or associative context)	Creative thinking—bisociation of data projected onto *more than one frame* of discernment (or bisociative contexts) at the same time.
Characteristics: • Association on one frame of discernment at a time • Critical, conscious activities • Repetition—applying well-understood frames of discernment • Conservative process—straightforward analysis and synthesis	Characteristics: • Bisociation on multiple frames of discernment simultaneously • Creative, subconscious activities • Novelty—exploring and testing new frames of discernment • Constructive and destructive—deep and repetitive analysis and synthesis cycles

Figure 5.3 Koestler's graphical representation of discovery.

orthogonal plane, the data can explain a sinusoid. Though undersampled, as much intelligence data is, the sinusoid represents a new and novel explanation that may remain hidden if the analyst does not explore more than the common, immediate, or simple interpretation.

In a similar sense, the inductive discovery by an intelligence analyst (aha!) may take on many different forms, following the simple geometric metaphor. For example:

- A subtle and unique correlation between the timing of communications (by traffic analysis) and money transfers of a trading firm may lead to the discovery of an organized crime operation.

- A single anomalous measurement may reveal a pattern of denial and deception to cover the true activities at a manufacturing facility in which many points of evidence, are, in fact deceptive data "fed" by the deceiver. Only a single piece of anomalous evidence (D5 in the figure) is the clue that reveals the existence of the true operations (a new plane in the figure). The discovery of this new plane will cause the analyst to search for additional supporting evidence to support the deception hypothesis.

Each frame of discernment (or plane in Koestler's metaphor) is a framework for creating a single or a family of multiple hypotheses to explain the evidence. The creative analyst is able to entertain multiple frames of discernment, alternatively analyzing possible "fits" and constructing new explanations, exploring the many alternative explanations. This is Koestler's *constructive-destructive* process of discovery.

Koestler's work, and Thomas Kuhn's classic, *The Structure of Scientific Revolutions* [11], both attempt to understand the point of creative inspiration that enables inductive discovery. Kuhn argued that scientific discovery resulted from the crisis when anomalies (e.g., experimental results that failed to fit the accepted paradigm) challenged the belief in the current paradigm. These crises caused searches for new explanations of the anomalous phenomena and resulted in the discovery of new all-encompassing paradigms. Kuhn referred to philosopher Michael Polanyi's (then) pioneering works [12] in tacit knowledge to explain how these crises led scientists to develop shared beliefs and internal "personal knowledge" (i.e., tacit knowledge), which led them to discovery. The essence of both scientific discovery and intelligence analysis depends on the creative conception of hypotheses and the subsequent testing of those hypotheses by either or both of two means [13]:

1. Confirmation of hypotheses;

2. Disconfirmation (or falsification) [14] of hypotheses.

Collaborative intelligence analysis (like collaborative scientific discovery) may produce a healthy environment for creative induction or an unhealthy competitive environment that stifles induction and objectivity. The goal of collaborative analysis is to allow alternative hypotheses to be conceived and objectively evaluated against the available evidence and to guide the tasking for evidence to confirm or disconfirm the alternatives.

5.2.3 Abductive Reasoning

Abduction is the informal or pragmatic mode of reasoning to describe how we "reason to the best explanation" in everyday life. Abduction is the practical description of the interactive use of analysis and synthesis to arrive at a solution or explanation creating and evaluating multiple hypotheses. Abduction incorporates both inductive (hypothesis-creating) and deductive (hypothesis-testing) operations. The reasoning process is expressed as a pragmatic syllogism in the following form [15]:

D is a collection of data.

Hypotheses H_1, H_2,...H_n all can explain D.

H_k *explains D best.*

∴ Therefore Accept Hypothesis H_k as the best explanation.

Unlike infallible deduction, abduction is fallible because it is subject to errors (there may be other hypotheses not considered or another hypothesis, however unlikely, may be correct). But unlike deduction, it has the ability to extend belief *beyond the original premises.* Peirce contended that this is the logic of discovery and is a formal model of the process that scientists apply all the time.

Consider a simple intelligence example that implements the basic abductive syllogism. Data has been collected on a foreign trading company, TraderCo, which indicates its reported financial performance is not consistent with (less than) its level of operations. In addition, a number of its executives have subtle ties with organized crime figures.

The operations of the company can be explained by at least three hypotheses:

Hypothesis (H_1)—TraderCo is a legitimate but poorly run business; its board is unaware of a few executives with unhealthy business contacts.

Hypothesis (H$_2$)—TraderCo is a legitimate business with a naïve board that is unaware that several executives who gamble are using the business to pay off gambling debts to organized crime.

Hypothesis (H$_3$)—TraderCo is an organized crime *front* operation that is trading in stolen goods and laundering money through the business, which reports a loss.

Hypothesis H$_3$ best explains the evidence.

∴ Therefore, Accept Hypothesis H$_3$ as the best explanation.

Of course, the critical stage of abduction unexplained in this set of hypotheses is the judgment that H$_3$ is the best explanation. The process requires a criteria for ranking hypotheses, a method for judging which is best, and a method to assure that the set of candidate hypotheses cover all possible (or feasible) explanations.

The stages of Peirce's abductive method of scientific investigation of a process (as in the empirical scientific method introduced earlier) include abduction, induction, and deduction in the following sequence:

1. Observe the process that is not explained; collect data.
2. Apply abduction to create feasible hypotheses that are able to explain the process.
3. Apply induction to test the hypotheses in experiments.
4. Apply deduction to confirm that the selected hypothesis is able to properly predict the process and new observed data.

5.2.3.1 Creating and Testing Hypotheses

Abduction introduces the competition among multiple hypotheses, each being an attempt to explain the evidence available. These alternative hypotheses can be compared, or competed on the basis of how well they explain (or fit) the evidence. Furthermore, the created alternative hypotheses provide a means of identifying three categories of evidence important to explanation:

- *Positive evidence.* This is evidence revealing the presence of an object or occurrence of an event in a hypothesis.
- *Missing evidence.* Some hypotheses may fit the available evidence, but the hypothesis "predicts" that additional evidence that should exist if the hypothesis were true is "missing." Subsequent searches and testing for this evidence may confirm or disconfirm the hypothesis.

- *Negative evidence.* Hypotheses that contain evidence of a nonoccurrence of an event (or nonexistence of an object) may confirm a hypothesis. This is the kind of "dog that didn't bark" evidence applied by Sherlock Holmes in the short story, *Silver Blaze* [16].

This process inherently demands a search for alternative hypotheses that extend beyond the hard evidence available. The U.S. Commission on Theater Ballistic Missile Threats has noted the importance of intelligence analysis exploring hypotheses that go beyond available evidence:

> Yet, in a large number of cases examined, Commissioners found analysts unwilling to make estimates that extended beyond the hard evidence they had in hand, which effectively precluded developing and testing alternative hypothesis about the actual foreign programs taking place [17].

5.2.3.2 Hypothesis Selection

Abduction also poses the issue of defining which hypothesis provides the *best* explanation of the evidence. The criteria for comparing hypotheses, at the most fundamental level, can be based on two principle approaches established by philosophers for evaluating truth propositions about objective reality [18]. The correspondence theory of the truth of a proposition p is true is to maintain that "p corresponds to the facts." For the intelligence analyst this would equate to "hypothesis h corresponds to the evidence"—it explains *all* of the pieces of evidence, with no expected evidence missing, all without having to leave out any contradictory evidence. The coherence theory of truth says that a proposition's truth consists of its fitting into a coherent system of propositions that create the hypothesis. Both concepts contribute to practical criteria for evaluating competing hypotheses (Table 5.5).

In the next chapter, we will introduce the practical implementation of abduction in the methodology of alternative competing hypotheses (ACH). We now turn to integrating these formal and informal methods of reasoning for practical analysis-synthesis in the intelligence problem-solving environment.

5.3   The Integrated Reasoning Process

The analysis-synthesis process combines each of the fundamental modes of reasoning to accumulate, explore, decompose to fundamental elements, and then fit together evidence. The process also creates hypothesized explanations of the evidence and uses these hypotheses to search for more confirming or refuting elements of evidence to affirm or prune the hypotheses, respectively. The previous section introduced the formal descriptions of the reasoning modes; here, we

Table 5.5
Hypothesis Evaluation Criteria

Basis of Truth	Hypothesis Testing Criteria	Application to Intelligence Analysis-Synthesis Criteria
Correspondence	The hypothesis corresponds to all of the data	1. Completeness—all expected data is present (e.g., there is no missing evidence)
		2. Exclusivity—all available data matches the hypothesis; no data contradicts the hypothesis
		3. Nonconflicting—there are no mutually exclusive hypotheses that also correspond to the data
Coherence	The hypothesis coheres to (is consistent with) all propositions that make up the hypothesis	1. Consistency of logic—the hypothesis-creating system that leads from evidence, relationships (e.g., casual, organizational, or behavioral), and processes (e.g., laws of physics or rules of behavior) to predicted outcomes is logical and consistent
		2. Consistency of hypotheses—all hypotheses follow the same consistent hypothesis-creating system

describe how the fundamental inference methods are notionally integrated into the intelligence analysis-synthesis process.

We can see the paths of reasoning in a simple flow process (Figure 5.4), which proceeds from a pool of evidence and a *question* (a query to explain the evidence) posed about the evidence. This process of proceeding from an evidentiary *pool* to detections, explanations, or discovery has been called *evidence marshaling* because the process seeks to marshal (assemble and organize) into a representation (a model) that:

- Detects the presence of evidence that match previously known premises (or patterns of data);

- Explains underlying processes that gave rise to the evidence;

- Discovers new patterns in the evidence—patterns of circumstances or behaviors not known before (learning).

The figure illustrates four basic paths that can proceed from the pool of evidence, our three fundamental inference modes and a fourth feedback path:

1. *Deduction.* The path of deduction tests the evidence in the pool against previously known patterns (or templates) that represent hypotheses of activities that we seek to detect. When the evidence fits the hypothesis template, we declare a *match*. When the evidence fits multiple

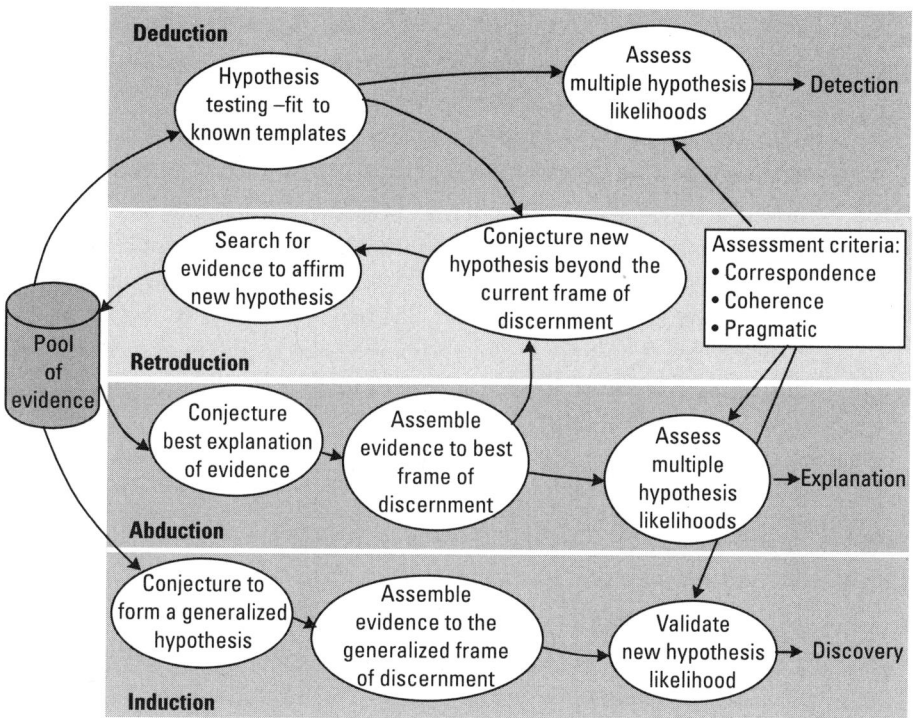

Figure 5.4 Integrating the basic reasoning flows.

hypotheses simultaneously, the likelihood of each hypothesis (determined by the *strength* of evidence for each) is assessed and reported. (This likelihood may be computed probabilistically using Bayesian methods, where evidence uncertainty is quantified as a probability and prior probabilities of the hypotheses are known.)

2. *Retroduction.* This feedback path, recognized and named by C.S. Peirce as yet another process of reasoning, occurs when the analyst conjectures (synthesizes) a new conceptual hypothesis (beyond the current frame of discernment) that causes a return to the evidence to seek evidence to match (or test) this new hypothesis. The insight Peirce provided is that in the testing of hypotheses, we are often inspired to realize new, different hypotheses that might also be tested. In the early implementation of reasoning systems, the *forward* path of deduction was often referred to as *forward chaining* by attempting to automatically fit data to previously stored hypothesis templates; the path of

retroduction was referred to as *backward chaining*, where the system searched for data to match hypotheses queried by an inspired human operator.

3. *Abduction.* The abduction process, like induction, creates explanatory hypotheses inspired by the pool evidence and then, like deduction, attempts to fit items of evidence to each hypothesis to seek the best explanation. In this process, the candidate hypotheses are refined and new hypotheses are conjectured. The process leads to comparison and ranking of the hypotheses, and ultimately the best is chosen as the explanation. As a part of the abductive process, the analyst returns to the pool of evidence to seek support for these candidate explanations; this return path is called *retroduction.*

4. *Induction.* The path of induction considers the entire pool of evidence to seek general statements (hypotheses) about the evidence. Not seeking point matches to the small sets of evidence, the inductive path conjectures new and generalized explanation of clusters of similar evidence; these generalizations may be tested across the evidence to determine the breadth of applicability before being declared as a new discovery.

Now we can examine how this process might flow in a typical intelligence process. Consider the case where a terrorist group ("ACQM") has attacked a facility of country A, and the analyst is posed with the I&W question: "Is there evidence that that the group has capabilities, plans, or operations to conduct other imminent attacks?" The flow of analytic activities (numbered 1–8) is sequentially illustrated in Figure 5.5:

1. *Deduction.* The analyst immediately checks all intelligence sources and pools the evidence about ACQM to determine if the evidence fits any known patterns of attack of other facilities. This hypothesis-testing process seeks to deduce attack capabilities, plans, or operations initiated; if deduction fails, it may be due to lack of evidence, lack of breadth of hypothesis templates (not robust enough), or insufficient templates to cover new categories of attack.

2. The analyst hypothesis tests the evidence against known patterns of attack. No matches to existing templates deduce that attacks (of known types or for known targets) are not borne out by the evidence.

3. *Retroduction.* The analyst conjectures that ACQM may be planning attacks on other targets (people, transportation, media) using the same

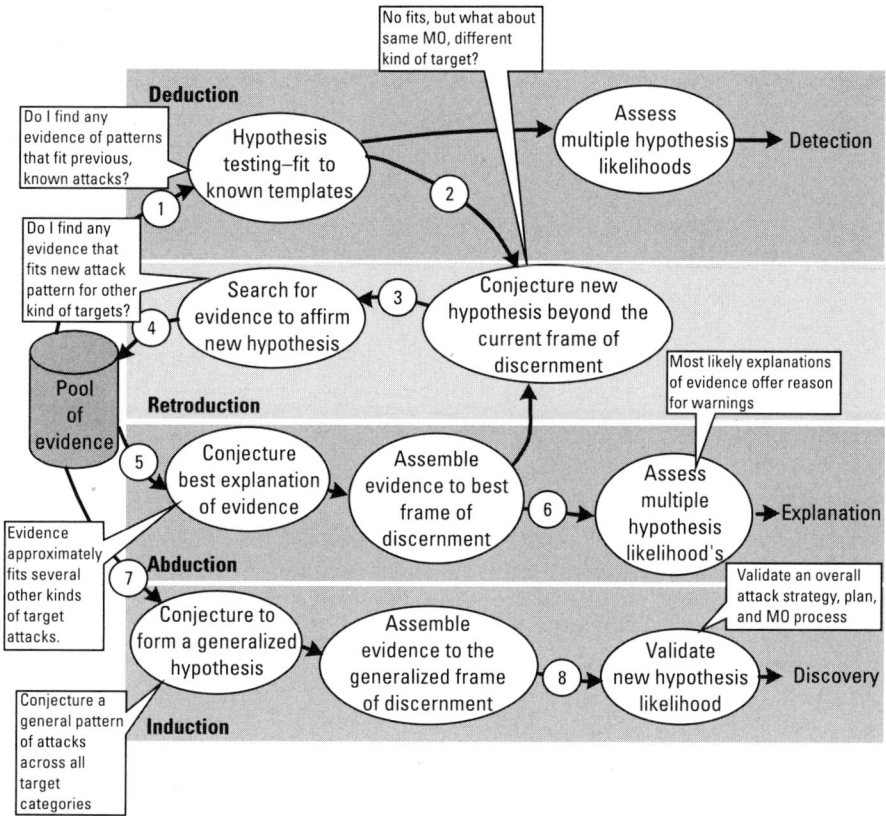

Figure 5.5 Representative intelligence problem search sequence.

modus operandi (MO). This new frame of discernment goes beyond the hypotheses that were considered within the deductive process.

4. This conjecture creates the basis for a search back through (retro) the evidence pool to explore other new patterns of attack that might target people, transportation, and media.

5. *Abduction.* Indeed, several hypotheses for new kinds of attacks on maritime transportation targets are quite feasible.

6. The evidence arrayed against these hypotheses is compared, additional collections of data are requested, and the results show that two target hypotheses (transportation around ports and rivers) are feasible and even likely. This provides a basis (the indication) for warning these categories of targets and an explanation for the warning.

7. *Induction.* Finally, as more evidence is accumulated over time, and the ACQM plans and conducts more attacks (some successes, some failures), the evidence shows a more general pattern of behavior of the group—characterized by special forms of financing, a hatred for certain cultural symbols, and special communications behaviors.

8. This generalized pattern is tested against all previous attacks and can be validated to provide a high-level template for future hypothesis testing in the deductive process.

This example illustrates one *thread* of many possible flows through the reasoning processes that analysts apply to iteratively analyze the growing pool of evidence and synthesize feasible hypotheses to be explored. The process also illustrates the validation of templates, created by induction, and their use in the deduction process. Once discovered by induction, these templates may be used for future attack detection by deduction.

5.4   Analysis and Synthesis As a Modeling Process

The fundamental reasoning processes are applied to a variety of practical analytic activities performed by the analyst.

- *Explanation and description.* Find and link related data to explain entities and events in the real world.

- *Detection.* Detect and identify the presence of entities and events based on known signatures. Detect potentially important deviations, including *anomaly detection* of changes relative to "normal" or "expected" state or *change detection* of changes or trends over time.

- *Discovery.* Detect the presence of previously unknown patterns in data (signatures) that relate to entities and events.

- *Estimation.* Estimate the current qualitative or quantitative state of an entity or event.

- *Prediction.* Anticipate future events based on detection of known indicators; extrapolate current state forward, project the effects of linear factors forward, or simulate the effects of complex factors to synthesize possible future scenarios to reveal anticipated and unanticipated (emergent) futures.

In each of these cases, we can view the analysis-synthesis process as an evidence-decomposing and model-building process. The objective of this process is to sort through and organize data (analyze) and then to assemble (synthesize), or *marshal* related evidence to create a hypothesis—an instantiated model that represents one feasible representation of the intelligence subject (target). The model is used to marshal evidence, evaluate logical argumentation, and provide a tool for explanation of how the available evidence best fits the analyst's conclusion. The model also serves to help the analyst understand what evidence is missing, what strong evidence supports the model, and where negative evidence might be expected. The terminology we use here can be clarified by the following distinctions:

- A real intelligence *target* is abstracted and represented by *models*.

- A model has descriptive and stated *attributes* or *properties*.

- A particular instance of a model, populated with evidence-derived and conjectured properties, is a *hypothesis*.

A target may be described by multiple models, each with multiple instances (hypotheses). For example, if our target is the financial condition of a designated company, we might represent the financial condition with a single financial *model* in the form of a spreadsheet that enumerates many financial *attributes*. As data is collected, the model is populated with data elements, some reported publicly and others estimated. We might maintain three instances of the model (legitimate company, faltering legitimate company, and illicit front organization), each being a competing explanation (or *hypothesis*) of the incomplete evidence. These hypotheses help guide the analyst to identify the data required to refine, affirm, or discard existing hypotheses or to create new hypotheses.

Inherent in this process is the *explicit* modeling of intelligence targets themselves, as well as multiple hypotheses regarding their description or state behavior. A collaborative intelligence analysis-synthesis process requires such explicit modeling. Tacit mental models in the minds of individual domain experts must be made explicit to be shared for collaborative analysis. Tacit mental models, exposed only as rationale for final intelligence judgments, are closed to independent scrutiny, while remaining vulnerable to errors of omission and cognitive biases of the owner. Explicit model representations provide a tool for collaborative construction, marshaling of evidence, decomposition, and critical examination. Mental and explicit modeling are complementary tools of the analyst; judgment must be applied to balance the use of both.

Former U.S. National Intelligence Officer for Warning (1994–1996) Mary McCarthy has emphasized the importance of the explicit modeling to analysis:

Rigorous analysis helps overcome mindset, keeps analysts who are immersed in a mountain of new information from raising the bar on what they would consider an alarming threat situation, and allows their minds to expand other possibilities. Keeping chronologies, maintaining databases and arraying data are not fun or glamorous. These techniques are the heavy lifting of analysis, but this is what analysts are supposed to do [19].

Though not glamorous, modeling provides the rigor that enables deeper (structured) and broader (collaborative) analysis: The model is an abstract representation that serves two functions:

1. *Model as hypothesis.* Based on partial data or conjecture alone, a model may be instantiated as a feasible proposition to be assessed, a hypothesis. In a homicide investigation, each conjecture for "who did it" is a hypothesis, and the associated model instance is a feasible explanation for "how they did it." The model provides a framework around which data is assembled, a mechanism for examining feasibility, and a basis for exploring data to confirm or refute the hypothesis. The model is often viewed as an abstract representation of an intelligence target: an organizational structure, a financial flow network, a military unit, a corporation, a trajectory of a submarine, or a computer-aided design (CAD) model of an adversary's weapon or a competitor's product.

2. *Model as explanation.* As evidence (relevant data that fits into the model) is assembled on the general model framework to form a hypothesis, different views of the model provide more robust explanations of that hypothesis. Narrative (story), timeline, organization relationships, resources, and other views may be derived from a common model. In a criminal investigation, the explanation seeks to prove the case, without a doubt—a case that is both coherent (all elements of the hypothesis are consistent with the evidence and are noncontradictory) and correspondent (all hypothesis expectations are consistent with and not contradicted by evidence from the real world).

The process of implementing data decomposition (analysis) and model construction-examination (synthesis) can be depicted in three process phases or spaces of operation (Figure 5.6):

1. *Data space.* In this space, data (relevant and irrelevant, certain and ambiguous) are indexed and accumulated. Indexing by time (of collection and arrival), source, content topic, and other factors is performed to allow subsequent search and access across many dimensions.

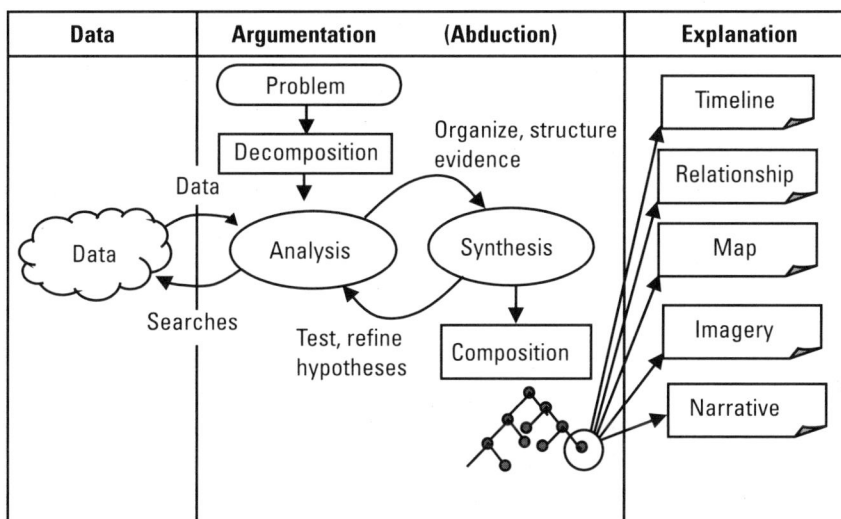

Figure 5.6 The model construction process.

2. *Argumentation space.* The data is reviewed; selected elements of potentially relevant data (evidence) are correlated, grouped, and assembled into feasible categories of explanations, forming a set (structure) of high-level hypotheses to explain the observed data. This process applies exhaustive searches of the data space, accepting some as relevant and discarding others. In this phase, patterns in the data are discovered, although all the data in the patterns may not be present; these patterns lead to the *creation* of hypotheses even though all the data may not exist. Examination of the data may lead to creation of hypotheses by conjecture, even though no data supports the hypothesis at this point. The hypotheses are examined to determine what data would be required to reinforce or reject each; hypotheses are ranked in terms of likelihood and needed data (to reinforce or refute). The models are tested and various excursions are examined. This space is the *court* in which the case is made for each hypothesis, and they are judged for completeness, sufficiency, and feasibility. This examination can lead to requests for additional data, refinements of the current hypotheses, and creation of new hypotheses.

3. Explanation space. Different "views" of the hypothesis model provide explanations that articulate the hypothesis and relate the supporting evidence. The intelligence report can include a single model and explanation that best fits the data (when data is adequate to assert the

single answer) or alternative competing models, as well as the supporting evidence for each and an assessment of the implications of each. Figure 5.6 illustrates several of the views often used: timelines of events, organization-relationship diagrams, annotated maps and imagery, and narrative story lines.

The form of the hypothesis-model is a function of the problem being addressed, and the model can have many views or perspectives of explanation. By a hypothesis or explanation, we can refer to a set of views or models that represent the single hypothesis. As in a criminal investigation, the model of a crime can be viewed from many perspectives as the evidence is fitted to a comprehensive explanation (e.g., the timeline of events, the path of the suspect on a map, or the spreadsheet of stolen assets matching evidence found in the suspect's home).

Figure 5.7 illustrates several of the common forms of models, where each may provide a different perspective on a subject of investigation: an entity, an event, a process, or a target object. Robert Clark has enumerated and explained practical analytic methods to quantify and synthesize descriptive and normative models for a wide range of intelligence applications in *Intelligence Analysis: Estimation and Prediction* [20]. Consider the range of analytic modeling activities that are required to answer the diverse questions posed by national and military intelligence consumers:

- *What is the gross domestic product of a closed foreign regime?* Economic questions regarding the gross domestic product of a closed foreign nation requires the development of a quantitative economic model, with inputs from measurement of crops, industrialized production, import-export activities, and other factors. The model provides a gross domestic product estimate and quantifies contributing factors and uncertainties.

- *What is the status of a foreign nation's weapon development?* Questions regarding the status of science and technology programs require a program schedule (timeline) model to be hypothesized, and milestones on the schedule must be evaluated against observations (e.g., weapons testing or facilities construction) [21].

- *What is the air order of battle of a foreign nation?* Order of battle questions require development of a model (the model is often a spreadsheet) that describes the force structure (organization) and enumerates the size of individual units (personnel and weapons).

- *How can a military facility target be functionally destroyed?* This targeting question requires the development of a functional model of the facility (e.g., the components that make up a radar installation, its electrical

Field or domain of analysis	Modeled intelligence target (subject)	Representative model forms (and typical visualization)	
Financial	Estimate a resource or budget Break down financial transactions and components	Spreadsheets Graphs Financial flows	
Organizational	Relationships and structure within human organizations	Organization charts Social networks Org directories	
Entity and event relationship	Entity-to-entity and entity-to-event linkages; noncausal relationships	Data linkage and network diagrams	
Event linkage, temporal sequence, operational process flows	Cause and effect dynamics, timing, sequence, and durations of events	Gantt charts Phase diagrams Flow charts Transaction sequences	
Physical objects	Physical structures, facilities, sites, and lines of communication	Annotated imagery 3D CAD models	
Geophysical areas, sites	Location of entities and events in geospatial context; cultural and contextual features	Annotated imagery, maps Geographic info system models	
Object tracking	Kinematic behavior of physical objects (e.g., aircraft, ships, and ground vehicles)	Kinematic model (e.g., alpha-beta or statistical filter) produces dynamic "track"	

Figure 5.7 Typical forms of intelligence models.

power source, and data links) and its operations (e.g., the personnel and heat and air conditioning). The targeting analysis evaluates which contributing functions or operations may be attacked to cease military functionality.

- *What are the intentions of a foreign leader?* The challenge of estimating the intentions, beliefs, and perceptions of human leaders (decision makers) is among the most difficult, yet most important, tasks posed to analysts. As noted by the U.S. Director of Central Intelligence, George Tenet:

From the mid-1960s on to the Soviet collapse, we knew roughly how many combat aircraft or warheads the Soviets had, and where. But why did they need that many or that kind? What did they plan to do with them? To this day, Intelligence is always much better at counting heads than divining what is going on inside them. That is, we are very good at gauging the size and location of militaries and weaponry. But for obvious reasons, we can never be as good at figuring out what leaders will do with them. Leadership analysis remains perhaps the most difficult of analytic specialties. Mikhail Gorbachev's rise to power in the Soviet Union—assessing his evolving thinking and policies, their implications and the chances for their success—posed huge analytical dilemmas. It is tough to divine leadership intentions in a secretive, centrally controlled society—particularly if that leadership, as was true under Gorbachev, ceases to be static. Assessing thinking *beyond* the leadership—identifying other societal forces at work and weighing their impacts, is even tougher [22].

For a single target under investigation, we may create and consider (or *entertain*) several candidate hypotheses, each with a complete set of model views. If, for example, we are trying to determine the true operations of the foreign company introduced earlier, TradeCo, we may hold several hypotheses:

1. H_1—The company is a legal clothing distributor, as advertised.
2. H_2—The company is a legal clothing distributor, but company executives are diverting business funds for personal interests.
3. H_3—The company is a front operation to cover organized crime, where hypothesis 3 has two subhypotheses:
   - H_{31}—The company is a front for drug trafficking.
   - H_{32}—The company is a front for terrorism money laundering.

In this case, H_1, H_2, H_{31}, and H_{32} are the four root hypotheses, and the analyst identifies the need to create an organizational model, an operations flow-process model, and a financial model for each of the four hypotheses—creating $4 \times 3 = 12$ models. The models help the analyst define the data needed to distinguish the hypotheses; the organization structure, financial flow, and operations behaviors combined give insight into the character of the true business. In practical application, three versions of the three basic model types are maintained, and evidence is fitted to the models to determine which hypothesis best fits the data.

5.5   Intelligence Targets in Three Domains

We have noted that intelligence targets may be objects, events, or dynamic processes—or combinations of these. The development of information operations has brought a greater emphasis on intelligence targets that exist not only in the physical domain, but in the realms of information (e.g., networked computers and information processes) and human decision making [23]. Information operations (IO) are those actions taken to affect an adversary's information and information systems, while defending one's own information and information systems [24]. The U.S. Joint Vision 2020 describes the Joint Chiefs of Staff view of the ultimate purpose of IO as "to facilitate and protect U.S. decision-making processes, and in a conflict, degrade those of an adversary" [25]. The JV2020 builds on the earlier JV2010 [26] and retains the fundamental operational concepts, two with significant refinements that emphasize IO. The first is the expansion of the vision to encompass the full range of operations (nontraditional, asymmetric, unconventional ops), while retaining warfighting as the primary focus. The second refinement moves information superiority concepts beyond technology solutions that deliver information to the concept of superiority in decision making. This means that IO will deliver increased information at all levels and increased choices for commanders. Conversely, it will also reduce information to adversary commanders and diminish their decision options. Core to these concepts and challenges is the notion that IO uniquely requires the coordination of intelligence, targeting, and security in three fundamental realms, or domains of human activities [27]. These are likewise the three fundamental domains of intelligence targets, and each must be modeled:

1. The *physical* domain encompasses the material world of mass and energy. Military facilities, vehicles, aircraft, and personnel make up the principal target objects of this domain. The *orders of battle* that measure military strength, for example, are determined by enumerating objects of the physical world.

2. The abstract *symbolic* domain is the realm of information. Words, numbers, and graphics all encode and represent the physical world, storing and transmitting it in electronic formats, such as radio and TV signals, the Internet, and newsprint. This is the domain that is expanding at unprecedented rates, as global ideas, communications, and descriptions of the world are being represented in this domain. The domain includes the *cyberspace* that has become the principal means by which humans shape their perception of the world. It interfaces the physical to the cognitive domains.

3. The *cognitive* domain is the realm of human thought. This is the ultimate locus of all information flows. The individual and collective thoughts of government leaders and populations at large form this realm. Perceptions, conceptions, mental models, and decisions are formed in this cognitive realm. This is the ultimate target of our

adversaries: the realm where uncertainties, fears, panic, and terror can coerce and influence our behavior.

These are not three arbitrary domains; even early philosophers have recognized them as the basic components of our knowledge. Aristotle, an empiricist philosopher, identified these three domains in his *Metaphysics,* written in 350 B.C. He distinguished physical objects and the abstractions (ideas) that the mind creates once the senses perceive the object. He further distinguished the words that the mind creates to symbolize or represent the abstractions of the mind. He described three processes of the intellect that manipulate these domains:

1. *Apprehension* is the process by which the mind perceives and understands the sensed physical object and creates a mental abstraction. (Physical-to-cognitive object mappings are formed.)

2. *Predication* is the process of making declarations or propositions about the object—characterizing the object and its behavior. (Cognitive-to-symbolic mappings are created.)

3. *Reasoning* is the process, then, of applying logical principles to the propositions to create new conclusions, or syllogisms. Here, Aristotle recognized the methods of deduction and induction. (Symbolic logic draws new conclusions about cognitive and physical objects.)

More recently, C.S. Peirce developed a mathematical theory of signs, or semiotics, that also embraces the three fundamental domains [28]. More explicitly than Aristotle, Peirce's logic distinguished a *triad* of relationships between the physical object, the symbolic sign that represents it, and the cognitive thought in the mind:

> Indeed, representation necessarily involves a genuine triad. For it involves a sign, or representamen, of some kind, inward or outward, mediating between an object and an interpreting thought [29].

The primary emphasis of military intelligence analysis to date has focused on the physical domain—physical military targets (aircraft, ships, ground vehicles, and personnel) and physical situations (the positions and courses of action of the physical targets.) Intelligence support to IO has emphasized the need to recognize that there also exist targets, target states, observable phenomena, and feasible detection and tracking methods in the symbolic and cognitive realms as well (Figure 5.8). It is these kinds of targets that are the focus of interest in the IO disciplines of computer network attack/defense (CNA/CND) and the

Domain	Physical	Symbolic	Cognitive
Target Objects	• Vehicles • Facilities	• Packets • Sessions	• Mental states • Ideas
Phenomena Domain	• Laws of Physics	• Network routing, stack, and protocols	• Human cognition
Sensors	• EO.IR, SAR, spectral sensors	• SIGINT, NETINT, intrusion sensors	• No direct sensors
States, Features, Complexities	• Physical components • Sensor perspective • Target articulations • Environmental signature variance	• Data components • ISO layer of target and form • Target transformation • Net environment signature variance • Data pattern	• Mental components • Mental states • Cultural, cognitive biases • Behavioral phenomena
Detection Methods	• Signature pattern matching; model matching	matching; model based matching	• Behavior pattern matching; cognitive model matching

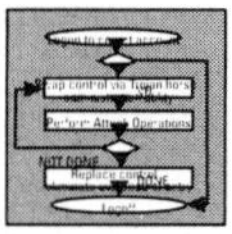

Figure 5.8 Representative targets, states, and observable phenomena in three domains.

perception management disciplines of psychological operations (PSYOPS)/ deception, respectively.

Current IO concepts have appropriately emphasized the targeting of the second domain—especially electronic information systems and their information content. The expansion of networked information systems and the reliance on those systems has focused attention on network-centric forms of warfare. Ultimately, though, IO must move toward a focus on the full integration of the cognitive realm with the physical and symbolic realms to target the human mind [30]. Recent studies within the DoD are moving toward this focus [31]. U.S. Joint Doctrine for Information Operations cites Liddell Hart's 1944 insightful assertion that: "The real target in war is the mind of the enemy commander, not the bodies of his troops" [32]. Yesterday's emphasis on physical military operations is giving way to today's emphasis on operations in the information realm. Future operations will target all three realms in an integrated fashion. PSYOPS and military deception operations have always targeted the minds of foreign populations and military units, respectively, but the disciplines have not yet achieved full integration with military operations, let alone preeminence. These disciplines, once fully integrated, will allow precision cognitive operations. IO operational concepts that target the human mind and its supporting information systems uniquely refocus the need for intelligence to model the other two domains beyond the physical: electronic information systems and the minds of decision makers [33]. Intelligence must understand and model the complete system or complex of the targets of IO: the interrelated systems of physical behavior, information perceived and exchanged, and the perception and mental states of decision makers.

Of importance to the intelligence analyst is the clear recognition that most intelligence targets exist *in all three domains,* and models must consider all three aspects. A terrorist organization, for example, includes:

1. Leaders and actors who perceive, believe, intend, plan, and decide in the cognitive domain;
2. Terrorist cells that communicate messages and finances in the symbolic domain of information;
3. Facilities, people, means of transport, weapons, and materials that exist and move within the physical domain.

The intelligence model of such an organization must include linked models of all three domains—to provide an understanding of how the organization perceives, decides, and communicates through a networked organization, as well as where the people and other physical objects are moving in the physical world. The concepts of detection, identification, and dynamic tracking of intelligence targets apply to objects, events, and processes in all three domains.

5.6 Summary

Intelligence analysis and synthesis is inherently an evidence decomposition and hypothesis assembly (or model-construction) process, where the model provides the framework around which evidence is marshaled. This framework forms the basis for structuring alternative hypotheses and supporting arguments to provide answers to the questions of intelligence consumers. In this chapter, we have developed the basic concepts of reasoning and approaches to explicitly model intelligence topics and targets, as well as the hypotheses regarding their description or behavior. A collaborative intelligence analysis-synthesis process requires such explicit modeling (versus unshared mental models—tacit target representations locked in the minds of an individual domain expert analysts). We have shown that the analysis-synthesis process proceeds from intelligence analysis to operations analysis and then to policy analysis. The knowledge-based intelligence enterprise requires the capture and explicit representation of such models to permit collaboration among these three disciplines to achieve the greatest effectiveness and sharing of intellectual capital.

In the next chapters, we consider the practical implementation of these principles in the knowledge-based intelligence organization.

Endnotes

[1] In this text, we use the terms *data* (a scientific term generally used to refer to quantitative items) and *evidence* (a legal term generally used to refer to qualitative facts used to substantiate a case) to refer to all forms of known facts, raw measurements, observations, or reports that provide the basis for analysis-synthesis. Evidence refers to data that is relevant to a problem at hand.

[2] Richard, A. O., *International Trafficking in Women to the United States: A Contemporary Manifestation of Slavery and Organized Crime*, U.S. Dept of State Bureau of Intelligence and Research, DCI Center for the Study of Intelligence Monograph, Washington, D.C.: CIA, November 1999.

[3] See, for example, Dulles, A., *The Craft of Intelligence*, New York: Harper and Row, 1963; Dulles, Alan, "The Role of Intelligence in Policy Making," Harvard Law School Forum (audio), December 13, 1963; Davis, J., "The Challenge of Managing Uncertainty: Paul Wolfowitz on Intelligence-Policy Relations," *CIA Studies in Intelligence,* Vol. 39, No. 5, 1996; and Ford, Harold P., *CIA and the Vietnam Policymakers: Three Episodes 1962-1968*, CIA Center for Studies in Intelligence, 1997.

[4] Cited in Davis, J., "The Kent-Kendall Debate of 1949," *CIA Studies in Intelligence,* Vol. 36, No. 5, 1992, p. 93.

[5] Helgerson, J. L., *CIA Briefings of Presidential Candidates*, Center for the Study of Intelligence, May 22, 1996. Quoted in Chapter 7, "Concluding Observations," section entitled "Keeping out of Politics."

[6] Bush, G. H. W., "Remarks at the Dedication Ceremony for the George Bush," Center for Intelligence, April 26, 1999.

[7] Figure adapted from Waltz, E., "Fundamentals of Reasoning and Multisensing," Figure 5, page 41, in Hyder, Shabazian, and E. Waltz (eds.), "Multisensor Data Fusion," Dordrecht, the Netherlands: Kluwer, 2002.

[8] Barwise, J., and J. Etchemendy, *Language, Proof, and Logic*, New York: Seven Bridges Press, 2000.

[9] Holland, J. H., et al., *Induction: Processes of Inference, Learning and Discovery,* Cambridge, MA: MIT Press, 1986.

[10] Koestler, A., *The Act of Creation*, New York: Macmillan, 1964, pp. 105–109. Also see his more recent work: Koestler, A., *Janus*, Chapters 6 and 7, New York: Random House, 1978.

[11] Kuhn, T. S., *The Structure of Scientific Revolutions*, (3rd ed.), Chicago: University of Chicago Press, 1996.

[12] Polanyi, M., *The Tacit Dimension*, Garden City, NY: Doubleday, 1966.

[13]   In practical intelligence analysis, efforts to both confirm and disconfirm are important methods of comparing, evaluating, and selecting hypotheses. It is important to note, however, that the philosophy of science has hotly debated these methods, with some viewing either one or the other as valid means of obtaining objective knowledge, but not both.

[14]   Philosopher Karl Popper (1902–1994) applied the term *falsification* to the process of gaining certain knowledge by disconfirming conjectures. Popper rejected the traditional logic of induction and confirmation as the basis for scientific discovery, asserting that certain knowledge is gained only through falsification.

[15]   Josephson, J. R., and S. G. Josephson (eds.), *Abductive Inference*, Cambridge, England: Cambridge University Press, 1996.

[16]   In the "Silver Blaze" episode, Sherlock Holmes realizes that the criminal was, in fact, the owner of the farm because no one heard the dog bark during the commitment of the crime. The nonoccurrence of the barking revealed the owner to be the only person that could be present at night without causing the dog to bark.

[17]   "Commission to Assess the Ballistic Missile Threat to the United States," Side Letter to the Rumsfeld Report, March 18, 1999. This unclassified letter was prepared subsequent to the 1998 formal report to specifically articulate the commission's concerns about intelligence analysis processes.

[18]   The coherence and correspondence theories of truth in epistemology are competing approaches to objective truth; both hold valuable insights into basic principles for evaluating intelligence data. Here we apply the basic principles for illustration in practical analysis but do not intend to apply the deeper philosophical implications of each theory.

[19]   McCarthy, M., "The Mission to Warn: Disaster Looms," *Defense Intelligence Journal*, Vol. 7, No. 2, 1998, p. 21.

[20]   Clark, R. M., *Intelligence Analysis: Estimation and Prediction*, Baltimore: American Literary Press, 1996.

[21]   See, for example, the *Executive Report of the Commission to Assess the Ballistic Missile Threat to the United States*, July 1998, accessed on-line on May 25, 2002, at http://www.fas.org/irp/threat/bm-threat.htm; see also *Foreign Missile Developments and the Ballistic Missile Threat to the United States Through 2015*, U.S. National Intelligence Council, September 1999.

[22]   Remarks of DCI George J. Tenet, "Opening Remarks," The Conference on CIA's Analysis of the Soviet Union, 1947–1991, Princeton University, March 8, 2001, accessed on-line on October 30, 2001, at http://www.odci.gov/cia/public_affairs/speeches/dci_speech_03082001.html. Also see Alberts, D. S. (et al.), *Understanding Information Age Warfare*, Washington, D.C.: CCRP, September 2001, accessed on-line on October 30, 2002, at http://www.dodccrp.org/Publications/pdf/UIAW.pdf.

[23]   This section is adapted from the author's paper: Waltz, E.," Data Fusion in Offensive and Defensive Information Operations," *Proc. of National Symposium of Sensor and Data Fusion*, San Antonio, TX, June 2000, Vol. 1, pp. 219–232.© Veridian, used by permission.

[24]   IO definition from DoD Joint Publication JP 3-13.

[25]   U.S. DoD Joint Chiefs of Staff J-5, *Joint Vision 2020,* Washington, D.C.: Government Printing Office, May 24, 2000, p. 28.

[26]   U.S. DoD Joint Chiefs of Staff, *Joint Vision 2010,* Washington, D.C.: Government Printing Office, 1977.

[27]   This concept of describing intelligence targets in three domains of reality was first introduced in Waltz, E., *Information Warfare Principles and Operations,* Norwood MA: Artech House, 1998, see Sections 1.2 and 5.2. For a more thorough discussion of this concept, see Waltz et al., "The Critical Role of Cognitive Models in Next Generation Intelligence Architectures," in *Proc. of 8th Annual AIPA Symp.,* Washington D.C., March 23–24, 1998. Also see Alberts, D. S. (et al.), *Understanding Information Age Warfare,* Washington, D.C.: CCRP, September 2001, accessed on-line on October 30, 2002 at http://www .dodccrp.org/Publications/pdf/UIAW.pdf.

[28]   The development of concepts of semiotics applied more generally to linguistics, and human interpretation is attributed to Peirce's contemporary, Swiss linguist Ferdinand de Saussiere (1857–1913). These works are applicable to the problems of perception management by the use of signs (symbolic objects) to influence thought (cognitive objects).

[29]   Peirce, C. S., *C.P. 1-480—The Logic of Mathematics,* 1896. In a manuscript a year later, Peirce further developed this triad, calling the cognitive object *the interpretant:* "A sign, or representamen, is something which stands to somebody for something in some respect or capacity. It addresses somebody, that is, creates in the mind of that person an equivalent sign or perhaps a more developed sign. That sign which it creates I call the interpretant of the first sign." Peirce, C. S., *C.P. 2-228—Division of Signs,* v. 1897.

[30]   The term *target* is used throughout this chapter to refer to the object of attention, rather than a specific target of offensive attack (though this is one function of IO). The mind of an individual or a group is targeted as an object to be understood, modeled, and explained by intelligence so actions can be taken. Actions to induce or coerce the mind of an individual or audience (group) also target the mind. For an early examination of this issue, see Szfranski, R., "Neocortical Warfare? The Acme of Skill," *Military Review,* November 1994, pp. 41–55. The article is also available in Arquilla, J., and D. Ronfeldt (eds.), *In Athena's Camp: Preparing for Conflict in the Information Age,* Santa Monica, CA: RAND, 1997.

[31]   Since Operation Allied Force, the U.S. DoD has considered refining the broad definition of IO to focus more narrowly on this cognitive aspect of IO, including perception management. See Verton, Dan, "DoD Redefining Information Operations," *Federal Computer Week,* May 29, 2000.

[32]   Quotation by Captain Sir B. H. Liddell Hart in *Thoughts on War* (1944), cited in *Joint Doctrine for Information Operations,* Joint Pub 3-13, October, 9, 1998, p. II-4. It is interesting to note that Liddell Hart observed that Sun Tzu had noted the same concept.

[33]   It should be noted that both domains could be considered to be *metaphysical,* though classical philosophers would likely object. Both the cognitive domain and the symbolic (entirely a product of human cognition, though represented in physical phenomena) are abstract in nature, transcend physical science, and concern the mind.

6

The Practice of Intelligence Analysis and Synthesis

Intelligence operations ranging in scale from small private-sector, competitive intelligence cells to large national intelligence organizations must implement similar process flows and address similar implementation considerations to integrate analysts with intelligence processes and tools. While the last chapter introduced the theoretical aspects of analysis and synthesis, this chapter addresses the practical implementation considerations unique to intelligence organizations. The chapter moves from high-level functional flow models toward the processes implemented by analysts. In Chapter 7, we will describe the detailed functional interactions between analysts and their automated KM systems.

While the last chapter dealt with intelligence analysis-synthesis from the perspective of rational and logical reasoning processes, here we describe the process from the perspective of the intelligence consumer and the implementers of enterprises of people, processes, and technologies to conduct analysis-synthesis. A practical description of the process by one author summarizes the perspective of the intelligence user:

> A typical intelligence production consists of all or part of three main elements: descriptions of the situation or event with an eye to identifying its essential characteristics; explanation of the causes of a development as well as its significance and implications; and the prediction of future developments. Each element contains one or both of these components: data, provided by knowledge and incoming information and assessment, or judgment, which attempts to fill the gaps in the data [1].

Consumers expect description, explanation, and prediction; as we saw in the last chapter, the process that delivers such intelligence is based on evidence (data), assessment (analysis-synthesis), and judgment (decision). We now describe the specific expectations of consumers and the practical implementation of solutions by analysts.

6.1   Intelligence Consumer Expectations

Several U.S. government reports have articulated the specific expectations of policymakers from analysis. In this section we cite two recent reports that describe the specific standards of methodology that decision makers expect from the analysis-synthesis process, in their own words.

The U.S. Government Accounting Office (GAO) noted the need for greater clarity in the intelligence delivered in U.S. national intelligence estimates (NIEs) in a 1996 report, enumerating five specific standards for analysis, from the perspective of policymakers.

Based on a synthesis of the published views of current and former senior intelligence officials, the reports of three independent commissions, and a CIA publication that addressed the issue of national intelligence estimating, an objective NIE should meet the following standards [2]:

- [G1]: quantify the certainty level of its key judgments by using percentages or *bettors' odds*, where feasible, and avoid overstating the certainty of judgments (note: bettors' odds state the chance as, for example, "one out of three");

- [G2]: identify explicitly its assumptions and judgments;

- [G3]: develop and explore *alternative futures:* less likely (but not impossible) scenarios that would dramatically change the estimate if they occurred;

- [G4]: allow dissenting views on predictions or interpretations;

- [G5]: note explicitly what the IC does not know when the information gaps could have significant consequences for the issues under consideration.

Two years later the Rumsfeld Commission to Assess the Ballistic Missile Threat to the United States specifically described the need for intelligence analysis of alternative hypotheses (introduced as abduction in the last chapter) and greater exploration of the unknowns in the analysis:

The Commission would urge that the [IC] adopt as a standard of its methodology that in addition to considering what they know, analysts consider as well what they know they don't know about a program and set about filling gaps in their knowledge by:

- [R1] taking into account not only the output measures of a program, but the input measures of technology, expertise and personnel from both internal sources and as a result of foreign assistance. The type and rate of foreign assistance can be a key indicator of both the pace and objective of a program into which the IC otherwise has little insight.

- [R2] comparing what takes place in one country with what is taking place in others, particularly among the emerging ballistic missile powers. While each may be pursuing a somewhat different development program, all of them are pursuing programs fundamentally different from those pursued by the US, Russia and even China. A more systematic use of comparative methodologies might help to fill the information gaps.

- [R3] employing the technique of alternative hypotheses. This technique can help make sense of known events and serve as a way to identify and organize indicators relative to a program's motivation, purpose, pace and direction. By hypothesizing alternative scenarios a more adequate set of indicators and collection priorities can be established. As the indicators begin to align with the known facts, the importance of the information gaps is reduced and the likely outcomes projected with greater confidence. The result is the possibility for earlier warning than if analysts wait for proof of a capability in the form of hard evidence of a test or a deployment. Hypothesis testing can provide a guide to what characteristics to pursue, and a cue to collection sensors as well.

- [R4] explicitly tasking collection assets to gather information that would disprove a hypothesis or fill a particular gap in a list of indicators. This can prove a wasteful use of scarce assets if not done in a rigorous fashion. But moving from the highly ambiguous absence of evidence to the collection of specific evidence of absence can be as important as finding the actual evidence [3].

The two reports cover the spectrum of intelligence issues, providing excellent guidelines for analysis. The GAO report addressed NIEs that produce broad conceptual estimates (e.g., nation-state capabilities and global threat assessments and projections) while the Rumsfield report addressed more focused hard-target problems where data is scarce and the subjects employ denial and deception measures. The essence of these nine recommendations can be summarized (Table 6.1) to reveal what kind of rigor is expected by policymakers.

Notice that intelligence consumers want more than estimates or judgments; they expect concise explanations of the evidence and reasoning processes behind judgments with substantiation that multiple perspectives, hypotheses, and consequences have been objectively considered. They expect a depth of

Table 6.1
Intelligence Methodology Standards

Process	Standard	Reference
Tasking	Create collection tasking based on alternative hypotheses; seek evidence to prove or disprove alternative hypotheses	[R3, R4]
Analysis-synthesis	Identify explicit assumptions for decomposing evidence and synthesizing models	[G1]
	Explicitly model processes, evaluating evidence of inputs and outputs to determine internal processes	[R1]
	Synthesize and compare alterative hypotheses (models of current processes or projected *futures*) to explain evidence	[R3], [G3]
	Encourage dissenting views in the process	[G4]
	Employ comparative models to enable comparison from target to target	[R2]
	Identify and task collection to seek negative evidence disprove hypotheses	[R4]
Reporting (dissemination)	Explicitly distinguish key assumptions (linchpins), assumptions, and judgments	[G2]
	Explicitly report information gaps; explain consequences on alternate hypotheses	[G5]
	Explicitly report uncertain judgments in quantified terms	[G1]

analysis-synthesis that explicitly distinguishes assumptions, evidence, alternatives, and consequences—with a means of quantifying each contribution to the outcomes (judgments). To meet these expectations, the analysis-synthesis process must be structured, explicit, and thorough. The intelligence tradecraft best practices described in Chapter 4 were produced to provide just such structure for analysis [4], and to provide the rigor required by national intelligence officers [5].

In the following sections, we address the practical procedures to implement this kind of structure.

6.2   Analysis-Synthesis in the Intelligence Workflow

Analysis-synthesis is one process within the intelligence cycle, the highest level abstract business model of intelligence, introduced in Chapter 2. It represents a process that is practically implemented as a continuum rather than a cycle, with all phases being implemented concurrently and addressing a multitude of different intelligence problems or targets. Further, the process integrates multiple INTs to deliver integrated products to consumers derived from all sources.

Several abstract models have been developed to describe the details of the process, each with a different perspective and focus (Figure 6.1) [6]. The figure is organized with increasing levels of model granularity moving down the chart. The first two models focus on command and control decision making for military action, while the second two models are focused on the delivery of intelligence. The models are all cyclic, including the feedback from results to actions that include sensor tasking to better observe a situation, or military response to change a situation.

The *stimulus-hypothesis-option-response* (SHOR) *model*, described by Joseph Wohl in 1986, emphasizes the consideration of multiple perception hypotheses to explain sensed data and assess options for response. The model detailed the considerations for commander decision making by making choices among alternative course of action [7]. The *observe-orient-decide-act* (OODA) loop, developed by Col. John Warden, is a high-level abstraction of the military command and control loop that considers the human decision-making role and its dependence on observation and orientation—the process of placing the observations in perceptual framework for decision making [8]. While the OODA model applies to the entire command and control process (in which intelligence provides the *observe* function), the entire loop may be applied to the intelligence control loop in which the *act* function governs tasking and collection. Both of these models focus on the military situation as the object of control; the next two models view

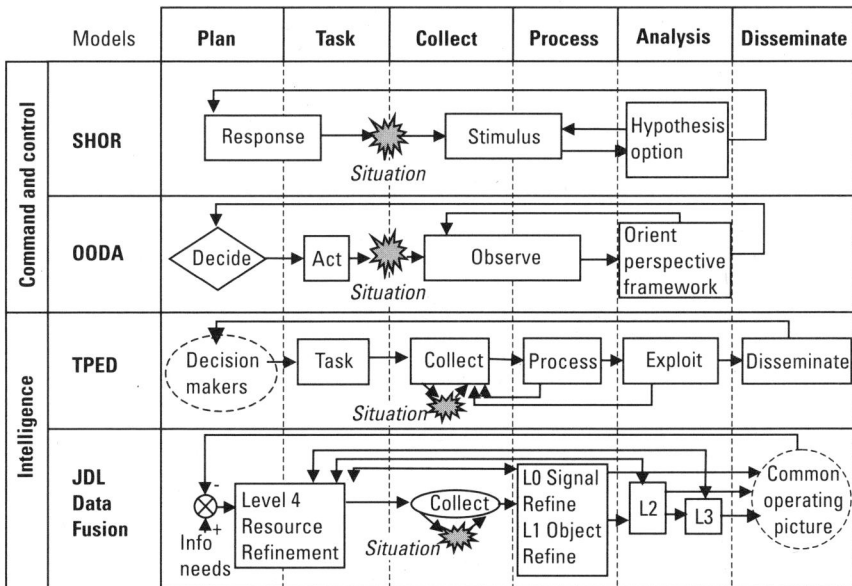

Figure 6.1 Comparison of models that describe the intelligence process.

the situation as an object of surveillance, where the control loop serves to better observe and understand the situation.

The *tasking, processing, exploitation, dissemination* (TPED) model used by U.S. technical collectors and processors [e.g., the U.S. National Reconnaissance Office (NRO), the National Imagery and Mapping Agency (NIMA), and the National Security Agency (NSA)] distinguishes between the processing elements of the national technical-means intelligence channels (SIGINT, IMINT, and MASINT) and the all-source analytic exploitation roles of the CIA and DIA. The TPED process has been applied to independent stovepipe intelligence channels, and concepts have been developed to implement wide-scale multi-INT TPED processes [9]. The model is a high-level organizational model that does not include planning *per se* because it includes policy-level activities organizationally above the processing chain.

The DoD Joint Directors of Laboratories (JDL) *data fusion model* is a more detailed technical model that considers the use of multiple sources to produce a common operating picture of individual objects, situations (the aggregate of objects and their behaviors), and the consequences or impact of those situations. The model includes a hierarchy of data correlation and combination processes at three levels (level 0: signal refinement; level 1: object refinement; level 2: situation refinement; level 3: impact refinement) and a corresponding feedback control process (level 4: process refinement) [10]. The JDL model is a functional representation that accommodates automated processes and human processes and provides detail within both the processing and analysis steps. The model is well suited to organize the structure of automated processing stages for technical sensors (e.g., imagery, signals, and radar).

The practical implementation of the processing and analysis stages in a typical intelligence workflow can be described using the JDL model to distinguish the characteristics of each stage (Figure 6.2). The processing stage is characterized by high-volume single-INT processing channels (stovepipes) to perform the JDL data fusion model level 0 and 1 functions:

- *Level 0: signal refinement* automated processing correlates and combines raw signals (e.g., imagery pixels or radar signals intercepted from multiple locations) to detect objects and derive their location, dynamics, or identity.
- *Level 1: object refinement* processing detects individual objects and correlates and combines these objects across multiple sources to further refine location, dynamics, or identity information.

These processing stages may also include cross-INT cueing to enable the detection of objects in one channel to cue the processing for confirming data or

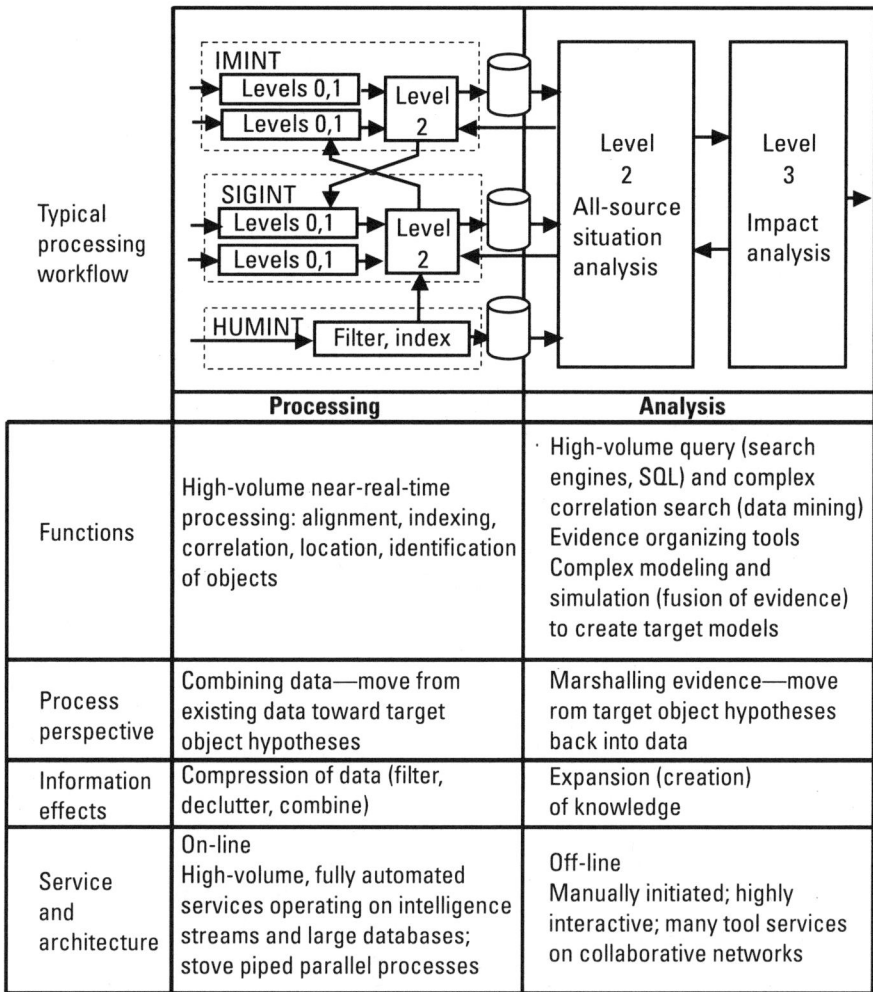

	Processing	Analysis
Functions	High-volume near-real-time processing: alignment, indexing, correlation, location, identification of objects	High-volume query (search engines, SQL) and complex correlation search (data mining) Evidence organizing tools Complex modeling and simulation (fusion of evidence) to create target models
Process perspective	Combining data—move from existing data toward target object hypotheses	Marshalling evidence—move rom target object hypotheses back into data
Information effects	Compression of data (filter, declutter, combine)	Expansion (creation) of knowledge
Service and architecture	On-line High-volume, fully automated services operating on intelligence streams and large databases; stove piped parallel processes	Off-line Manually initiated; highly interactive; many tool services on collaborative networks

Figure 6.2   Processing-analysis workflow.

data to resolve object identity in other channels. The stage may also correlate and combine this data across channels to perform limited level 2 situation assessments (e.g., correlation and identification of a cluster of tanks as an armored unit on the move toward a likely target). This stage may be implemented as an integration of high-volume processing and analysts (e.g., an IMINT chain of image processors and imagery analysts who enter images and linked textual analysis reports into an IMINT database for subsequent analysis). In this case, the processing chain includes processing and single-INT analysis by specialists.

The output of this stage is a set of heterogeneous databases (e.g., imagery, video, text, or audio) or a data warehouse for subsequent all-source analysis.

The analysis stage in the figure performs the analysis-synthesis functions described in Chapter 5 for higher level understanding of situations and their consequences:

- *Level 2: situation refinement analysis* correlates and combines the detected objects across all sources within the background context to produce estimates of the situation—explaining the aggregate of static objects and their behaviors in context to derive an explanation of activities with estimated status, plans, and intents.

- *Level 3: impact refinement analysis* estimates the consequences of alternative courses of action.

Figure 6.2 illustrates the general contrast in the processing and analysis stages; the processing stage is on-line, processing near-real-time, high-volume single-INT data channels while the all-source analysis stage is off-line, focused on selecting only the required data to solve consumer problems. The processing stage is data driven, processing data as it is collected to produce intermediate products for large databases, while the analysis stage is goal driven, responding to queries for intelligence answers from consumers (e.g., targeting, I&W, or order of battle or national capability estimates). The analysis stage employs semiautomated detection and discovery tools to access the data in large databases produced by the processing stage. In general, the processing stage can be viewed as a factory of processors, while the analysis stage is a lower volume shop staffed by craftsmen—the analytic team.

The level 4 process refinement flows are not shown in the figure, though all forward processing levels can provide inputs to refine the process to: focus collection or processing on high-value targets, refine processing parameters to filter unwanted content, adjust database indexing of intermediate data, or improve overall efficiency of the production process. The level 4 process effectively performs the KM business intelligence functions introduced in Section 3.7.

The practical implementation of this workflow, whether in a large national or military intelligence organization or in a small corporate competitive intelligence cell, requires a structural model of the workflow processes, policies, and procedures that move from raw data to finished intelligence products. Later, in Chapter 9, we illustrate the process to translate such a workflow into an enterprise functional design. The following sections focus on the critical role of the human analyst and integration with the automation components of the enterprise.

6.3 Applying Automation

Automated processing has been widely applied to level 1 object detection (e.g., statistical pattern recognition) and to a lesser degree to level 2 situation recognition problems (e.g., symbolic artificial intelligence systems) for intelligence applications. The problem space for which automated technologies may be applied to analysis-synthesis has been structured by artificial intelligence pioneer Marvin Minsky and adapted in Figure 6.3 [11]. The space is defined by two dimensions that describe the complexity of the intelligence subject. Minsky's two dimensions correspond to the two factors of complex situations introduced in Chapter 4, Section 4.4.2. Viewing these dimensions as the number of nodes (causes) and number of interactions (influencing the scale of effects) in a dynamic system, the problem space depicts the complexity of the situation being analyzed:

- *Causal diversity.* The first dimension relates to the number of causal factors, or actors, that influence the situation behavior.

- *Scale of effects.* The second dimension relates to the degree of interaction between actors, or the degree to which causal factors influence the behavior of the situation.

As both dimensions increase, the potential for nonlinear behavior increases, making it more difficult to model the situation being analyzed. This

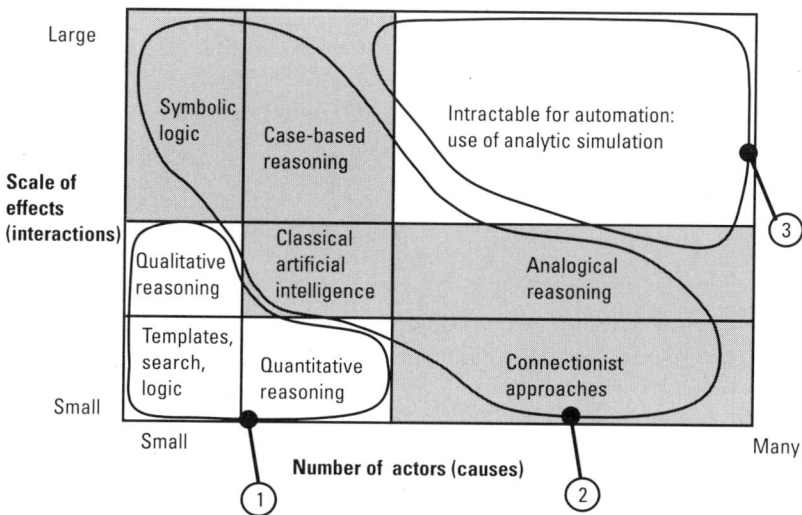

Figure 6.3 Three categories of intelligence problems and automated reasoning technology solutions [12].

problem space includes nine general areas of automation technologies, with the general intelligence problem categories overlaid (See Table 6.2 for example intelligence problems that correspond to the three categories in Figure 6.3). The problem categories move in increasing situation complexity from lower left (category 1) to upper right (category 3) and can be generally related to the levels of the JDL data fusion model. The simplest level 0 and 1 problems (lower left corner) with few causal factors and small linear effects can be solved with simple template matching (matched filter detection), applying search processes to exhaustively search for matching patterns and logic to match the current situation with prior patterns. These problems include the detection of straightforward objects in images, content patterns in text, and emitted signal matching. More difficult problems still in this category include dynamic situations with moderately higher numbers of actors and scales of effects that require qualitative (propositional logic) or quantitative (statistical modeling) reasoning processes These include those problems where small dimension deterministic or stochastic models can accurately represent the situations for comparison with collected data, such as the kinematic tracking of physical targets with statistical Kalman filter models.

The middle band of more complicated (category 2) problems addresses higher dimensional and highly dynamic situations; automated processes resort to higher order reasoning. Approaches to deal with large numbers of actors with small and moderate scales of effects apply connectionist solutions (e.g., Bayesian and neural networks) and reasoning by analogy. Where smaller numbers of

Table 6.2
Representative Intelligence Problem Categories

Problem Category	Example Intelligence Problems
1. Pattern detection and tracking	Simple content pattern recognition in text
	Military vehicle target tracking; unit tracking
	Automatic target recognition; change detection
2. Complex patterns and dynamic behavioral recognition	Relationship and novelty discovery in large databases
	Military order of battle and operations analysis
	Contextual pattern recognition in multimedia
	Financial transactional analysis
3. Complex situation recognition and prediction	Leadership analysis
	Foreign political, social, and economic analysis
	Foreign covert missile program analysis
	Regional nation-state analysis
	Global futures alternatives analysis

actors are involved but the scale of effects are greater, case-based reasoning is applied to compare the present situation with stored cases. Classical heuristic expert systems solve situations in the middle regions, but they are limited as the situations exhibit nonlinearity and emergent behaviors that exceed the representations of knowledge in limited heuristic models and create intractable search demands on the system.

The most difficult category 3 problems, intractable to fully automated analysis, are those complex situations characterized by high numbers of actors with large-scale interactions that give rise to emergent behaviors. Supportive simulation tools, described in the next chapter, can provide support to analysts for tackling these kinds of problems.

The implementation of these automated processes to support knowledge externalization-internalization and combination are described in Chapters 7 and 8, respectively.

6.4　The Role of the Human Analyst

The analyst applies tacit knowledge to search through explicit information to create tacit knowledge in the form of mental models and explicit intelligence reports for consumers. This creative process includes two complementary activities that move from data to knowledge (Figure 6.4):

1. *Reasoning,* the more explicit and reductionist form of analysis-synthesis, creates explicit models of situations. In Chapter 3, we introduced reasoning as the process resulting from the Western emphasis on logic and

	Reasoning processes Explicit knowledge	Sensemaking processes Tacit knowledge
Knowledge	• Focus: objective intellect • Result: explicit models integrated from linked data; multiple hypotheses • Model components: perceptions, beliefs, explanations	• Focus: subjective character • Result: mental models or mindsets integrated from clustered feelings; multiple "hunches" • Model components: imagination, intuition, insight, values
Information	Organized and related data (explicit multimedia databases)	Clusters of experience and feelings (metaphors, ideas, "images")
Data	Text, imagery, audio, video, scientific measurements	Experiences, narrative stories, feelings

Figure 6.4　Reasoning-sensemaking distinctions.

dualism (René Descartes) to decompose problems and create explicit abstractions of truth that can be articulated and shared; knowledge emphasis is on the intellect.

2. *Sensemaking,* the more tacit and holistic form of analysis-synthesis, which creates mental models or mindsets, has been studied in the cognitive sciences. In contrast with reasoning, this mode has been emphasized in the East, where holistic intuition and oneness (humanity-nature oneness, mind-body oneness, and self-other oneness) has been embraced. Knowledge emphasis is placed on the *action* of truth (character). The term sensemaking refers to our "deep understanding" or "feel" of a situation and includes the components of a prior tacit knowledge of a situation, a rich contextual awareness, and a tacit perception of alternatives, futures, and consequences [13].

The analysis process requires the analyst to integrate the cognitive reasoning and more emotional sensemaking processes with large bodies of explicit information to produce explicit intelligence products for consumers. To effectively train and equip analysts to perform this process, we must recognize and account for these cognitive and emotion components of comprehension. The complete process includes the automated workflow, which processes explicit information, and the analyst's internal mental workflow, which integrates the cognitive and emotional modes (Figure 6.5). Research in the cognitive sciences is exploring the relationship between these two modes of creating knowledge, which produce an analyst's integrated understanding of a subject. Antonio Damasio, in *Descartes Error: Emotion, Reason and the Human Brain,* has offered one explanation of this interaction that is illustrated in the figure [14]. Stimulus to the analyst—new explicit information from the intelligence pipeline—is matched to the current mental model or mindset. This is the analyst's "distillation of the intelligence analyst's cumulative factual and conceptual knowledge into a framework for making estimative judgments on a complex subject" [15]. The products of reasoning and sensemaking, the analyst's mental models, influence the perception of new information—filtering, focusing, and even distorting the incoming stimuli. The shortcomings of these distortions are addressed in the next section.

The coordinated mind-body (cognition-emotion) interactions according to Damasio include the placement of the stimuli in an emotional framework to assess the feelings about the new information, applying subconscious tacit knowledge to reduce the alternative space for the reasoning process. Complementary logical and emotional frameworks are based on the current mental model of beliefs and feelings and the new information is compared to these frameworks; differences have the potential for affirming the model (agreement), learning and refining the model (acceptance and model adjustment), or rejecting the new

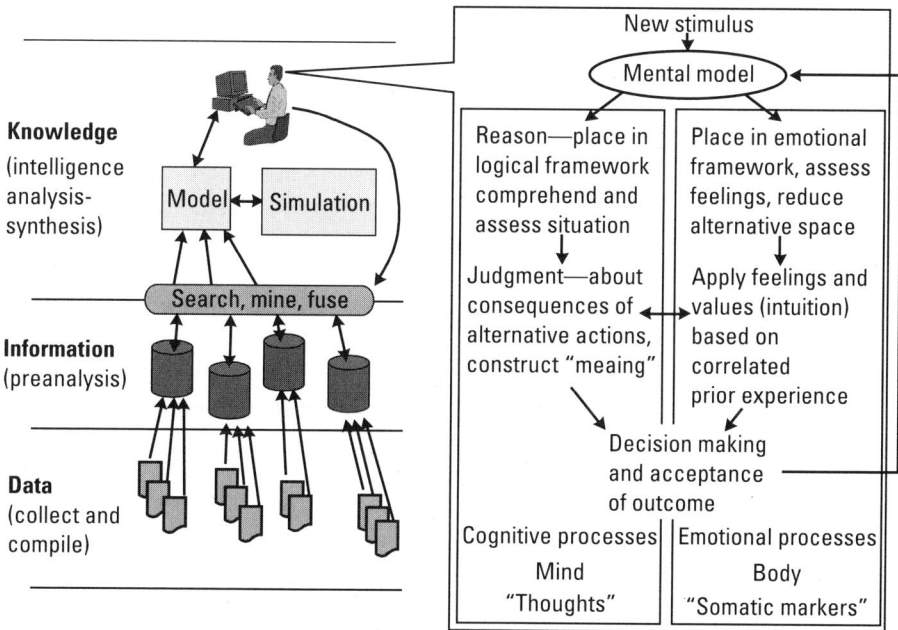

Figure 6.5 Mental model formation within the intelligence workflow.

information. Judgment integrates feelings about consequences and values (based on experience) with reasoned alternative consequences and courses of action that construct the *meaning* of the incoming stimulus. Decision making makes an intellectual-emotional commitment to the impact of the new information on the mental model (acceptance, affirmation, refinement, or rejection).

6.5　Addressing Cognitive Shortcomings

The intelligence analyst is not only confronted with ambiguous information about complex subjects, but is often placed under time pressures and expectations to deliver accurate, complete, and predictive intelligence. Consumer expectations often approach infallibility and omniscience. In this situation, the analyst must be keenly aware of the vulnerabilities of human cognitive shortcomings and take measures to mitigate the consequences of these deficiencies. The natural limitations in cognition (perception, attention span, short- and long-term memory recall, and reasoning capacity) constrain the objectivity of our reasoning processes, producing errors in our analysis. Veteran analyst and author Richards Heuer, Jr., has carefully enumerated these shortcomings and the implications for intelligence officers. In *The Psychology of Intelligence*

Analysis, Heuer has identified the major biases we exhibit when evaluating evidence, attributing causality in explaining relationships, and in estimating relative probabilities (Table 6.3) [16]. To these biases in analytic reasoning, researcher Thomas Gilovich has added the bane of the common subjective biases of motivation: believing self-serving beliefs (e.g., those the intelligence chief believes), imagined agreement (exaggerating the agreement of other analysts and colleagues), and inaccurate narratives (distortion of tacit knowledge in narrative stories) [17]. In each case, the analyst is inclined to create mental models that distort perception of the significance of evidence or derived inferences, then attribute undeserved support for those models. In "Combatting

Table 6.3
Cognitive Shortcomings to be Addressed by the Analyst

Area of Bias	Categories of Cognitive Shortcomings
Evaluating Evidence	Vividness bias—vivid, concrete, personal information is biased (preferred) over pallid, abstract information
	Missing evidence bias—absent evidence is often ignored or not factored into analytic judgments; present evidence is biased over missing gaps
	Consistency bias—a small body of consistent evidence is biased over a larger body of less consistent evidence
	Persistent impression bias—prior uncertain impressions persist (are preferred) even after the evidence on which they are based is discredited
Attributing Causality	Causality bias—evidence that falls into an orderly causal pattern is preferred
	Centralized direction bias—evidence that fits centralized coherent nation-state control explanations are biased over more random, accidental explanations
	Internal factor bias—bias that increases influence of internal (beliefs attitudes) over external (constraints) factors in decision making
	Own importance bias—due to analysts' greater knowledge of own-nation actions, those influences are biased over influences less well understood
Estimating Probabilities	Availability bias—probability estimates are biased toward evidence that is more available and against missing evidence
	Anchor bias—probability estimates are anchored by a natural starting point that becomes preferred, then are adjusted (erroneously) in response to new information
Succumbing to Social Factors	Motivation bias—propensity to accept self-serving beliefs
	Imagined agreement bias—the exaggerated belief in the agreement of others
	Secondhand story bias—the distortion of tacit knowledge in secondhand narrative stories

Mind-Set," respected analyst Jack Davis has noted that analysts must recognize the subtle influence of *mindset,* the cumulative mental model that distills analysts' beliefs about a complex subject and "find[s] strategies that simultaneously harness its impressive energy and limit[s] the potential damage" [18].

Davis recommends two complementary strategies:

1. *Enhancing mindset.* Creating explicit representation of the mindset—externalizing the mental model—allows broader collaboration, evaluation from multiple perspectives, and discovery of subtle biases.

2. *Ensuring mind-set.* Maintaining multiple explicit explanations and projections and opportunity analyses provides insurance against single-point judgments and prepares the analyst to switch to alternatives when discontinuities occur.

While these shortcomings address the problem of understanding the *subject* of an analysis, Davis has also cautioned analysts to beware the *paradox of expertise* phenomenon that can distract attention from the *purpose* of an analysis. This error occurs when discordant evidence is present and subject experts tend to be distracted and focus on situation analysis (solving the discordance to understand the subject situation) rather than addressing the impact on the analysis of the *consequences* of the discrepancy. In such cases, the analyst must focus on providing value added by addressing what action alternatives exist for alternatives and their consequences in cost-benefit terms [19].

Heuer emphasized the importance of supporting tools and techniques to overcome natural analytic limitations [20]: "Weaknesses and biases inherent in human thinking processes can be demonstrated through carefully designed experiments. They can be alleviated by conscious application of tools and techniques that should be in the analytical tradecraft toolkit of all intelligence analysts." These tools and techniques support the kind of critical thinking introduced in earlier chapters; the practical methods for marshaling evidence, structuring argumentation, and evaluating hypotheses are introduced in the next section.

6.6 Marshaling Evidence and Structuring Argumentation

In the *Thinker's Toolkit,* former analyst Morgan Jones distinguishes between our common-sense instinctive analysis and structured analysis [21]. Instinctive analysis focuses on a single or limited range of alternatives, moves on a path to satisfy minimum needs (*satisficing,* or finding an acceptable explanation), and is performed implicitly using tacit mental models. Structured analysis follows the principles of critical thinking introduced in Chapter 4, organizing the problem to consider all reasonable alternatives, systematically and explicitly representing the alternative solutions to comprehensively analyze all factors. Though we recognize the objective benefits of structured reasoning, intelligence is all too often plagued by analyses that are more instinctive than structured.

Intelligence organizations have recognized the importance of instilling the value of structured thinking in the analytic community, and structured processes are emphasized in analyst training [22]. The desired discipline is to ensure that analysts will synthesize alternative hypotheses, marshal evidence to affirm the hypotheses, and then objectively evaluate the alternatives. The ability to explain the hypothesis is referred to as *argumentation*, where the hypothesis provides a means of structuring the argument for presentations to decision makers. In this section, we introduce the concepts of synthesizing structured hypotheses and then marshaling evidence around competing alternatives. In the next section, we describe how competing hypotheses are compared.

6.6.1   Structuring Hypotheses

To illustrate the structure of hypotheses in practical intelligence problems, consider the intelligence conclusion in a critical 1964 U.S. CIA intelligence report estimating the likelihood of the location and timing of China's first nuclear test. The report, written in August 1964, concluded:

> On the basis of new overhead photography, we are now convinced that the previously suspect facility at Lop Nor in Western China is a nuclear test site that could be ready for use in about two months. On the other hand the weight of available evidence indicates the Chinese will not have sufficient fissionable material for a test of a nuclear device in the next few months. Thus, the evidence does not permit a very confident estimate of the chances of a Chinese Communist nuclear detonation in the next few months. Clearly the possibility of such a detonation before the end of this year cannot be ruled out—the test may occur during this period. On balance, however, we believe that it will not occur until sometime after the end of 1964 [23].

The simple hierarchical structure of the hypothesis for this report is depicted in Figure 6.6. The basic binary hypothesis set, $H = \{H_0, H_1\}$, includes the hypothesis that the Lop Nor is a test site, H_1 or the complement H_0 that it is not. Furthermore, if the site is a nuclear test site, the subset of hypotheses, $H_2 = \{H_{21}, H_{22}\}$ deal with the imminence of a test. The analysts' conclusion and argumentation is:

- Hypothesis H_1—Lop Nor is indeed a nuclear test site, based on the site characteristics recognized in overhead photography, and could be ready for a test within 2 months.

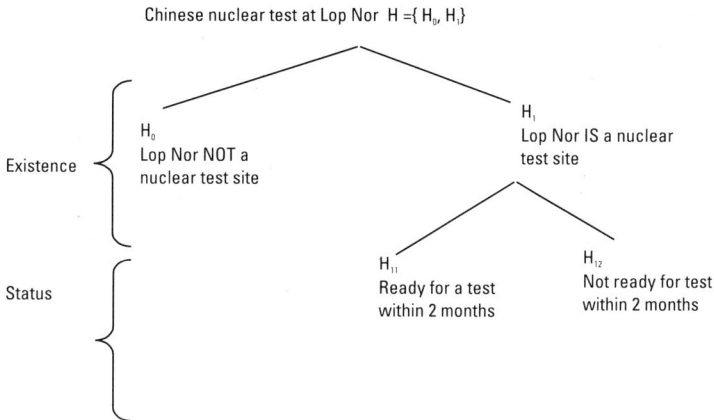

Figure 6.6 Hypothesis structure.

- Hypothesis H_{11}—there is insufficient evidence of available fissional material to conclude that a test can occur within 2 months, but a test within the next 4 months cannot be ruled out.

- Hypothesis H_{12}—the likelihood of a test increases beyond 2 months and more likely beyond 4 months (in 1965).

Notice that in the text, the analyst clearly qualifies the confidence in the conclusion by stating, "the evidence does not permit a very confident estimate of the chances of a Chinese Communist nuclear detonation in the next few months." That statement carefully identifies equivocation (the presence of uncertainty) in the judgments being made. The actual Chinese nuclear test was conducted at Lop Nor on October 16, just 2 months after this report.

6.6.2 Marshaling Evidence and Structuring Arguments

There exist a number of classical approaches to representing hypotheses, marshaling evidence to them, and arguing for their validity. Argumentation structures propositions to move from premises to conclusions. Three perspectives or disciplines of thought have developed the most fundamental approaches to this process (Table 6.4):

1. Rhetoric has historically contributed to the disciplined structuring of informal oral or written arguments to provide accuracy of thought, clarity of communication, and strength of persuasion. Aristotle emphasized three modes of persuasive appeal (proof): *logos* appeals to

Table 6.4
Evidence Marshaling and Hypothesis Argumentation Structures

Approach		Implementation of Knowledge Representation and Inference Process	Intelligence Application
Structured inferential argumentation (informal logic)		Rational, practical, structured organization of argumentation and inference in near-natural language, which distinguishes data, evidence, inferential reasoning principles, and rules that lead from data to conclusions, or assertions of hypotheses	All-source analysis (across multiple unstructured sources) Natural language explanation of analytic results
Formal logic	Propositional logic	Logic that combines assertions of truth (propositions) to deduce combined propositions	
	Predicate logic	Logic that allows the assignment of attributes (quantifiers) to entities and therefore permits the combination of assertions about the properties of entities	
	Fuzzy logic	Logical representation of attributes and hypotheses about entities as *fuzzy* functions; fuzzy inference is performed by an algebra that combines uncertain data to derive uncertain deductions	Inferential networks to implement automated data fusion
Mathematical statistics	Bayesian inference	Mathematical representation of evidence and possible states (hypotheses) in probabilities; Bayesian inference permits mathematical computation of *posterior* hypothesis probabilities from *prior* probabilities and current evidence	Database evidence cross-correlation and linking
	Dempster-Shafer evidential reasoning	More general that Bayesian; represents evidence in terms of belief functions and performs mathematical inference by computing accumulated mass of belief for any given hypothesis	

reason (explicit knowledge), *pathos* to emotion (tacit knowledge), and *ethos* to the character (truth). The appeal to reason has emphasized the careful structure of natural language to accurately and clearly explain the basis for arguing from evidence and premises to conclusion.

2. Philosophy has developed formal logic to structure and combine simple propositions (assertions) such that judgments of truth can be made about the validity of more complex propositions inferred from the combination of simple propositions.

3. Mathematics has contributed probabilistic methods to describe uncertainty and quantitatively perform the inference process. These methods impose greater structure on both evidence and hypothesis and provide a quantified method of reasoning that can be automated, presuming evidence and belief can be quantified. (Applications of these automated methods are described in Chapter 8.)

Each discipline has contributed methods to represent knowledge and to provide a structure for reasoning to infer from data to relevant evidence, through intermediate hypotheses to conclusion. The term *knowledge representation* refers to the structure used to represent data and show its relevance as evidence, the representation of rules of inference, and the asserted conclusions. In the following paragraphs we survey these approaches and their contributions to the analysis-synthesis process.

6.6.3   Structured Inferential Argumentation

Philosophers, rhetoricians, and lawyers have long sought accurate means of structuring and then communicating, in natural language, the lines of reasoning, that lead from complicated sets of evidence to conclusions. Lawyers and intelligence analysts alike seek to provide a clear and compelling case for their conclusions, reasoned from a mass of evidence about a complex subject. Although less formal that the logic briefly introduced in our discussion of deduction in Chapter 5, we will introduce here three approaches to structuring natural-language arguments.

We first consider the classical forms of argumentation described as informal logic, whereby the argument connects premises to conclusions. The common forms include:

1. *Linked.* Multiple premises, when taken together, lead to but one conclusion. For example: The radar at location A emits at a high pulse repetition frequency (PRF); when it emits at high PRF, it emits on frequency (F) → the radar at A is a fire control radar.

2. *Convergent.* Multiple premises independently lead to the same conclusion. For example: The radar at A is a fire control radar. Also Location A stores canisters for missiles. → A surface to air missile (SAM) battery must be at location A.

3. *Serial.* A single premise leads to but one conclusion, for example: A SAM battery is located at A the battery at A → must be linked to a command and control (C2) center.

4. *Divergent.* A single premise can support more than one conclusion. For example: The SAM battery could be controlled by the C2 center at golf, or The SAM battery could be controlled by the C2 center at hotel.

These four basic forms may be combined to create complex sets of argumentation, as in the simple sequential combination and simplification of these examples:

- The radar at A emits at a high PRF; when it emits at high PRF, it emits on frequency F, so it must be a fire control radar. Also, location A stores canisters for missiles, so there must be a SAM battery there. The battery at A must be linked to a C2 center. It could be controlled by the C2 centers at golf or at hotel.

The structure of this argument can be depicted as a *chain* of reasoning or argumentation (Figure 6.7) using the four premise structures in sequence.

Next, consider Toulmin's practical approach to structuring argumentation, which attempted to refine the classical approach with a predefined reasoning sequence (argument form) and a careful distinction between evidence and the logic or principles of inference. Toulmin distinguished six elements of all arguments [24]:

1. Data (D), at the beginning point of the argument, are the explicit elements of data (relevant data, or evidence) that are observed in the external world.
2. Claim (C), is the assertion of the argument.
3. Qualifier (Q), imposes any qualifications on the claim.

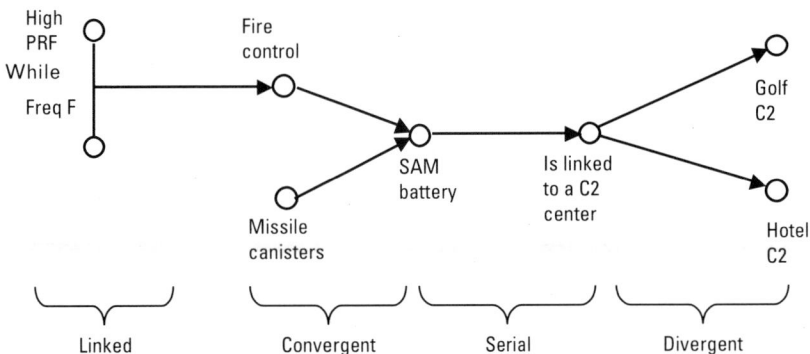

Figure 6.7 A simple informal logic chain.

4. Rebuttals (R) are any conditions that may refute the claim.

5. Warrants (W) are the implicit propositions (rules, principles) that permit inference from data to claim.

6. Backing (B) are assurances that provide authority and currency to the warrants.

Applying Toulmin's argumentation scheme requires the analyst to distinguish each of the six elements of argument and to fit them into a standard structure of reasoning—see Figure 6.8(a)—which leads from datum (D) to claim (C). The scheme separates the domain-independent structure from the warrants and backing, which are dependent upon the field in which we are working (e.g., legal cases, logical arguments, or morals).

The general structure, described in natural language then proceeds from datum (D) to claim (I) as follows:

- The datum (D), supported by the warrant (W), which is founded upon the backing (B), leads directly to the claim (C), qualified to the degree (Q), with the caveat that rebuttal (R) is present.

In Figure 6.8(b), we insert the elements of the Chinese nuclear test argument (used earlier in this section) into the Toulmin schema to illustrate how the schema forces structure to the analyst's argument. Such a structure requires the analyst to identify all of the key components of the argument—and explicitly report if any components are missing (e.g., if rebuttals or contradicting evidence is not existent).

The benefits of this scheme are the potential for the use of automation to aid analysts in the acquisition, examination, and evaluation of natural-language arguments. As an organizing tool, the Toulmin scheme distinguishes data (evidence) from the warrants (the universal premises of logic) and their backing (the basis for those premises). Notice that in the previous informal logic example, data (the radar at location A emits at a high PRF) and warrants (so there must be a SAM battery located there) were not distinguished; warrants and data are equally treated as premises. It must be noted that formal logicians have criticized Toulmin's scheme due to its lack of logical rigor and ability to address probabilistic arguments. Yet, it has contributed greater insight and formality to developing structured natural-language argumentation.

6.6.4 Inferential Networks

Moving beyond Toulmin's structure, we must consider the approaches to create network structures to represent complex chains of inferential reasoning. While

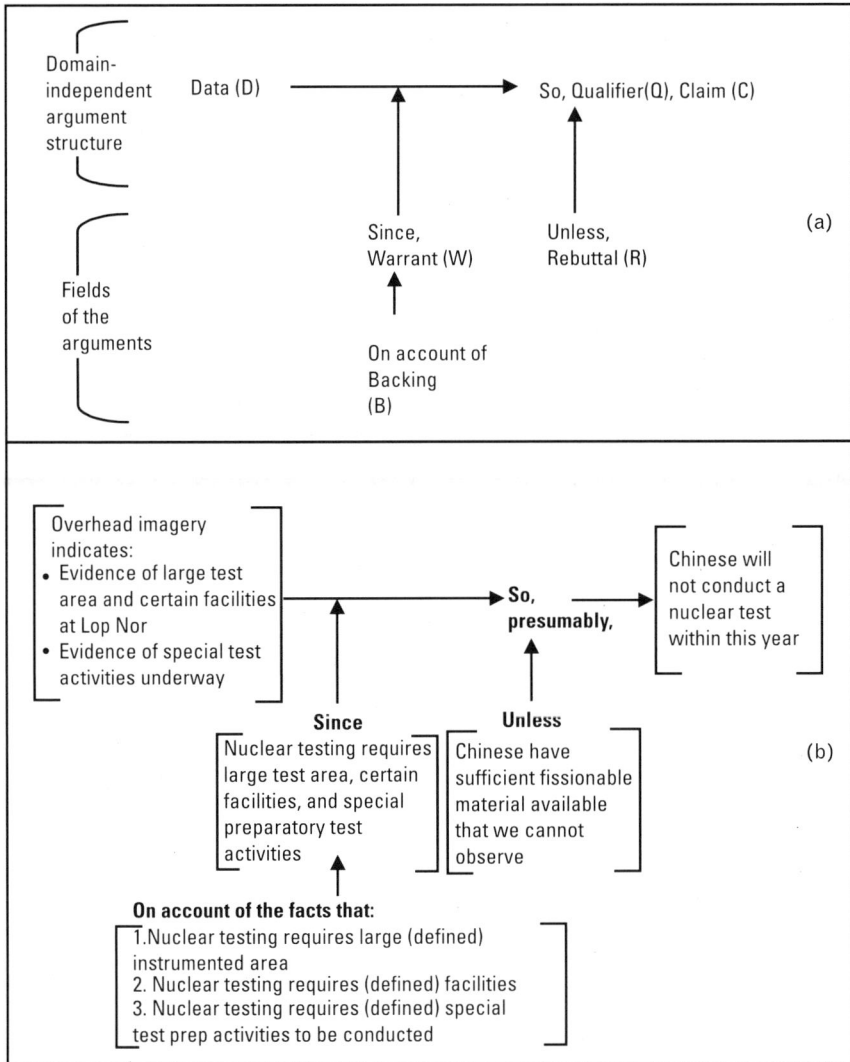

Figure 6.8 (a) Toulmin's argument structure, and (b) populated argument structure example.

the Toulmin structure allowed us to represent arguments with a handful of data (evidence), warrant, and backing elements, the single-thread structure becomes cumbersome (and unable to fully represent all factors) when many elements exist and there are interactions between those elements. The development of graphical approaches proceeds from the legal graphs of evidence introduced by Whigmore in *The Science of Judicial Proof* (1937) to the directed graph representations of inferential networks currently used to logically and mathematically structure complex arguments. We illustrate these networks in the following discussion using the directed acyclic graph forms introduced by Schum in his foundational work, *Evidence and Inference for the Intelligence Analyst* [25], and his subsequent exhaustive text, *The Evidential Foundations for Probabilistic Reasoning* [26].

The use of graph theory to describe complex arguments allows the analyst to represent two crucial aspects of an argument:

- *Argument structure.* The directed graph represents evidence (E), events, or intermediate hypotheses inferred by the evidence (i), and the ultimate, or final, hypotheses (H) as graph nodes. The graph is *directed* because the lines connecting nodes include a single arrow indicating the single direction of inference. The lines move from a source element of evidence (E) through a series of inferences (i_1, i_2, i_3, ... i_n) toward a terminal hypothesis (H). The graph is *acyclic* because the directions of all arrows move from evidence, through intermediate inferences to hypothesis, but not back again: there are no closed-loop cycles.

- *Force of evidence and propagation.* In common terms we refer the *force, strength,* or *weight* of evidence to describe the relative degree of contribution of evidence to *support* an intermediate inference (i_n), or the ultimate hypothesis (H). The graph structure provides a means of describing supporting and refuting evidence, and, if evidence is quantified (e.g., probabilities, fuzzy variables, or other belief functions), a means of propagating the accumulated weight of evidence in an argument.

Like a vector, evidence includes a direction (toward certain hypotheses) and a magnitude (the inferential force). The basic categories of argument can be structured to describe four basic categories of evidence combination (illustrated in Figure 6.9):

1. *Direct.* The most basic serial chain of inference moves from evidence (E) that the event E occurred, to the inference (i_1) that E did in fact occur. This inference expresses belief in the evidence (i.e., belief in the veracity and objectivity of human testimony). The chain may go on serially to further inferences because of the belief in E.

2. *Consonance.* Multiple items of evidence may be synergistic resulting in one item *enhancing* the force of another; their joint contribution provides more inferential force than their individual contributions. Two items of evidence may provide *collaborative* consonance; the figure illustrates the case where ancillary evidence (E_2) is favorable to the

Evidence form	Inferential structure	Inferential effect
Direct		Evidence E of an event directly infers the intermediate hypothesis that the event occurred (i_1) hypotheses (i_1).
Consonant — Corroborative		Auxiliary evidence E_2 supports the credibility of evidence E_1; it therefore enhances the inferential force of E_1.
Consonant — Convergent		E_1 and E_2 provide evidence of the occurrence of different events (i_1 and i_2), when occurring together, favor a common subsequent inference. The effect is a possible enhancement of evidential force.
Redundant — Corroborative		Two or more sources supply identical evidence (E_1 and E_2) of a common event inference (i_1); one may diminish the force of the other to avoid "doublecounting" the force of the redundant evidence.
Redundant — Cumulative		Redundant E_1 and E_2, although inferring intermediate hypotheses (i_1, i_2), lead to a common hypothesis (i_3); This redundant contribution to (i_3) diminishes the contribution of inferential force from E_2
Dissonant — Contradictive		E_1 and E_2 report, mutually exclusively, that the event E did occur and did not occur, respectively. The force of each must be considered.
Dissonant — Conflicting		E_1 and E_2 report two separate events i_1 and i_2 (both of which may have occurred, but not jointly), but these events favor mutually exclusive hypotheses at i_3.

Figure 6.9 Basic directed graph evidence and inference forms.

credibility of the source of evidence (E_1), thereby increasing the force of E_1. Evidence may also be *convergent* when E_1 and E_2 provide evidence of the occurrence of different events, but those events, together, favor a common subsequent inference. The enhancing contribution (i_1) to (i_2) is indicated by the dashed arrow.

3. *Redundant.* Multiple items of evidence (E_1, E_2) that redundantly lead to a common inference (i_1) can also *diminish* the force of each other in two basic ways. *Corroborative* redundancy occurs when two or more sources supply identical evidence of a common event inference (i_1). If one source is perfectly credible, the redundant source does not contribute inferential force; if both have imperfect credibility, one may diminish the force of the other to avoid *double counting* the force of the redundant evidence. *Cumulative* redundancy occurs when multiple items of evidence (E_1, E_2), though inferring different intermediate hypotheses (i_1,i_2), respectively, lead to a common hypothesis (i_3) farther up the reasoning chain. This redundant contribution to (i_3), indicated by the dashed arrow, necessarily reduces the contribution of inferential force from E_2.

4. *Dissonance.* Dissonant evidence may be *contradictory* when items of evidence E_1 and E_2 report, mutually exclusively, that the event E did occur and did not occur, respectively. *Conflicting* evidence, on the other hand, occurs when E_1 and E_2 report two separate events i_1 and i_2 (both of which may have occurred, but not jointly), but these events favor mutually exclusive hypotheses at i_3.

Note that these four forms elaborate on the four classical forms (introduced earlier) as serial, linked, convergent, and divergent, respectively. These basic components can be assembled into complex argument structures to carefully represent the many transitional inferences and supporting relationships that lead from evidence to a final hypothesis.

A military deception example illustrates the complexity of even a simple argument. It also illustrates the many hidden or unspoken inferential components generally overlooked when the argument is simply structured in natural language.

Consider the binary hypothesis, H = {H_0, H_1} , where:

H_0 = The soap factory is a concealed military vehicle depot.

H_1 = The soap factory is not a concealed military vehicle depot.

Six elements of direct evidence are received regarding the factory:

E1 = Imagery sources report that the factory ceased soap production 6 months ago.

E2 = A human source S1 said soap production was ongoing 3 months ago.

E3 = The company advertises in open reports that soap production is ongoing.

E4 = A human source S2 says military officers reside in the factory office buildings.

E5 = UAV 1 sensors reported signal K emissions from the factory.

E6 = UAV 2 sensors also reported signal K emissions from the factory at the same time UAV 1 issued its report.

In addition, three items of auxiliary evidence are applied:

E7 = Evidence that backs the credibility of the imagery analysis process.

E8 = Evidence that supports the accuracy of the methods of observation employed by certain clandestine human observers.

E9 = Evidence that certain signals are unique to military C2 vehicles.

We now consider the directed graph (Figure 6.10) that represents the structure of this example from evidence to inference, using the basic forms. The tree structure moves from six elements of direct evidence at the bottom to the single binary hypothesis at the top. On the left side are elements of auxiliary evidence that determine the strength of inferential links; in Toulmin's terms, these are the backing to inferential warrants. The graph moves from bottom to top in the following sequence:

1. Direct evidence at the bottom;

2. Evidence credibility inferences are the first row above evidence, inferring the veracity, objectivity, and sensitivity of the source of evidence;

3. Relevance inferences move from credibility-conditioned evidence through a chain of inferences toward final hypothesis;

4. The final hypothesis is at the top.

The three principle chains of reasoning supporting H_1, the hypothesis that the factory is a concealed military vehicle facility, can readily be described:

Chain one: Factory is a cover for non-soap-making operation.

1. The directly observed imagery evidence from imagery analysis, E1, leads to the inference i1 that the factory is no longer producing phenomena associated with soap production and has not been for 6 months. Auxiliary evidence E7 that backs the credibility of the imagery analysis process used to assess soap production supports i2 that soap production has stopped within the period.

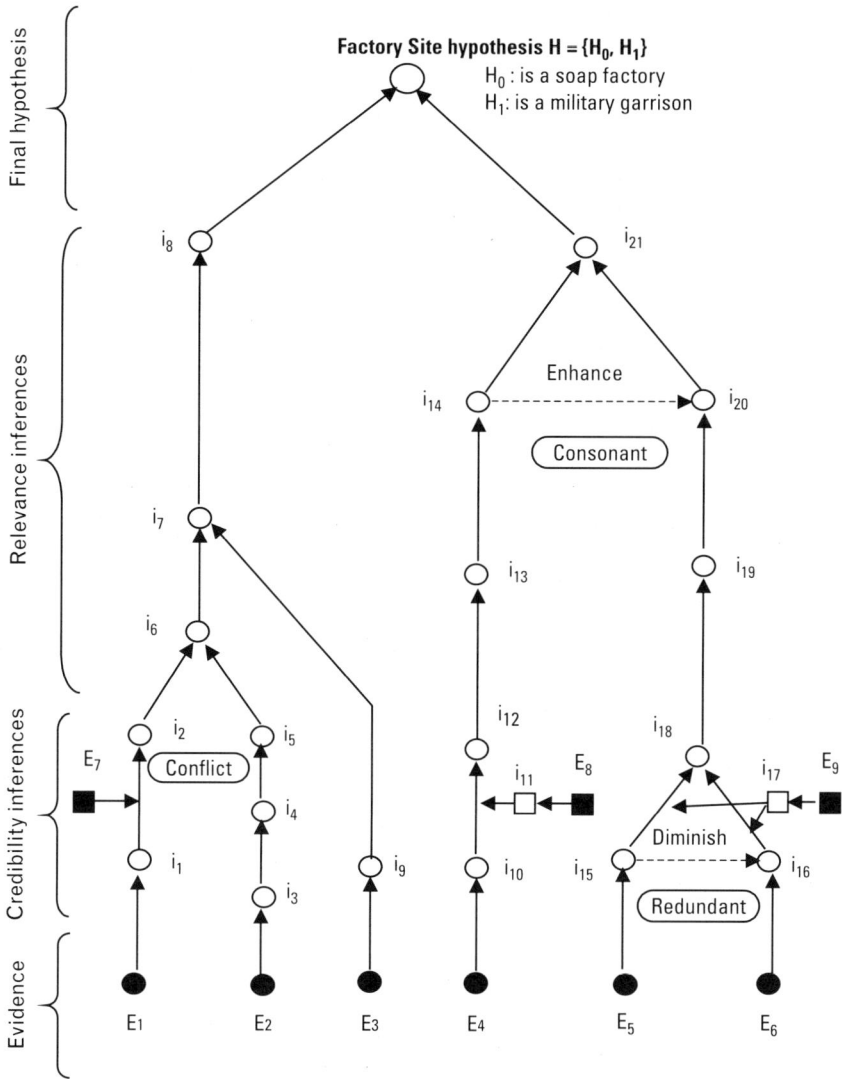

Figure 6.10 A simple inference network for the example military deception hypothesis.

2. Evidence E2 from a human source, stating that production was observed only 3 months ago, leads to i3 that the source believes soap production was ongoing as recent as 3 months ago (source's veracity). This leads to the subsequent i4 that the source actually detected production (source's observational sensitivity and accuracy), and subsequently i5, that production occurred 3 months ago.

3. *Conflicting* dissonant i2 and i5 merge at the intermediate hypothesis (i6) that the factory is no longer producing soap.

4. Evidence E3 that the state-operated company publicly states that the factory is in production leads to the inference that the statement is a credible assertion of the company (i9).

5. Inference i6 infers (from i6 and i9) that the company is inconsistent in its performance and its public statement.

6. Inference i8 infers intent; the company is deliberately deceptive about its factory operations and denies information on the real activities at the site.

Chain two: Factory conducts military operations.

1. A reliable human source reported evidence E4 that military personnel have been seen almost daily in the factory office buildings in recent months; i10 infers the source believes the report to be true (veracity) and that the observation was accurate (i12). Auxiliary evidence E8 supports the accuracy of the sources observation means, and i11 infers that the source is a very accurate observer, supporting i12.

2. This evidence leads to a serial chain of inferences: i12 leads to the inference that the military personnel are conducting business at the factory (i13), leading to i14, that the offices are being used to conduct military operations.

Chain three: Factory houses military vehicles

1. *Redundant* evidence (repeated measurements from two UAVs at the same time periods over the past four months) E5 and E6 detect military C2 vehicle signal emissions from the factory. Inferences i15 and i16 (that the signal was emitted from a source at the factory) contain redundancy and therefore one diminishes the full force of the other.

2. Both i15 and i16 lead to the common inference that military vehicles were located at the factory (i18). These inferences are supported by auxiliary evidence E9 that the signals are unique to military C2 vehicles, and i17 that the signal is uniquely associated with military C2 vehicles.

3. This leads to the i19, that military vehicles are often at the factory, and subsequently i20, that the factory is a location that is intended to house military C2 vehicles.

Final hypothesis

The final hypothesis $H = \{H_0, H_1\}$ weights the inferential force from the three chains. *Consonant* inferences i14 and i20 from chains 2 and 3 (i14 enhances i20) lead to i21, that the factory is a military vehicle garrison (military personnel are conducting operations where military vehicles are stored). This inference and i8 (the company is conducting denial and deception) provide the combined inferential force for H. If the accumulated evidential force is sufficient, the analyst makes the judgment H_1 that the former factory provides cover, concealment, and deception (CCD) for a military garrison.

Some may wonder why such rigor is employed for such a simple argument. This relatively simple example illustrates the level of inferential detail required to formally model even the simplest of arguments. It also illustrates the real problem faced by the analyst in dealing with the nuances of redundant and conflicting evidence. Most significantly, the example illustrates the degree of care required to accurately represent arguments to permit machine-automated reasoning about all-source analytic problems.

We can see how this simple model demands the explicit representation of often-hidden assumptions, every item of evidence, the entire sequence of inferences, and the structure of relationships that leads to our conclusion that H_1 is true.

Inferential networks provide a logical structure upon which quantified calculations may be performed to compute values of inferential force of evidence and the combined contribution of all evidence toward the final hypothesis. In these cases, evidence, intermediate inferences, and hypotheses (E, i, H) are expressed as random variables using probabilities or other expressions to represent the inferential force. The most common approaches to apply quantitative measures of *uncertainty* to evidence and to compute the inferential combination of uncertain evidence are summarized in Table 6.5. In addition to Schum [26], standard texts on multisensor data fusion and reasoning in uncertainty develop the mathematics of these approaches [27].

6.7 Evaluating Competing Hypotheses

Heuer's research indicated that the single most important technique to overcome cognitive shortcomings is to apply a systematic analytic process that allows objective comparison of alternative hypotheses:

> The ideal is to generate a full set of hypotheses, systematically evaluate each hypothesis, and then identify the hypothesis that provides the best fit to the data … The simultaneous evaluation of multiple, competing hypotheses permits a more systematic and objective analysis than is possible when an

Table 6.5
Quantitative Approaches to Inference Computation

Evidential Representation	Inferential Methodology	Inference Computation
Probabilities	Evidence is represented in terms of prior and conditional probabilities	Bayes Rule
	Bayesian networks implement directed acyclic graphs to compute inferential force in terms of forward conditional probabilities	
	Inferential force is represented as posterior probabilities	
Fuzzy variables	Membership functions represent imprecise evidence in terms of fuzzy set theory	Fuzzy algebra
	Fuzzy logic combines evidential membership functions using fuzzy logical functions	
	Inferential force is a fuzzy variable	
Belief functions	Belief functions represent the total evidential force for any hypothesis	Dempster's Rule of Combination
	Total belief for any hypothesis is computed as the mass of all belief that supports the hypothesis	
	Inferential force is a mass function	

analyst focuses on a single, most-likely explanation or estimate. The simultaneous evaluation of multiple, competing hypotheses entails far greater cognitive strain than examining a single, most-likely hypothesis [28].

The logical process of reasoning to the best explanation (abduction) was introduced in the last chapter, and a number of approaches to explicitly structure evidence to support such reasoning have been offered to aid the analyst or investigator. The acyclic graphs introduced in the last section link evidence through inference structures to hypotheses to permit mathematical computation of hypothesis likelihoods using methods such as Bayesian networks. Wigmore diagrams, named after the nineteenth century legal scholar, provide a relatively complex symbolic methodology to array evidence and annotate inferences of causality and relationship to hypothesized legal case explanations [29].

In this section, we introduce the method of analysis of competing hypotheses (ACH), a straightforward process that structures a matrix to compare alternative hypotheses that was introduced by Heuer to visualize the basis for an analyst's judgments.

The approaches are complementary, not competitive. Inferential networks are useful at the detail level, where evidence is rich and the ACH approach is useful at the higher levels of abstraction and where evidence is sparse. Networks are valuable for automated computation; ACH is valuable for collaborative analytic reasoning, presentation, and explanation. The ACH approach provides a methodology for the *concurrent competition* of multiple explanations, rather than the focus on the currently most plausible. The methodology focuses on explicit representation and objective evaluation to overcome many of the biases introduced in the previous section. The ACH structure approach described by Heuer uses a matrix to organize and describe the relationship between evidence and alternative hypotheses [30]. The sequence of the analysis-synthesis process (Figure 6.11) includes:

1. *Hypothesis synthesis.* A multidisciplinary team of analysts creates a set of feasible hypotheses, derived from imaginative consideration of all possibilities before constructing a complete set that merits detailed consideration.

2. *Evidence analysis.* Available data is reviewed to locate relevant evidence and inferences that can be assigned to support or refute the hypotheses. Explicitly identify the assumptions regarding evidence and the arguments of inference. Following the processes described in the last chapter, list the evidence-argument pairs (or chains of inference) and identify, for each, the intrinsic value of its contribution and the potential for being subject to denial or deception (D&D).

3. *Matrix synthesis.* Construct an ACH matrix that relates evidence-inference to the hypotheses defined in step 1.

4. *Matrix analysis.* Assess the diagnosticity (the significance or diagnostic value of the contribution of each component of evidence and related inferences) of each evidence-inference component to each hypothesis. This process proceeds for each item of evidence-inference *across* the rows, considering how each item may contribute to each hypothesis. An entry may be supporting (consistent with), refuting (inconsistent with), or irrelevant (not applicable) to a hypothesis; a contribution notation (e.g., +, −, or N/A, respectively) is marked within the cell. Where possible, annotate the likelihood (or probability) that this evidence would be observed if the hypothesis is true. Note that the diagnostic significance of an item of evidence is reduced as it is consistent with multiple hypotheses; it has no diagnostic contribution when it supports, to any degree, all hypotheses.

5. *Matrix synthesis (refinement).* Evidence assignments are refined, eliminating evidence and inferences that have no diagnostic value.

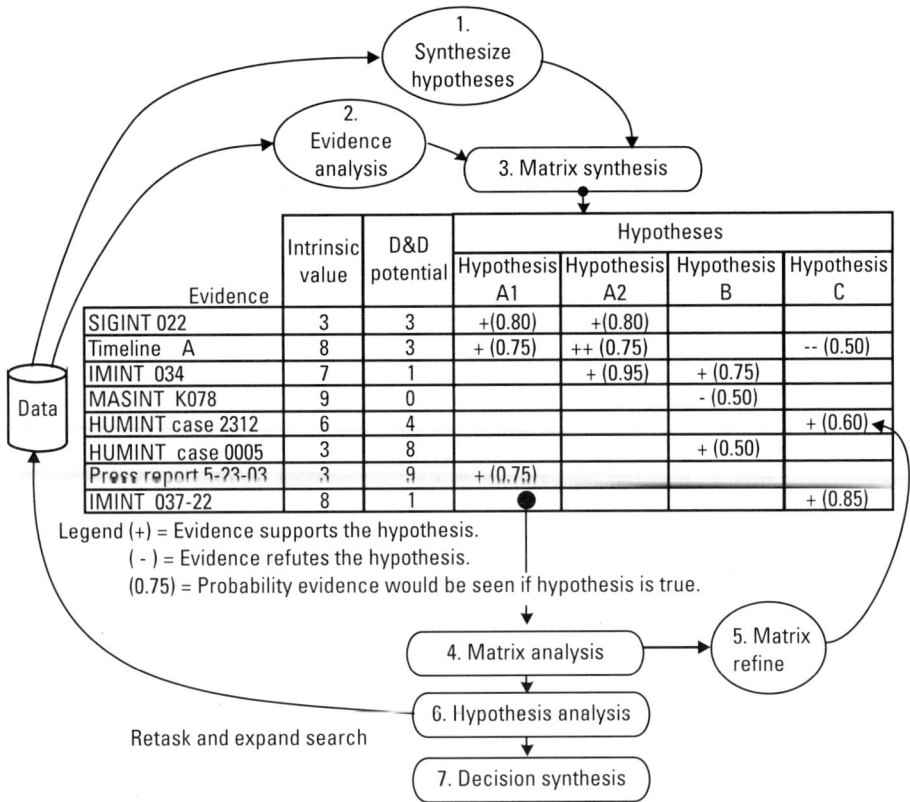

Figure 6.11 ACH process flow.

6. *Hypotheses analysis.* The analyst now proceeds to evaluate the likelihood of each hypothesis, by evaluating entries *down* the columns. The likelihood of each hypothesis is estimated by the characteristics of supporting and refuting evidence (as described in the last chapter). Inconsistencies and gaps in expected evidence provide a basis for retasking; a small but high-confidence item that refutes the preponderance of expected evidence may be a significant indicator of deception. The analyst also assesses the sensitivity of the likely hypothesis to contributing assumptions, evidence, and the inferences; this sensitivity must be reported with conclusions and the consequences if any of these items are in error. This process may lead to retasking of collectors to acquire more data to support or refute hypotheses and to reduce the sensitivity of a conclusion.

7. *Decision synthesis (judgment).* Reporting the analytic judgment requires the description of all of the alternatives (not just the most

likely), the assumptions, evidence, and inferential chains. The report must also describe the gaps, inconsistencies, and their consequences on judgments. The analyst must also specify what should be done to provide an update on the situation and what indictors might point to significant changes in current judgments.

Notice that the ACH approach deliberately focuses the analyst's attention on the contribution, significance, and relationships of evidence to hypotheses, rather than on building a case for any one hypothesis. The analytic emphasis is, first, on evidence and inference across the rows, before evaluating hypotheses, down the columns.

We can now illustrate an example structured analytic flow that leads to the ACH process. The example process integrates a variety of services to explore and determine the characteristics, behavior, and locations of a criminal organization from a massive volume of disparate and unstructured data. In this case, an unstructured database may include entities and events (Table 6.6) provided by collectors of financial, communication transactions, and known organizational relationships.

The stages of the structured analysis-synthesis methodology (Figure 6.12) are summarized in the following list:

- *Organize.* A data mining tool (described in Chapter 8, Section 8.2.2) automatically clusters related data sets by identifying linkages (relationships) across the different data types. These linked clusters are visualized using link-clustering tools used to visualize clusters and linkages to allow the analyst to consider the meaningfulness of data links and discover potentially relevant relationships in the real world.

- *Conceptualize.* The linked data is translated from the abstract relationship space to diagrams in the temporal and spatial domains to assess

Table 6.6
Typical Criminal Database Categories

Data Category	Entities	Events
Organizational	People, positions (roles), organizations	Decisions, commands (orders), statements
Financial	Accounts, banks, owners	Open/close accounts, transactions
Communications	Senders, recipients, messages, media (channels)	Message transactions
Travel	Travelers, agencies, flight numbers, airlines, payments	Time and date of transit and clearance through customs

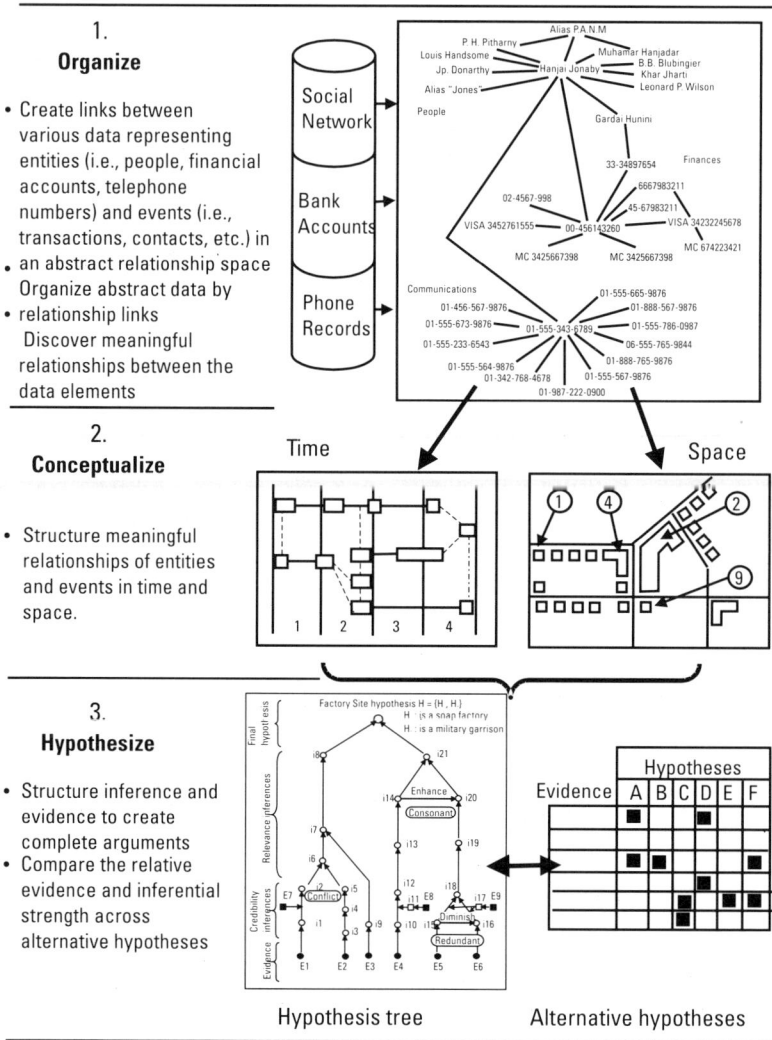

Figure 6.12 An example of an organizational link analysis process flow.

real-world implications of the relationships. These temporal and spatial models allow the analyst to conceptualize alternative explanations that will become working hypotheses. Analysis in the time domain considers the implications of sequence, frequency, and causality, while the spatial domain considers the relative location of entities and events.

- *Hypothesize.* The analyst synthesizes hypotheses, structuring evidence and inferences into alternative arguments that can be evaluated using the method of alternative competing hypotheses. In the course of this process, the analyst may return to explore the database and linkage diagrams further to support or refute the working hypotheses.

Notice that this process moves from an abstract domain (links of various data types) to the time and space domain, where the analyst considers feasibility of explanations, and then back to the abstract domain of hypothetical evidence and inference relationships.

6.8 Countering Denial and Deception

Because the targets of intelligence are usually high-value subjects (e.g., intentions, plans, personnel, weapons or products, facilities, or processes), they are generally protected by some level of secrecy to prevent observation. The means of providing this secrecy generally includes two components:

1. *Denial.* Information about the existence, characteristics, or state of a target is denied to the observer by methods of concealment. Camouflage of military vehicles, emission control (EMCON), operational security (OPSEC), and encryption of e-mail messages are common examples of denial, also referred to as *dissimulation* (hiding the real).

2. *Deception.* Deception is the insertion of false information, or *simulation* (showing the false), with the intent to distort the perception of the observer. The deception can include misdirection (m-type) deception to reduce ambiguity and direct the observer to a simulation—away from the truth—or ambiguity (a-type) deception, which simulates effects to increase the observer's ambiguity or understanding about the truth [31].

D&D methods are used independently or in concert to distract or disrupt the intelligence analyst, introducing distortions in the collection channels, ambiguity in the analytic process, errors in the resulting intelligence product, and misjudgment in decisions based on the product. Ultimately, this will lead to distrust of the intelligence product by the decision maker or consumer. Strategic D&D poses an increasing threat to the analyst, as an increasing number of channels for D&D are available to deceivers [32]. Six distinct categories of strategic D&D operations (Table 6.7) have different target audiences, means of implementation, and objectives.

Propaganda or psychological operations (PSYOP) target a general population using several approaches. *White propaganda* openly acknowledges the

Table 6.7
Categories of Strategic Deception Operations

	Strategic Denial and Deception Operations					
	Propaganda (PSYOP)			Denial and Deception		
	White	Gray	Black	Leadership Deception	Intelligence Deception	Denial (OPSEC)
Objective	Influence a general belief to an audience with an interest			Induce a specific belief to an audience with focused interest on a given topic (or target)		Deny access to information about intent and capabilities
Target Audience	Population at large			National or military leadership	Intelligence collectors or analysts	
Deception Methods and Objectives	Use declared sources and organizations to influence target audiences to accept general beliefs	Use undeclared sources and organizations to influence target audiences to accept general beliefs	Use false sources and organizations to influence target audiences to accept general beliefs	Use diplomatic channels and sympathetic influences to induce beliefs Use open news sources and channels to induce beliefs	Deceive and defeat human and technical collectors	Minimize the signature of entities and activities

source of the information, *gray propaganda* uses undeclared sources. *Black propaganda* purports to originate from a source other its actual sponsor, protecting the true source (e.g., clandestine radio and Internet broadcast, independent organizations, or agents of influence [33]). Coordinated white, gray, and black propaganda efforts were strategically conducted by the Soviet Union throughout the Cold War as *active measures* of disinformation:

> … for the Soviet Union, active measures constitute a dynamic and integrated array of overt and covert techniques for influencing events and behavior in, and the actions of, foreign countries. These measures are employed to influence the policies of other governments, undermine confidence in the leaders and institutions of these states, disrupt the relations between various nations, and discredit and weaken major opponents. This

frequently involves attempts to deceive the target, and to distort the target's perception of reality [34].

PSYOP activities are doctrinally distinct from the following deception operations. *Leadership deception* targets leadership or intelligence consumers, attempting to bypass the intelligence process by appealing directly to the intelligence consumer via other channels. Commercial news channels, untrustworthy diplomatic channels, suborned media, and personal relationships can be exploited to deliver deception messages to leadership (before intelligence can offer D&D cautions) in an effort to establish mindsets in decision makers. The literature examining the examples and principles of D&D employed by military leadership through history and particularly during the Second World War is extensive [35]. The effects of leadership deception in international politics have been described by Jervis [36].

Intelligence deception specifically targets intelligence collectors (technical sensors, communications interceptors, and humans) and subsequently analysts by combining denial of the target data and by introducing false data to disrupt, distract, or deceive the collection or analysis processes (or both processes). The objective is to direct the attention of the sensor or the analyst away from a correct knowledge of a specific target. Military deception is directed at an adversary's surveillance and reconnaissance sensors, seeking to misdirect sensors away from knowledge of true force movements, capabilities, and intentions. The successful covert preparation and conduct of a nuclear test by India in 1998 was studied extensively by the U.S. to determine the effectiveness of India's intelligence deception activities to counter U.S. national intelligence. The study concluded that countering D&D required greater rigor and more collaborative cross-INT analysis:

> More rigor needs to go into analysts' thinking when major events take place. Two mechanisms would help: A) bring in outside substantive experts in a more systematic fashion, so that we work against this "everybody thinks like us" mindset. And, B) bring in experts in the process of analysis when the IC faces a transition on a major intelligence issue, like the [Indian political party] BJP election, and like other things that you can think of. Look at establishing effective mechanisms to guarantee stronger integration of the analysis and greater collaboration and coordination of intelligence agencies and disciplines. So that instead of looking up at each of these stovepipes, we look at the product and the interaction between the stovepipes [37].

Denial operations by means of OPSEC seek to deny access to true intentions and capabilities by minimizing the signatures of entities and activities.

The cognitive shortcomings noted in the prior section can contribute to self deception on the part of the analyst. Earnest Mays summarized the three basic vulnerabilities of the intelligence analysts:

[Analysts] are vulnerable in the first place because they follow an almost unavoidable rule of trying to fit the evidence they have into some coherent, rational whole. … They are vulnerable in the second place because, partly perhaps from awareness of the power of prejudice and preconception, they have a preference for data that is quantifiable and therefore appear comparatively objective. … And thirdly they are vulnerable to deception because, after having to judge hard issues, they are prone to look for confirming rather than disconfirming evidence [38].

Two primary categories of countermeasures for intelligence deception must be orchestrated to counter either the simple deception of a parlor magician or the complex intelligence deception program of a rogue nation-state. Both collection and analysis measures (Table 6.8) are required to provide the careful observation and critical thinking necessary to avoid deception. Improvements in collection can provide broader and more accurate coverage, even limited penetration of some covers. Because of this, there is often a tendency to focus on collection improvements over analysis, but Richards Heuer has noted that "Any systematic counterdeception program must focus primarily on problems of analysis, only

Table 6.8
Countermeasures to D&D

	Collection	Analysis
Deceiver's Objective	Dissimulation—deny the observation and detection of true target phenomena Simulation—insert false signals to simulate a false situation that is expected; draw attention of collection assets away from true target	Reinforcement—reinforce analyst's mental sets and expectations; condition the analyst to reduce sensitivity to changes and target phenomena Integration—bring together a combination of signals to create false interpretation; reduce probability of correct interpretation or beliefs assigned to correct hypotheses
Counterdeception Measures	Increase sensor spatial resolution, phenomenological domains (multi-INT) Increase sensor sampling rates: revisit rate and dwell time on target Evaluate voracity of human collection sources	Create awareness of a target's potential D&D capabilities and activities Collect and index potential D&D indicators (e.g., incongruities) Consider and evaluate D&D hypotheses in analysis Estimate the opponent's D&D plan as an intelligence target itself

secondarily on collection" [39]. Deception is an effect of the analyst's mind; countermeasures to deception must likewise aid the analyst's reasoning process to consider the alternative D&D hypothesis and the available, missing, and negative evidence. Deceived analysts are susceptible to surprise about the occurrence of events of significant consequence for one of several reasons: they are unaware of the possibility of an event—a failure to envision a possibility or predict; they lack sufficient evidence and inferences to predict an event—a failure to detect; or both. The problem of mitigating intelligence surprise, therefore, must be addressed by considering both *large numbers of models or hypotheses (analysis)* and *large sets of data* (collection, storage, and analysis) [40].

In his classic treatise, *Strategem*, Barton Whaley exhaustively studied over 100 historical D&D efforts and concluded, "Indeed, this is the general finding of my study—that is, the deceiver is almost always successful regardless of the sophistication of his victim in the same art. On the face of it, this seems an intolerable conclusion, one offending common sense. Yet it is the irrefutable conclusion of historical evidence" [41]. In spite of this dire conclusion, Whaley offered hope for the intelligence analyst to counter the threat of a well-orchestrated intelligence D&D operation. First, he notes that exhortations to awareness (i.e., D&D awareness training for the analyst) and the study of examples of D&D are necessary but insufficient to protect against deception. Second, he states that the analyst should be supported by a decision model "designed to analyze the signals of stratagem, rather than the one designed to synthesize their false signals" [42]. By this, Whaley indicated that the opponent's deception plan, itself, must be an active target of intelligence analysis; attempting to filter out deceptive signals alone is insufficient. Michael Handel agreed:

> Analytic awareness provided by training and simulation is not only insufficient, but some caution that it is paradoxically dangerous because increasing the level of alertness raises our skepticism of all evidence and causes us to consider even the noise—potentially increasing our susceptibility to deception [43].

The components of a rigorous counter D&D methodology, then, include the estimate of the adversary's D&D plan as an intelligence subject (target) and the analysis of specific D&D hypotheses as alternatives (Figure 6.13). Incorporating this process within the ACH process described earlier amounts to assuring that reasonable and feasible D&D hypotheses (for which there may be no evidence to induce a hypothesis) are explicitly considered as alternatives. The process maintains a knowledge base of the adversary's past patterns of D&D (and their estimated current capabilities) and of the vulnerabilities of one's own current intelligence capabilities (e.g., current vulnerable channels). This process has

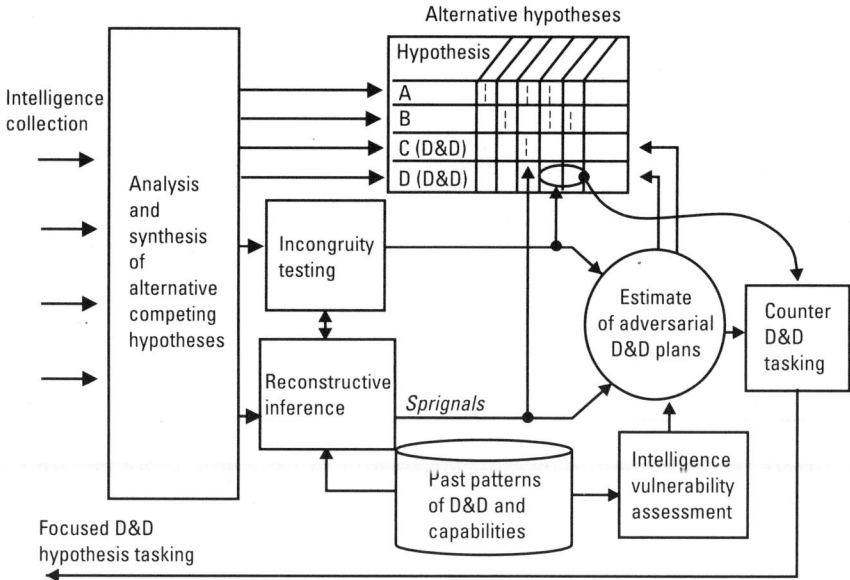

Figure 6.13 Components of counter D&D analysis process.

been described by Harris in a study on countering D&D and includes two active searches for evidence to support, refute, or refine the D&D hypotheses [44]:

1. *Reconstructive inference.* This deductive process seeks to detect the presence of spurious signals (Harris call these *sprignals*) that are indicators of D&D—the faint evidence predicted by conjectured D&D plans. Such sprignals can be strong evidence confirming hypothesis A (the simulation), weak contradictory evidence of hypothesis C (leakage from the adversary's dissimulation effort), or missing evidence that should be present if hypothesis A were true.

2. *Incongruity testing.* This process searches for inconsistencies in the data and inductively generates alternative explanations that attribute the incongruities to D&D (i.e., D&D explains the incongruity of evidence for more than one reality in simultaneous existence).

These processes should be a part of any rigorous alternative hypothesis process, developing evidence for potential D&D hypotheses while refining the estimate of the adversaries' D&D intents, plans, and capabilities. The processes also focus attention on special collection tasking to support, refute, or refine current D&D hypotheses being entertained.

6.9 Summary

Central to the intelligence cycle, analysis-synthesis requires the integration of human skills and automation to provide description, explanation, and prediction with explicit and quantified judgments that include alternatives, missing evidence, and dissenting views carefully explained. The challenge of discovering the hidden, forecasting the future, and warning of the unexpected cannot be performed with infallibility, yet expectations remain high for the analytic community. The U.S. director of central intelligence (DCI) has described these expectations:

> What, then, if not infallibility, should our national leaders, and ultimately the American public, expect of our analysts?
>
> • First and foremost, they should expect our analysts to deliver intelligence that is objective, pulls no punches, and is free from political taint.
> • Next, they should expect that our analysts think creatively, constantly challenging the conventional wisdom and tapping expertise wherever it lies—inside the IC or in the private sector and academia.
> • They should expect that our analysts always act with the highest standards of professionalism.
> • They should expect that they take risks—analytic risks—and make the tough calls when it would be easier to waffle.
> • They should expect that they respond to the President's and other decision makers' needs on demand—juggling analytic priorities and capabilities to meet the most urgent missions.
> • And, finally, they should expect that our analysis not only tell policymakers about what is uppermost on their minds, but also alert them to things that have not yet reached their in boxes [45].

The practical implementation of collaborative analysis-synthesis requires a range of tools to coordinate the process within the larger intelligence cycle, augment the analytic team with reasoning and sensemaking support, overcome human cognitive shortcomings, and counter adversarial D&D. In the next two chapters, we introduce the integration of the analyst's tradecraft with KM technologies for internalization-externalization (Chapter 7) and combination (Chapter 8).

Endnotes

[1]   Kam, E., *Surprise Attack*, Boston: Harvard University Press, 1988, p. 120.

[2]   "Foreign Missile Threats: Analytic Soundness of Certain National Intelligence Estimates,"
      U.S. Government Accounting Office, B-274120, August 30, 1996, accessed   on-line
      in December 2001 at http://www.house.gov/hasc/openingstatementsandpressreleases/
      104thcongress/gaonie.pdf.

[3]   "Commission to Assess the Ballistic Missile Threat to the United States," Side Letter to
      the Rumsfeld Commission Report, March 18, 1999. This unclassified letter was prepared
      subsequent to the 1998 formal report to specifically articulate the commission's concerns
      about intelligence analysis processes.

[4]   *A Compendium of Analytic Tradecraft Notes, Volume I (Notes 1–10)*, Washington, D.C.:
      CIA, 1995. Note 3 addresses the means of articulating assumptions, note 4 addresses the
      methods to articulate alternative outcomes (hypotheses), and note 5 addresses the methods
      to depict facts and sourcing in intelligence judgments.

[5]   McCarthy, M., "The Mission to Warn: Disaster Looms," *Defense Intelligence Journal*,
      Vol.7, No.2, 1998, p. 21.

[6]   Note that the figure includes six steps at the top while the intelligence cycle introduced in
      Chapter 2 has five steps. The planning and direction step has been divided into planning
      and tasking in the table to facilitate discussion of the models.

[7]   Wohl, J. G., "Force Management Decision Requirements for Air Force Tactical Com-
      mand and Control," *IEEE Trans. Systems, Man, and Cybernetics*, Vol. SMC-11, No. 9,
      September 1981, pp. 618–639. For a description of the application of the SHOR model
      in automated intelligence data fusion, see, Waltz, E. L., and D. M. Buede, "Data Fusion
      and Decision Support for Command and Control," *IEEE Transactions on Systems, Man,
      and Cybernetics*, Vol. SMC-16, No. 6 November–December 1986), pp. 865–879.

[8]   Boyd, J. R., "The Essence of Winning and Losing," unpublished briefing, January 1996.
      For an overview of command models, see "Appendix: Alternative Models of Command
      and Control," *Command Concepts: A Theory Derived from the Practice of Command and
      Control*, MR-775-OSD, RAND, 1999. For commentaries on the contributions of John
      Boyd, see: Hammond, G. T., *The Mind of War: John Boyd and American Security*, Wash-
      ington D.C.: Smithsonian Institute Press, 2001; and Coram, R., *Boyd: The Fighter Pilot
      Who Changed the Art of War*, Boston: Little, Brown, 2002.

[9]   This model has also been called TCPED to include collection. For a discussion of the
      Multi-INT TPED process, see Section 14 in *The Information Edge: Imagery Intelligence
      and Geospatial Information In an Evolving National Security Environment*, Report of the
      Independent Commission on the National Imagery and Mapping Agency, Washington,
      D.C., January 9, 2001.

[10] The JDL model is described in further detail in Chapter 8.

[11] The concept and figure is adapted from Minsky, M., "Common-Sense Based Interfaces,"
      Communications of the ACM, Vol. 43, No. 8, p. 71. Minsky first published a version of
      this chart in July 1992 in *Toshiba Review*, Vol. 47, No. 7, accessed on-line on August 14,
      1998, at http://minsky,www.media.mit.edu/people/minsky/papers/CausalDiversity/html.

[12] Figure adapted from Minsky, M. "Common-Sense Based Interfaces," *Communications of
      the ACM*, Vol. 43, No. 8. Used by permission from ACM.

[13] Weick, K., *Sensemaking in Organizations*, Thousand Oaks, CA: Sage, 1995.

[14]   Damasio, A. R., *Descartes' Error: Emotion, Reason and the Human Brain,* New York: Putnam, 1994.

[15]   Davis, J., "Combating Mind-Set," *Studies in Intelligence,* Vol. 36, No. 5, 1992, p. 33.

[16]   Heuer Jr., R. J., *Psychology of Intelligence Analysis,* Washington D.C.: CIA Center for the Study of Intelligence, 1999. This table summarizes the biases described in Chapters 10, 11, and 12.

[17]   Gilovich, T., *How We Know What Isn't So,* New York: Free Press, 1991.

[18]   Davis, J., "Combating Mind Set," *Studies in Intelligence,* Vol. 36, No. 5, 1992, pp. 33–38.

[19]   See "Symposium on the Psychology of Intelligence," in *Bulletin of the Center for the Study of Intelligence,* Issue 11, Summer 2000, p. 1.

[20]   Heuer Jr., R. J., *Psychology of Intelligence Analysis,* Chapter 1, Washington D.C.: CIA Center for the Study of Intelligence, 1999.

[21]   Jones, M., *Thinker's Toolkit,* New York: Three Rivers Press, 1995, pp. 12–46.

[22]   The U.S. Joint Military Intelligence College, for example, emphasizes the importance of applying structured methodologies. See, Brei, W*., Getting Intelligence Right: The Power of Logical Procedure,* Occasional Paper 2, Joint Military Intelligence College, Washington D.C., January 1996, and Folker, R. D., *Intelligence Analysis in Theater Joint Intelligence Centers: An Experiment in Applying Structured Methods,* Occasional Paper 7, Joint Military Intelligence College, Washington D.C., January 2000.

[23]   "The Chances of an Imminent Communist Chinese Nuclear Explosion," Special National Intelligence Estimate, SNIE-13-4-64, August 26, 1964, in Rufner, K. C. (ed.), *Corona: America's First Satellite Program,* Washington D.C.: CIA Center for the Study of Intelligence, 1995, p. 239.

[24]   Toulmin, S. E., *The Uses of Argument,* Cambridge, England: Cambridge University Press, 1958.

[25]   Schum, D. A., *Evidence and Inference for the Intelligence Analyst,* Vols. I and II, Lanham, MD: University Press of America, 1987; this text was authored while Schum was a scholar in residence at the CIA.

[26]   Schum, D. A., *The Evidential Foundations for Probabilistic Reasoning,* Evanston IL: Northwestern University Press, 2001. The brief introduction to inferential networks in this section is based on Schum's exhaustive treatment, but does not approach the many critical nuances of the theory developed by Schum. The reader is encouraged to turn to Schum's works for the details necessary to implement inferential nets.

[27]   See, for example: Waltz, E., and J. Llinas, *Multisensor Data Fusion,* Norwood MA: Artech, 1990; Hall, D. L., *Mathematical Techniques in Multisensor Data Fusion,* Artech House: Boston, 1992; Antony, R., *Principles of Data Fusion Automation,* Boston: Artech House, 1995; Pearl, J., *CAUSALITY: Models, Reasoning, and Inference,* Cambridge, England: Cambridge University Press, 2000; Hall, D. L. and J. Llinas (eds.), *Handbook of Multisensor Data Fusion,* Boca Raton: CRC Press, 2001.

[28]   Heuer, R. J., Jr., *Psychology of Intelligence Analysis,* Chapter 4, "Strategies for Analytic Judgment," Washington D.C.: CIA Center for the Study of Intelligence, 1999.

[29]   See Anderson, T., and William T., *Analysis of Evidence: How to Do Things with Facts Based on Wigmore's Science of Judicial Proof,* Evanson, IL: Northwestern University Press, 1998. Wigmore's original presentation was in: Wigmore, J.H., *The Science of Judicial Proof,* Boston: Little Brown, 1937.

[30]   This process is adapted from the eight-step process in Heuer, R. J., Jr., *Psychology of Intelligence Analysis,* Chapter 8: "Analysis of Competing Hypotheses." See also Sawka, K., "Competing Hypothesis Analysis," *Competitive Intelligence,* Vol. 2, No. 3, July–Sept. 1999, pp. 37–38.

[31]   Daniel, D. C., and K. L. Herbig, "Propositions on Military Deception," in *Strategic Military Deception,* Daniel, Donald, C., and Herbig, K. L., (eds.), New York: Pergamon, 1982, p. 5.

[32]   For a discussion of new D&D challenges, see Wirtz, J. J., and R. Godson, *Strategic Denial and Deception: The 21st Century Challenge,* New Brunswick, NJ: Transaction Publishers, 2002.

[33]   Agent-of-influence operations carry out the subornation (knowing or unwitting) of a person who will use their position, influence, power, or credibility to promote the objectives of a foreign power.

[34]   Schultz, R. H., and R. Goodson, *Dezinformatsia: Active Measures in Soviet Strategy,* Washington D.C.: Pergamon-Brasseys, 1984, p.16; see also Bittman, Ladislav, *The KGB and Soviet Disinformation,* Washington D.C.: Pergamon-Brassey's, 1985.

[35]   See Herbig, D., D., and K. Herbig, *Strategic Military Deception,* New York: Pergamon Press, 1982; Wohlstetter, R., *Pearl Harbor: Warning and Decision,* Palo Alto, CA: Stanford University Press, 1962; and Hesketh, R. (with foreword by Nigel West), *Fortitude: The D-Day Deception Campaign,* London: St Ermin's Press, 1999.

[36]   Jervis, R., *Perception and Misperception in International Politics,* Princeton, NJ: Princeton University Press, 1976.

[37]   Quotation by Adm. Jeremiah in news conference on recommendations from the Study of U.S. Intelligence Community Performance in Detecting Indian Nuclear Testing, June 2, 1998, released CIA Public Affairs Staff on June 4, 1998.

[38]   May, E. R., "Capabilities and Proclivities," in *Knowing One's Enemies: Intelligence Assessment before the Two World Wars,* May, Ernest R. (ed.), Princeton, NJ: Princeton University Press) 1984, pp. 537–538.

[39]   Heuer, R., "Cognitive Factors on Deception and Counterdeception," in *Strategic Military Deception,* Daniel, Donald C., and Katherine L. Herbig (eds.), New York: Pergamon, 1982, p. 61. A prevision version of the chapter was published as "Strategic Deception and Counterdeception," *International Studies Quarterly,* Vol. 25, No. 2, June 1981, pp. 294–327.

[40]   Waltz, E., "Employing Data Fusion Tools within Intelligence Community Analysis," in *Proc. of National Symp. on Sensor and Data Fusion,* August 2002.

[41]   Whaley, B., *Strategem: Deception and Surprise in War,* unpublished monograph, MIT, 1969.

[42]   Whaley, B., *Strategem: Deception and Surprise in War*, unpublished monograph, MIT, 1969, p. 147.

[43]   Handel, M., I., "Intelligence and Deception," in Gooch, J., and A. Perlmutter, *Military Deception and Strategic Surprise*, London: Frank Cass, 1982, p. 144.

[44]   Harris, W. R., *On Countering Strategic Deception*, R-1230-ARPA, November 1973, pp. 33–50.

[45]   Remarks of DCI George J. Tenet, "Opening Remarks," The Conference on CIA's Analysis of the Soviet Union, 1947–1991, Princeton University, March 8, 2001, accessed on-line on August 21, 2001, at http://www.odci.gov/cia/public_affairs/speeches/dci_speech_03082001.html.

7

Knowledge Internalization and Externalization

The process of conducting knowledge transactions between humans and computing machines occurs at the intersection between tacit and explicit knowledge, between human reasoning and sensemaking, and the explicit computation of automation. The processes of externalization (tacit-to-explicit transactions) and internalization (explicit-to-tacit transactions) of knowledge, however, are not just interfaces between humans and machines; more properly, the intersection is between human thought, symbolic representations of thought, and the observed world. When an analyst writes, sketches, types on a keyboard, or explains some experience or conceptual mental model in the symbols of language, externalization is taking place. When collected sensor data, message traffic, a foreign television interview, or a computer visualization of multidimensional data is viewed and mentally absorbed by an analyst, internalization is taking place. Both processes conduct a transaction of information between the human mind (and emotion) and another explicit symbolic abstraction that represents the real world. In this chapter, we examine externalization and internalization in the intelligence workflow and the tools that support the processes.

7.1 Externalization and Internalization in the Intelligence Workflow

The knowledge-creating spiral described in Chapter 3 introduced the four phases of knowledge creation. In Chapter 4, we described the role of *collaborative* services to enable the socialization phase, where analysts socialize to exchange tacit knowledge. The externalization, combination, and internalization activities of the spiral deal with analyst interaction with explicit information; each activity has interactions between the analyst and supporting *cognitive* services (or tools) unique to each phase. The workflow in these three phases of the spiral includes distinguishable tacit activities in the analyst's mind and complementary tools or services to support those activities (Figure 7.1).

Externalization

Following social interactions with collaborating analysts, an analyst begins to explicitly frame the problem. The process includes the decomposition of the intelligence problem into component parts (as described in Section 2.2) and explicit articulation of essential elements of information required to solve the problem. The tacit-to-explicit transfer includes the explicit listing of these essential elements of information needed, candidate sources of data, the creation of searches for relevant SMEs, and the initiation of queries for relevant knowledge within current holdings and collected all-source data. The primary tools to interact with all-source holdings are query and retrieval tools that search and retrieve information for assessment of relevance by the analyst. The analyst also uses organizing tools to link relevant data (evidence) to the corresponding information requirements that they address. This process externalizes the

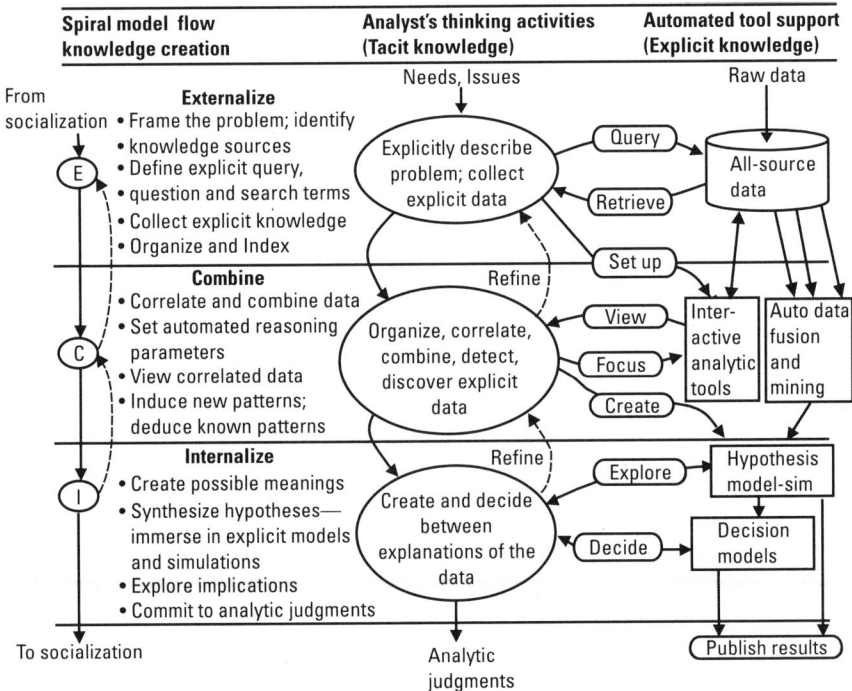

Figure 7.1   Analyst-computing services workflow interactions.

decomposed problem, collection requirements, and organized evidence, but the analyst's emerging (tacit) mental models of explanations are not yet externalized.

Combination

This explicit-explicit transfer process correlates and combines the collected data in two ways:

1. *Interactive analytic tools.* The analyst uses a wide variety of analytic tools (discussed further in this chapter) to compare and combine data elements to identify relationships and marshal evidence against hypotheses.

2. *Automated data fusion and mining services.* Automated data combination services (discussed further in Chapter 8) also process high-volume data to bring detections of known patterns and discoveries of "interesting" patterns to the attention of the analyst.

While the analyst is using these interactive tools and automated services to combine explicit data, the analyst also observes the results and continues to create and modify emerging tacit mental models that explain and represent meaning contained in the data.

Internalization

The analyst integrates the results of combination in two domains: external hypotheses (explicit models and simulations) and decision models (like the alternative competing hypothesis decision model introduced in the last chapter) are formed to explicitly structure the rationale between hypotheses, and internally, the analyst develops tacit experience with the structured evidence, hypotheses, and decision alternatives. Internalization refines the mental models and their explicit counterparts that can be shared with other analysts (for further collaborative socialization) and with consumers. In this explicit-to-tacit transfer phase, the analyst begins to develop confidence and form judgments about the meaning of the evidence; the internalization phase involves subjective sensemaking as well as the more objective cognitive reasoning about the hypotheses.

Studies of the intelligence workflow have identified the discrete tasks and associated cognitive processes to enhance analyst training and use of supporting tools [1]. This workflow describes the *operational* architecture of an intelligence enterprise; the structure of logical and physical components to implement the workflow is the *system* architecture. A typical intelligence enterprise system architecture, organized in three distinct tiers, is illustrated in Figure 7.2. Services in the *data tier* capture incoming data from processing pipelines (e.g., imagery

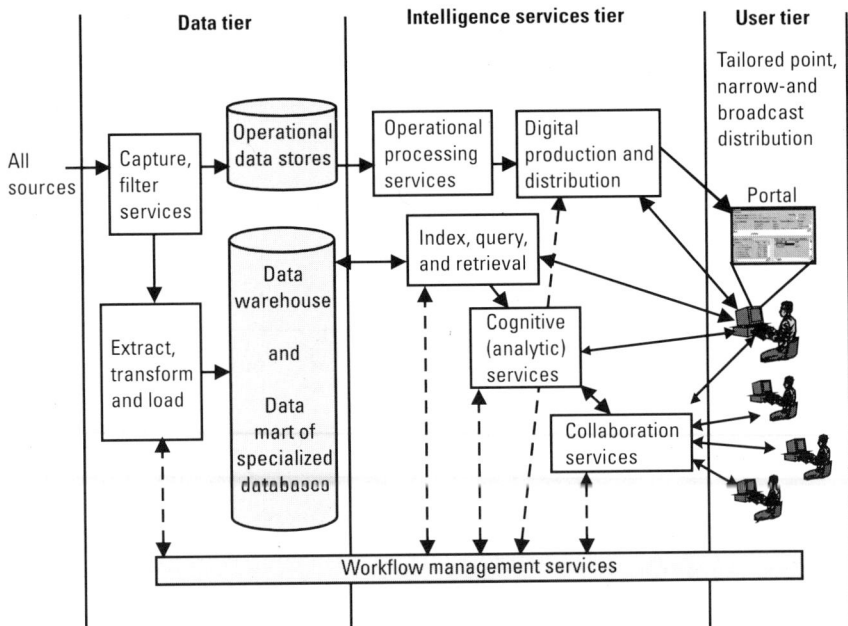

Figure 7.2   N-tier system architecture.

and signals producers), reporting sources (news services, intelligence reporting sources), and open Internet sources being monitored. Content appropriate for immediate processing and production, such as news alerts, indications, and warning events, and critical change data are routed to the operational storage for immediate processing. All data are indexed, transformed, and loaded into the long-term data warehouse or into specialized data stores (e.g., imagery, video, or technical databases). *The intelligence services tier* includes six basic service categories:

1. *Operational processing.* Information filtered for near-real-time critical-ity are processed to extract and tag content, correlate and combine with related content, and provide updates to operational watch offi-cers. This path applies the automated processes of data fusion and data mining to provide near-real-time indicators, tracks, metrics, and situa-tion summaries.

2. *Indexing, query, and retrieval.* Analysts use these services to access the cumulating holdings by both automated subscriptions for topics of interest to be pushed to the user upon receipt and interactive query and retrieval of holdings.

3. *Cognitive (analytic) services.* The analysis-synthesis and decision-making processes described in Chapters 5 and 6 are supported by cognitive services (thinking-support tools).

4. *Collaboration services.* These services, described in Chapter 4, allow synchronous and asynchronous collaboration between analytic team members.

5. *Digital production services.* Analyst-generated and automatically created dynamic products are produced and distributed to consumers based on their specified preferences.

6. *Workflow management.* The workflow is managed across all tiers to monitor the flow from data to product, to monitor resource utilization, to assess satisfaction of current priority intelligence requirements, and to manage collaborating workgroups.

Subsequent sections in this chapter describe the search and retrieval tools that are key to the externalization process (Section 7.2), as well as the analytic services that support interactive combination, hypothesis modeling, and decision making used in the internalization process (Section 7.3). Digital production processes are described in Section 7.4. The means of tacit-explicit interaction between the human and analyst and these services is described in Section 7.5.

7.2 Storage, Query, and Retrieval Services

At the center of the enterprise is the knowledge base, which stores explicit knowledge and provides the means to access that knowledge to create new knowledge. Before describing the means of query and retrieval, it is important to briefly distinguish the alternative database structures that make up the knowledge base.

7.2.1 Data Storage

Intelligence organizations receive a continuous stream of data from their own tasked technical sensors and human sources, as well as from tasked collections of data from open sources. One example might be Web *spiders* that are tasked to monitor Internet sites for new content (e.g., foreign news services), then to collect, analyze, and index the data for storage. The storage issues posed by the continual collection of high-volume data are numerous:

- *Diversity.* All-source intelligence systems require large numbers of independent data stores for imagery, text, video, geospatial, and special technical data types. These data types are served by an equally high number of specialized applications (e.g., image and geospatial analysis and signal analysis).

- *Legacy.* Storage system designers are confronted with the integration of existing (legacy) and new storage systems; this requires the integration of diverse logical and physical data types.

- *Federated retrieval and analysis.* The analyst needs retrieval, application, and analysis capabilities that span across the entire storage system.

These storage and application integration challenges are faced by most business enterprises and require storage structure and application integration trade-offs that influence performance, scalability, and system cost.

- *Storage structure alternatives.* Heterogeneous, or data-oriented, databases maintain independent, physically and logically separate databases for each unique data type; these generally collect raw intelligence data for initial processing. Data marts are information-oriented stores of processed information; they are application-specific, tailored to specific intelligence analysis domain needs (e.g., target development). Data marts are not easily scaled in breadth or expanded to other applications, and they may be a subset of a larger data warehouse. Large-scale data warehouses serve the strategic intelligence enterprisewide applications; they are centralized, shared (application independent), and scalable. Large intelligence enterprises employ all three categories of storage structures: Heterogeneous databases maintain all raw data sources for a short period for initial processing and analysis, selected data sets are transformed and loaded to data marts in a common format for midterm all-source analysis, and data warehouses retain long-term archives of high-value, high-quality data and finished intelligence products.

- *Application integration alternatives.* Three alternatives are considered to integrate the many analytic applications across heterogeneous data stores (Figure 7.3). Integration at the data level, or data federation, maintains independent data stores (e.g., IMINT and SIGINT) and specialized processing applications for each, but adds a data-level federation application to allow searches and analysis across the data sets (e.g., association of objects in imagery databases and emitter locations within signal databases). When the number of heterogeneous data stores grows large, an intermediate store (data mart or large data warehouse) can be created to maintain only the data necessary for federated processing. Data from the heterogeneous stores are extracted, transformed to a common format, and loaded onto the data mart or warehouse for

Approach	Structure	Characteristics
Data federation	Data fed	• Maintains local control, providing global viewing • Integrates many independent systems
Data marts and warehouses		• Extract-transform-load to warehouse • Historical archival analysis
Enterprise application integration	Middleware applications	• Synchronized updates • Unified view from multiple system reads • Not very scalable

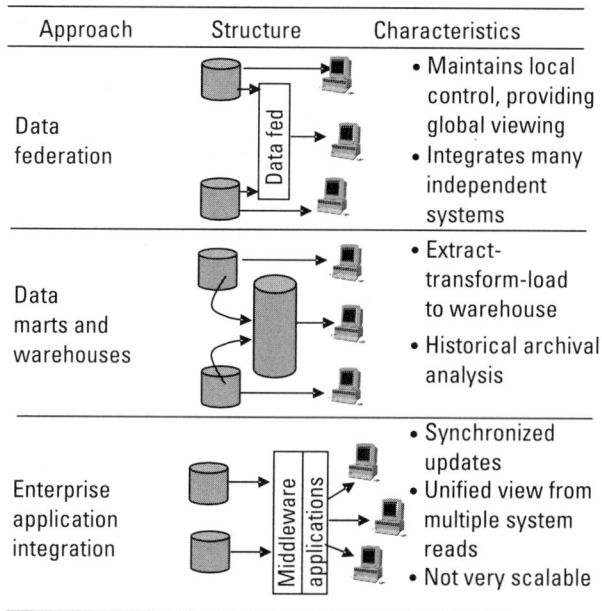

Figure 7.3 Alternative enterprise integration system architectures.

long-term archived analysis across all sources. Integration at the application level, or enterprise application integration (EAI), adds a layer of custom middleware that translates, synchronizes, and updates data from all heterogeneous stores across a layer of common applications.

7.2.2 Information Retrieval

Information retrieval (IR) is formally defined as "... [the] actions, methods and procedures for recovering stored data to provide information on a given subject" [2]. Two approaches to query and retrieve stored data or text are required in most intelligence applications:

1. *Data query and retrieval* is performed on structured data stored in relational database applications. Imagery, signals, and MASINT data are generally structured and stored in structured formats that employ structured query language (SQL) and SQL extensions for a wide variety of databases (e.g., Access, IBM DB2 and Informix, Microsoft SQL Server, Oracle, and Sybase). SQL allows the user to retrieve data by context (e.g., by location in data tables, such as date of occurrence) or by content (e.g., retrieve all record with a defined set of values).

2. *Text query and retrieval* is performed on both structured and unstructured text in multiple languages by a variety of natural language search engines to locate text containing specific words, phrases, or general concepts within a specified context.

Data query methods are employed within the technical data processing *pipelines* (IMINT, SIGINT, and MASINT). The results of these analyses are then described by analysts in structured or unstructured text in an analytic database for subsequent retrieval by text query methods. In this section, we briefly introduce the IR query and retrieval (or question and answer) approaches that deal with structured and unstructured text. Moldovan and Harabagiu have defined a five-level taxonomy of Q&A systems (Table 7.1) that range from the common keyword search engine that searches for relevant content (class 1) to reasoning systems that solve complex natural language problems (class 5) [3]. Each level requires increasing scope of knowledge, depth of linguistic understanding, and sophistication of reasoning to translate relevant knowledge to an answer or solution.

The first two levels of current search capabilities locate and return relevant content based on keywords (content) or the relationships between clusters of words in the text (concept). The performance of such retrieval is measured in terms of *precision* (the ratio of relevant content retrieved to the total content retrieved) and *recall* (the ratio of relevant content retrieved to the total relevant content available). These IR search engines offer the use of Boolean expressions, proximity searches to reduce the specificity around a word or concept, and weighted searching to specify different weights on individual search terms. The retrieved corpus may be analyzed to present the user with a taxonomy of the retrieved content to aid the user to select the most relevant content. While class 1 capabilities only match and return content that matches the query, class 2 capabilities integrate the relevant data into a simple response to the question.

Class 3 capabilities require the retrieval of relevant knowledge and reasoning about that knowledge to deduce answers to queries, even when the specific answer is not explicitly stated in the knowledge base. This capability requires the ability to both reason from general knowledge to specific answers and provide rationale for those answers to the user.

Class 4 and 5 capabilities represent advanced capabilities, which require robust knowledge bases that contain sophisticated knowledge representation (assertions and axioms) and reasoning (mathematical calculation, logical inference, and temporal reasoning). The DARPA High-Performance Knowledge Base and Rapid Knowledge Formation research programs and the Cycorp CYC® knowledge base and inference engine target these classes of Q&A performance. These problem-solving classes perform knowledge retrieval, rather than IR.

Table 7.1
A Taxonomy of Query and Retrieval Capabilities

Q&A Class	Level of Processing	Example Query and Answer
1. Dictionary search and find	Simple, heuristic pattern matching from query to relevant content Simple semantic expansion to locate related keywords (nouns)	Q: What forces are in Kirabiu? A: "… and the Kirabu Peninsula has mechanized brigades stationed …"
2. Ontology-based conceptual search	Use of ontology to reason about semantic relationships, actions (verbs) that makeup context Reasoning across multiple matching sets of content to integrate response	Q: How did the Kirabu brigade establish air defenses? A: Kirabu brigades used SA-6 units in 1995; Kirabu brigades trained with Condo radars in Brigauton in 1996; Kirabu terrain has three likely surveillance radar locations
3. Advanced natural language reasoning	Advanced natural language understanding Reasoning across multiple sets of facts to create new knowledge using deductive logic with uncertainty management	Q: Do SA-6 batteries defend Kirabu? A: No batteries are observed; condo radars emit from Kirabu; only SA-6 batteries employ condo; it is likely that Kirabu has camouflaged SA-6 batteries
4. Domain-specific high performance	Very large domain-specific (narrow) knowledge base with axiomatic knowledge Deductive and inductive reasoning	Q: What is the likely course of action for the Kirabu forces? A: Three COAs are … ; current weather conditions dictate that the likely COA is … because…
5. General problem solver	Very large, yet broad knowledge base with axiomatic knowledge across multiple domains Integrated deductive-inductive reasoning and the ability to model or simulate situations	Q: What should the Kandugan regime do to remain stable and economically viable if its neighbors annex the Kirabu Peninsula before winter? A: The solution is to …

7.3 Cognitive (Analytic Tool) Services

Cognitive services support the analyst in the process of interactively analyzing data, synthesizing hypotheses, and making decisions (choosing among alternatives). These interactive services support the analysis-synthesis activities described in Chapters 5 and 6. Alternatively called thinking tools, analytics, knowledge discovery, or analytic tools, these services enable the human to transform and view data, create and model hypotheses, and compare alternative hypotheses and consequences of decisions. The U.S. IC Strategic Investment Plan for Analysis has noted the importance of such services:

The development of analytic and data integration tools will be one of the most important and expensive areas for the analytic production community. Without such tools, the shrinking analytic workforce will have no hope of managing the flood of new intelligence information or shifting smoothly from one crisis or issue area to another. To achieve progress, we must develop: 1) An automated analytic workflow process relying on advanced analytic tools, such as visualization, search, processing, KM and dynamic updating, 2) New tools that reveal connections, facilitate analytic insights and deductions and streamline search by prioritizing information, automatically populating databases, and integrating data [4].

With similar emphasis, businesses have invested in OLAP tools to analyze numerical data, text retrieval tools for market and competitive analysis, and decision-support tools for management policymaking. Fuld and Company has published a survey of the growing line of commercial software tools suitable for competitive intelligence analysis [5].

A basic taxonomy of these interactive analytic tools (Table 7.2) organizes the tools into four basic categories by their use in the workflow.

Exploration tools allow the analyst to interact with raw or processed multimedia (text, numerical data, imagery, video, or audio) to locate and organize content relevant to an intelligence problem. These tools provide the ability to search and navigate large volumes of source data; they also provide automated taxonomies of clustered data and summaries of individual documents. The information retrieval functions described in the last subsection are within this category. The product of exploration is generally a relevant set of data/text organized and metadata tagged for subsequent analysis. The analyst may drill down to detail from the lists and summaries to view the full content of all items identified as relevant.

Reasoning tools support the analyst in the process of correlating, comparing, and combining data across all of the relevant sources. These tools support a wide variety of specific intelligence target analyses:

- *Temporal analysis.* This is the creation of timelines of events, dynamic relationships, event sequences, and temporal transactions (e.g., electronic, financial, or communication).

- *Link analysis.* This involves automated exploration of relationships among large numbers of different types of objects (entities and events).

- *Spatial analysis.* This is the registration and layering of 3D data sets and creation of 3D static and dynamic models from all-source evidence. These capabilities are often met by commercial geospatial information system and computer-aided design (CAD) software.

Table 7.2
A Basic Cognitive Service Taxonomy

1. Exploration	2. Reasoning	3. Sensemaking	4. Decision, Judgment
Search, navigate, organize, query, and explore (browse) data	Query for knowledge, create and structure hypothesis arguments; test hypotheses against data	Explore, evaluate, and compare alternative hypotheses; assign meaning	Evaluate COAs and consequences of decisions; weigh decision alternatives
Objects: data/text; massive volume $> 10^{10}$	Objects: information; information volume $> 10^{5}$	Objects: hypotheses; $\sim$ 10 hypotheses	Objects: decisions; $\sim$ 4 decision alternatives
Tools:	Tools:	Tools:	Decision support tools:
IR	Data/text mining (pattern discovery)	Modeling and simulation for immersion and exploration	Modeling and simulation for COA and consequence comparison
Ontology creation	Data/text fusion (pattern detection and content tracking)	Trend and forecast analysis	Risk analysis
Extraction (content, concepts, and relationships)	Change detection	Structured argumentation	Utility analysis
Conversion (content translation)	Link analysis	Alternative hypothesis comparison	Alternative decision comparison
Data/text clustering	Problem-solving knowledge retrieval	Creativity support	
Summarize, abstract, and categorize	Temporal-spatial mapping and analysis		
Filter, monitor database or Web site changes			
Visualize and interact with high-dimensionality data	Visualize and interact with organized information	Visualize and interact with arguments	Visualize and interact with decisions

- *Functional analysis.* This is the analysis of processes and expected observables (e.g., manufacturing, business, and military operations, social networks and organizational analysis, and traffic analysis).

While dealing with large data volumes, the tools in this category are interactive (in contrast with the fully automated processes described in the next chapter), allowing the analyst to manipulate data, processed results, and views of detected and discovered relationships in the data. These tools aid the analyst in five key analytic tasks:

1. *Correlation:* detection and structuring of relationships or linkages between different entities or events in time, space, function, or

interaction; association of different reports or content related to a common entity or event;

2. *Combination*: logical, functional, or mathematical joining of related evidence to synthesize a structured argument, process, or quantitative estimate;

3. *Anomaly detection*: detection of differences between expected (or modeled) characteristics of a target;

4. *Change detection*: detection of changes in a target over time—the changes may include spectral, spatial, or other phenomenological changes;

5. *Construction*: synthesis of a model or simulation of entities or events and their interactions based upon evidence and conjecture.

Sensemaking tools support the exploration, evaluation, and refinement of alternative hypotheses and explanations of the data. Argumentation structuring, modeling, and simulation tools in this category allow analysts to be immersed in their hypotheses and share explicit representations with other collaborators. This immersion process allows the analytic team to create shared meaning as they experience the alternative explanations. These tools and those that enhance creative thought, explore trends, and compare alternative hypotheses all support inductive thinking to discover subtle patterns and generalizations. They also support conjecture that leads to retroductive analysis (returning to the data to affirm or falsify conjectures).

Decision support (judgment) tools assist analytic decision making by explicitly estimating and comparing the consequences and relative merits of alternative decisions. These tools include models and simulations that permit the analyst to create and evaluate alternative COAs and weigh the decision alternatives against objective decision criteria. Decision support systems (DSSs) apply the principles of probability to express uncertainty and decision theory to create and assess attributes of decision alternatives and quantify the relative utility of alternatives. Normative, or decision-analytic DSSs, aid the analyst in structuring the decision problem and in computing the many factors that lead from alternatives to quantifiable attributes and resulting utilities [6]. These tools often relate attributes to utility by influence diagrams and compute utilities (and associated uncertainties) using Bayes networks [7].

The tools progressively move from data as the object of analysis (for exploration) to clusters of related information, to hypotheses, and finally on to decisions, or analytic judgments. Notice also that these tool categories move from high-volume, high-breadth tasks (exploration tools deal with massive volumes of raw data) to lower volume task dealing with great analytic depth (decision support tools). In addition to these analytic tools, intelligence workflow management software can provide a means to organize the process by providing the following functions:

- *Requirements and progress tracking:* maintains list of current intelligence requirements, monitors tasking to meet the requirements, links evidence and hypotheses to those requirements, tracks progress toward meeting requirements, and audits results;

- *Relevant data linking:* maintains ontology of subjects relevant to the intelligence requirements and their relationships and maintains a database of all relevant data (evidence);

- *Collaboration directory:* automatically locates and updates a directory of relevant subject matter experts as the problem topic develops.

We can illustrate one practical intelligence process spiral in a representative counter-drug analysis (though many spirals are generally involved) and the use of tools organized around a common knowledge base that accumulates the evidence and hypotheses about the problem at hand (Figure 7.4; compare this to Figure 7.1). The knowledge base also provides the current knowledge that may be digitally published to update consumers on progress in the analysis. In this example, an intelligence consumer has requested specific intelligence on a drug cartel named "Zehga" to support counter-drug activities in a foreign country. The sequence of one analyst's use of tools in the example include:

1. The process begins with synchronous collaboration with other analysts to discuss the intelligence target (Zehga) and the intelligence requirements to understand the cartel organization structure, operations, and finances. The analyst creates a peer-to-peer collaborative workspace that contains requirements, essential elements of information (EEIs) needed, current intelligence, and a directory of team members before inviting additional counter-drug subject matter experts to the shared space.

2. The analyst opens a workflow management tool to record requirements, key concepts and keywords, and team members; the analyst will link results to the tool to track progress in delivering finished intelligence. The tool is also used to request special tasking from technical collectors (e.g., wiretaps) and field offices.

3. Once the problem has been externalized in terms of requirements and EEIs needed, the sources and databases to be searched are selected (e.g., country cables, COMINT, and foreign news feeds and archives). Key concepts and keywords are entered into IR tools; these tools

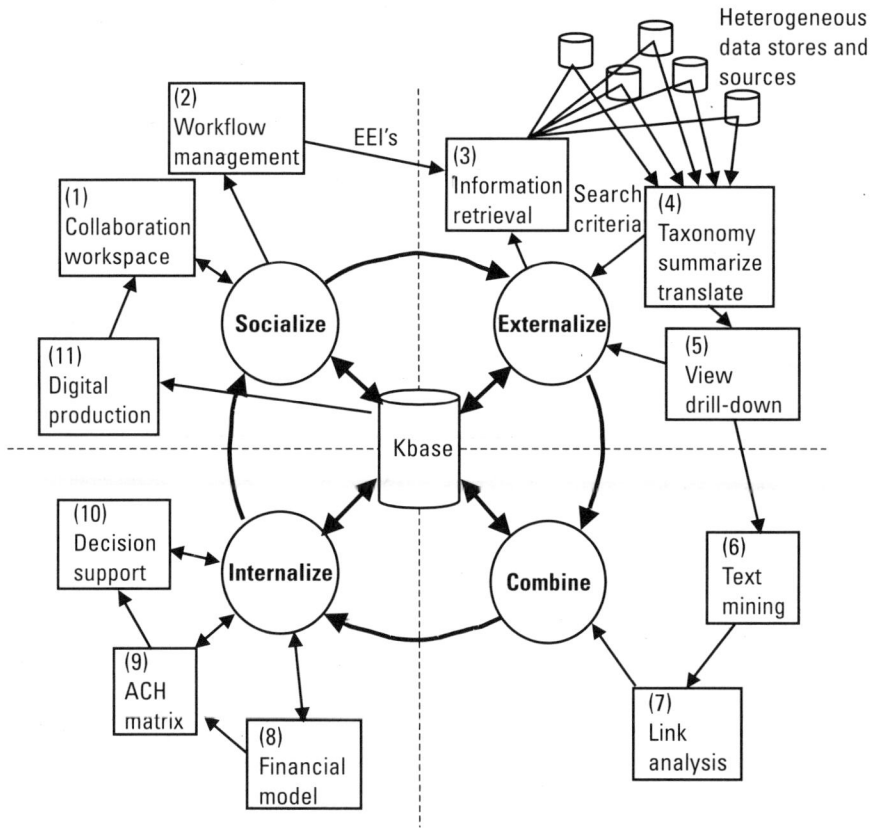

Figure 7.4 Example use of tools in a single spiral.

search current holdings and external sources, retrieving relevant multi-media content. The analyst also sets up monitor parameters to continually check certain sources (e.g., field office cables and foreign news sites) for changes or detections of relevant topics; when detected, the analyst will be alerted to the availability of new information.

4. The IR tools also create a taxonomy of the collected data sets, structuring the *catch* into five major categories: Zehga organization (personnel), events, finances, locations, and activities. The taxonomy breaks each category into subcategories of clusters of related content. Documents located in open-source foreign news reports are translated into English, and all documents are summarized into 55-word abstracts.

5. The analyst views the taxonomy and drills down to summaries, then views the full content of the most critical items to the investigation.

Selected items (or hyperlinks) are saved to the shared knowledge base for a local repository relevant to the investigation.

6. The retrieved catch is analyzed with text mining tools that discover and list the multidimensional associations (linkages or relationships) between entities (people, phone numbers, bank account numbers, and addresses) and events (meetings, deliveries, and crimes).

7. The linked lists are displayed on a link-analysis tool to allow the analyst to manipulate and view the complex web of relationships between people, communications, finances, and the time sequence of activities. From these network visuals, the analyst begins discovering the Zehga organizational structure, relationships to other drug cartels and financial institutions, and the timeline of explosive growth of the cartel's influence.

8. The analyst internalizes these discoveries by synthesizing a Zehga organization structure and associated financial model, filling in the gaps with conjectures that result in three competing hypotheses: a centralized model, a federated model, and a loose network model. These models are created using a standard financial spreadsheet and a network relationship visualization tool. The process of creating these hypotheses causes the analyst to frequently return to the knowledge base to review retrieved data, to issue refined queries to fill in the gaps, and to further review the results of link analyses. The model synthesis process causes the analyst to internalize impressions of confidence, uncertainty, and ambiguity in the evidence, and the implications of potential missing or negative evidence. Here, the analyst ponders the potential for denial and deception tactics and the expected subtle "sprignals" that might appear in the data.

9. An ACH matrix is created to compare the accrued evidence and argumentation structures supporting each of the competing models. At any time, this matrix and the associated organizational-financial models summarize the status of the intelligence process; these may be posted on the collaboration space and used to identify progress on the workflow management tool.

10. The analyst further internalizes the situation by applying a decision support tool to consider the consequences or implications of each model on counter-drug policy courses of action relative to the Zehga cartel.

11. Once the analyst has reached a level of confidence to make objective analytic judgments about hypotheses, results can be digitally published to the requesting consumers and to the collaborative workgroup

to begin socialization—and another cycle to further refine the results. (The next section describes the digital publication process.)

A wide variety of commercial off the shelf (COTS) and government-developed (GOTS) tools are being applied to intelligence analysis; the tools are often viewed as components of a tool suite, although interoperability among tools remains an integration challenge for enterprise designers. Tool suites such as the National Ground Intelligence Center's Pathfinder and AFRL WebTas integrate a variety of tools around a common database and exchange protocol [8]. The Defense Intelligence Agency's Joint Intelligence Virtual Architecture is integrating a wide variety of multimedia, multilingual, and multiparty (collaborative) tools for its analytic workforce [9]. Commercial tool suites such as Wincite's eWincite, Wisdom Builder's Wisdombuilder, and Cipher's Knowledge.Works similarly integrate text-based tools to support the competitive intelligence analysis. Table 7.3 summarizes the wide range of representative tool capabilities in the four categories and provides examples of existing GOTS and COTS tools. The list does not include IMINT imagery analysis tools, nor does it include special SIGINT and MASINT tools dedicated to technical analysis.

It is important to recognize that all of these analytic services must be capable of shared applications if they are to be effectively used for collaborative analysis. The collaborative-analysis process requires that all team members have access to retrieved data, emerging hypotheses, and decision analysis. The simple structure of two collaborating analysts (Figure 7.5) illustrates the function of collaborative tools to synchronize the socialization of two knowledge creation spirals and the collaborative contribution of access to common tools and a shared knowledge base. Tacit capture and collaborative filtering monitors the activities of all users on the network and uses statistical clustering methods to identify the emergent clusters of interest that indicate communities of common practice. Such filtering could identify and alert these two analysts to other analysts that are converging on a common suspect from other directions (e.g., money laundering and drug trafficking).

7.4   Intelligence Production, Dissemination, and Portals

The externalization-to-internalization workflow results in the production of digital intelligence content suitable for socialization (collaboration) across users and consumers. This production and dissemination of intelligence from KM enterprises has transitioned from static, hardcopy reports to dynamically linked digital softcopy products presented on portals. Publication has moved from point-in-time delivery (the day the report is printed on paper) to continuous delivery on a dynamic portal that is continually updated with recent intelligence

Table 7.3
Analytic Tools Summary

Category	Tool Descriptions	Representative GOTS Tools	Representative COTS Tools
Exploration	Information retrieval—retrieve relevant text (Section 7.2.2)	Pathfinder Query	Brio Intelligence Verity Inc. Verity
	Ontology creation—automated taxonomy generation for corpus of documents; automated metatagging		Veridian ThemeLink Semio Sequoia
	Query and report—manual OLAP query		Business Objects WebIntelligence
	Extraction/conversion—parses text, extracts and tags entities and events	MITRE Multimedia Extract MITRE Alembic Pathfinder X-Tractor	LockheedMartin NLToolset Memex Textract
	Summarization—summarize content and context of document	MITRE IntelGazette	Megaputer TextAnalyst Cartia ThemeScape
	Search, filter, and monitor—search for data; monitor sources for changes	Pathfinder DB Monitor	C-4-U Scout Caesius WebQL
Reasoning support	Data mining—perform automated multidimensional clustering to discover patterns and correlated data		SPSS Clementine SAS Enterprise Miner IBM intelligent Miner
	Text mining—perform automated linking of conceptually and semantically related textual content	Pathfinder Matrix	RetrievalWare Autonomy
	Data fusion—correlate and combine data to locate, identify, and track	U.S. Army ASAS tools DoD GCCS tools	Autometric InSight SAIC KnowledgeBoard
	Text fusion—correlate and combine text corpus to detect and track topics	NIST topic detection tracking tools	Commercial summarization and tracking tools
	Change detection—detect anomalies or temporal changes in content	Custom tools	NextLabs TrackEngine

Table 7.3 (continued)

Category	Tool Descriptions	Representative GOTS Tools	Representative COTS Tools
Reasoning support	Link analysis—perform automated discovery of re-lationships between multi-media data and text objects	Pathfinder MATRIX	i2 Analyst's Notebook Visual Analytics VisuaLinks Orion OrionMagic
	Temporal mapping—map events and entities to chronological timelines	AFRL WebTas	i2 Analyst's Notebook WisdomBuilder
	Spatial mapping—map physical objects to spatial database; register and overlay spatial information	MITRE GeoNODE Pathfinder MapViewer DoD Joint Mapping Toolkit	ESRI ArcView, ArcInfo Lockheed Martin ATAS Autometric EDGE
	Knowledge retrieval—performs automated Q&A deduction on large-scale knowledge base	SMU LASSO DARPA/GMU Disciple	CyCorp CYC
Sensemaking support	Creativity support—promotes inductive discovery; enables problem exploration		DiscoveryMachine Axion Idea Processor
	Static domain structure modeling—models functional hypotheses	Pathfinder CAMEO Custom Bayes Net Tools	Wincite eWincite
	Dynamic simulation, trend analysis—models time-dynamic process hypotheses; forecasts behaviors	MITRE Social indicator Analysis Model (SIAM)	Docere Market Analyzer
	Structured argumentation—construct and test logical arguments	Veridian CIM SRI SEAS	Knowledge Industries DXpress
Decision support	Modeling and simulation—apply simulation to evaluate alternative COAs	Military simulations: JWARS, NETWARS	Miya Viz CoMotion HPS iThink and Stella
	Alternative decision comparison—performs quantitative sensitivity and comparative analysis of risk and utility	Tools to display ACHs	Expert Choice Lumina Analytica

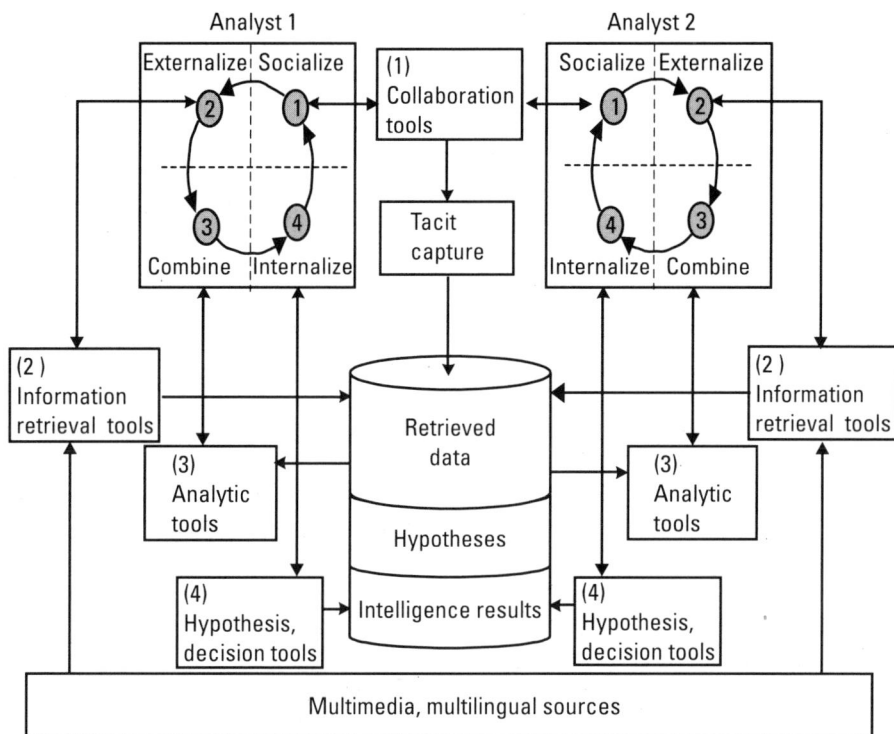

Figure 7.5 Collaborative knowledge creation workflow.

data and judgments. This trend follows the third-wave transition (described in Chapter 1) that moves from mass production of a standard report to all consumers, to mass customization of products; intelligence consumer portals may be customized to tailor intelligence delivery to each individual consumer's interests.

Digital production processes employ *content technologies* that index, structure, and integrate fragmented components of content into deliverable products. In the intelligence context, content includes:

1. Structured numerical data (imagery, relational database queries) and text [e.g., extensible markup language (XML)-formatted documents] as well as unstructured information (e.g., audio, video, text, and HTML content from external sources);

2. Internally or externally created information;

3. Formally created information (e.g., cables, reports, and imagery or signals analyses) as well as informal or ad hoc information (e.g., e-mail, and collaboration exchanges);

4. Static or active (e.g., dynamic video or even interactive applets) content.

Production content management services (Table 7.4) organize, dynamically assemble, and distribute diverse content elements into personalized intelligence products tailored to unique consumer requirements.

The key to dynamic assembly is the creation and translation of all content to a form that is understood by the KM system. While most intelligence data is transactional and structured (e.g., imagery, signals, MASINT), intelligence and open-source documents are unstructured. While the volume of open-source content available on Internet and closed-source intelligence content grows exponentially, the content remains largely unstructured. Content technology provides the capability to transform all-sources to a common structure for dynamic integration and personalized publication. The XML offers a method of embedding content descriptions by *tagging* each component with descriptive information that allows automated assembly and distribution of multimedia content [10]. XML allows metadata (data about the data including security, creation data, version, or type), text, imagery and graphics (format), and other content

Table 7.4
Production Content Management Service Categories

Services	Functions Performed
Content Creation	Creation and update of content taxonomies (knowledge ontologies)
	Structured authoring of content per standard content descriptions
	Analyst-appending metadata tags to describe external content
	Automated search and retrieval of relevant external content
Content Translation	Paper and electronic document capture, extraction, and conversion (decomposition) to content components
	Multilingual translation to common-language format
	Translation of technical data to tagged, structured forms
Content Integration	Manual to automated integration of content components into custom, timely products per user and group profiles
Personalized Portal Services	Integrated access to content tailored to user and desktop profiles
	Subscription (*push*) services—organization and delivery of custom news content per customer subscription profile
	Access (*pull*) services—library query and retrieval across all content; topical indices tailored to user profile
	Collaboration services—access to interact with communities of practice relevant to user and group profiles and current activity

categories to be uniquely tagged. The U.S. IC established a metadata and metadata markup working group to establish a communitywide mandated XML model to support interoperability of intelligence content across intelligence producers and consumers [11]. Intelligence standards being developed include an intelligence information markup language (ICML) specification for intelligence reporting and metadata standards for security, specifying digital signatures (XML-DSig), security/encryption (XML-Sec), key management (XML-KMS), and information security marking (XML-ISM) [12]. Such tagging makes the content interoperable; it can be reused and automatically integrated in numerous ways:

- Numerical data may be correlated and combined.

- Text may be assembled into a complete report (e.g., target abstract, targetpart1, targetpart2, …, related targets, most recent photo, threat summary, assessment).

- Various formats may be constructed from a single collection of contents to suit unique consumer needs (e.g., portal target summary format, personal digital assistant format, or pilot's cockpit target folder format).

In a typical intelligence application, SQL queries may be used to collect SIGINT results from a relational database, and then a standard document type description (DTD) is used to define the structure of an XML document. Figure 7.6 illustrates a simple XML SIGINT content example, with tagged data and metadata that can be read, processed, and integrated with other XML

```
<?xml version="1.0" ?>
<!DOCTYPE siginttype096report SYSTEM "st096.dtd">
<siginttype096report>
        <collect>
                <collect-source>Vandal033</collect-source>
                <collect-time-zulu year="2001" month="10" day="15"
                        hour="7" minute="44" second="12.35" />
                <platform>Hawk055</platform>
                <mission>Condor Talon</mission>
                <metadata-text>Excellent 10.3 second capture of emitter
1045-P during data transfer</metadata-text>
                <signal data parameterz1="1032.67" parameterz2="12.345"
                        parameterz3="10.15" parameterz4="0.00" />
                <singallibrarylocation url="vandal.033.sig0234.23.2" />
                <userlist>Simplex, Honcho5, Pico299, Gala</userlist>
                <targetname>Quintexiplex</targetname>
                <processors preproc="K25" Postproc="Vector33" />
        </collect>
                . . .
</siginttype096report>
```

Figure 7.6 Example structured XML SIGINT content.

content [13]. Finally, a document object model (DOM) tree can be created from the integrated result to transform the result into a variety of formats (e.g., HTML or PDF) for digital publication.

The analysis and single-source publishing architecture adopted by the U.S. Navy Command 21 K-Web (Figure 7.7) illustrates a highly automated digital production process for intelligence and command applications [14]. The production workflow in the figure includes the processing, analysis, and dissemination steps of the intelligence cycle:

1. *Content collection and creation (processing and analysis).* Both quantitative technical data and unstructured text are received, and content is extracted and tagged for subsequent processing. This process is applied to legacy data (e.g., IMINT and SIGINT reports), structured intelligence message traffic, and unstructured sources (e.g., news reports and

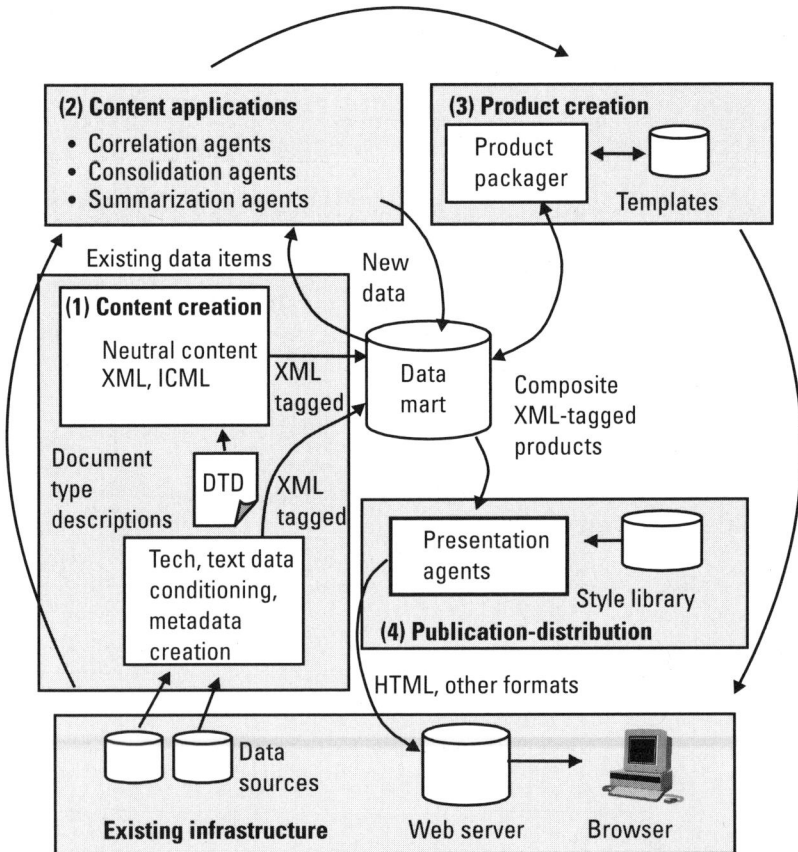

Figure 7.7 A single-source intelligence publishing workflow.

intelligence e-mail). Domain experts may support the process by creating metadata in a predefined XML metadata format to append to audio, video, or other nontext sources. Metadata includes source, pedigree, time of collection, and format information. New content created by analysts is entered in standard XML DTD templates.

2. *Content applications.* XML-tagged content is entered in the data mart, where data applications recognize, correlate, consolidate, and summarize content across the incoming components. A correlation agent may, for example, correlate all content relative to a new event or entity and pass the content on to a consolidation agent to index the components for subsequent integration into an event or target report. The data (and text) fusion and mining functions described in the next chapter are performed here.

3. *Content management-product creation (production).* Product templates dictate the aggregation of content into standard intelligence products: warnings, current intelligence, situation updates, and target status. These composite XML-tagged products are returned to the data mart.

4. *Content publication and distribution.* Intelligence products are personalized in terms of both style (presentation formats) and distribution (to users with an interest in the product). Users may explicitly define their areas of interests, or the automated system may monitor user activities (through queries, collaborative discussion topics, or folder names maintained) to implicitly estimate areas of interest to create a user's personal profile. Presentation agents choose from the style library and user profiles to create distribution lists for content to be delivered via e-mail, pushed to users' custom portals, or stored in the data mart for subsequent retrieval. The process of content syndication applies an information and content exchange (ICE) standard to allow a single product to be delivered in multiple styles and to provide automatic content update across all users.

The user's single entry point is a personalized portal (or Web portal) that provides an organized entry into the information available on the intelligence enterprise.

The portal provides an organized single-point access to a broad array of services (Table 7.5) that are tailored to the user's personal profile [15].

The U.S. IC chief information officer (CIO) introduced a pilot project in 2000 to provide automated digital production of an Intel Gazette product that may be personalized to individual consumers to deliver daily intelligence articles that are categorized, clustered (by topic similarity), and summarized [16].

Table 7.5
Intelligence Portal Service Categories

Service Category	Accessible Services
Collaboration, productivity, production	Synchronous and asynchronous collaboration tools
	Workspaces in which the user is a participant
	E-mail, chat, and forums (communities of practice)
	Scheduling and tasking (teams and personal agents)
	Digital production tools (XML authoring tools), office tools (e.g., word processor and spreadsheet)
Enterprise transactions	Customer (intelligence consumer) relationship management
	Enterprise resource management services
	Supply chain management services
Analysis-synthesis process	Information retrieval services
	Cognitive services (see Table 7.3)
	Decision support services
Personalized news	Current intelligence news, tailored to individual needs (by subscription), filtered, clustered and summarized, and pushed
	Advertisements (offerings) of available enterprise services
Personal knowledge base	Personal collection of raw data, linked information
	Working storage; in-process intelligence products

7.5 Human-Machine Information Transactions and Interfaces

In all of the services and tools described in the previous sections, the intelligence analyst interacts with explicitly collected data, applying his or her own tacit knowledge about the domain of interest to create estimates, descriptions, explanations, and predictions based on collected data. This interaction between the analyst and KM systems requires efficient interfaces to conduct the transaction between the analyst and machine. The following subsections consider two major forms of human-machine interaction employed in analysis—visual display of information in the spatial domain and intellectual interaction in the symbolic domain between analysts and their intelligent agents.

7.5.1 Information Visualization

Edward Tufte introduced his widely read text *Envisioning Information* with the prescient observation that, "Even though we navigate daily through a perceptual world of three dimensions and reason occasionally about higher dimensional arena with mathematical ease, the world portrayed on our information displays is caught up in the two-dimensionality of the flatlands of paper and video screen" [17]. Indeed, intelligence organizations are continually seeking technologies that will allow analysts to escape from this flatland. The U.S. IC conducted a broad study of information visualization technologies that "offer an enhanced method of analysis that enables discovery, understanding and presentation of situation assessment through the effective use of computer graphics and the interactive interface between analysts and their information sources" [18]. The study noted emerging research that has recognized visual perception as a thinking process that is not independent from, but integrated with, cognition. The essence of visualization is to provide multidimensional information to the analyst in a form that allows immediate understanding by this visual form of thinking [19].

A wide range of visualization methods are employed in analysis (Table 7.6) to allow the user to:

- Perceive patterns and rapidly grasp the essence of large complex (multidimensional) information spaces, then navigate or rapidly browse through the space to explore its structure and contents;

- Manipulate the information and visual dimensions to identify clusters of associated data, patterns of linkages and relationships, trends (temporal behavior), and outlying data;

- Combine the information by registering, mathematically or logically jointing (fusing), or overlaying.

The information in these visual spaces may include representations of the content text documents, technical data, imagery, video, or other forms. Often, the visualization process transforms high-dimensional information into a synthetic three-dimensional representation that allows the analyst to view an abstraction as if it were a physical artifact. The SPIRE and Themescape™ tools, for example, map the topical contents of a large corpus of text into a virtual 3D contour map; the map identifies the major topics within the corpus and their relationships. Major topics are displayed as mountains whose heights are determined by the amount of content; the relative distance between peaks is determined by the association between information sources contributing to the peaks [20].

7.5.2 Analyst-Agent Interaction

Intelligent software agents tailored to support knowledge workers are being developed to provide autonomous automated support in the information retrieval and exploration tasks introduced throughout this chapter. These *collaborative information agents,* operating in multiagent networks, provide the

Table 7.6
Information Visualization Applications

Visualization Methods	Functions	Example Intelligence Applications
Text visualization	View large corpus of documents to identify and track topics, authors, and sources	Review and track topics in and across multiple news feeds Locate topic, author, and relationships within scientific literature
Statistical data visualization	Graphical display of statistical metrics derived from high-dimensional data	Visualize multidimensional SIGINT and MASINT data sets Computer network analysis
Imagery and video visualization	Roam, zoom, and enhance image data; register and overlay imagery to maps	Imagery analysis for functional assessment, targeting, battle damage assessment
Geospatial information visualization	Create registered and layered geospatial (map) data sets of terrain, features, and annotation	Mapping of intelligence area of interest; overlay locations of sites, lines of communication, locations of events, and hypothesized routes
Synthetic modeling and simulation	Create synthetic models of intelligence targets: organizations, information flows, doctrinal processes, and facilities	Link analysis diagrams of network communications; social network diagrams of organizations Manufacturing or military process simulations
Virtual reality	Immerse the analyst in a 3D virtual world of physical places and abstract information spaces that include motion and animation	High volume multidimensional data immersion Virtual target world simulation for analysis or training

potential to amplify the analyst's exploration of large bodies of data, as they search, organize, structure, and reason about findings before reporting results [21]. Information agents are being developed to perform a wide variety of functions, as an autonomous collaborating community under the direction of a human analyst, including:

- *Personal information agents* (PIMs) coordinate an analyst's searches and organize bookmarks to relevant information; like a team of librarians, the PIMs collect, filter, and recommend relevant materials for the analyst [22].

- *Brokering agents* mediate the flow of information between users and sources (databases, external sources, collection processors); they can also act as sentinels to monitor sources and alert users to changes or the availability of new information.

- *Planning agents* accept requirements and create plans to coordinate agents and task resources to meet user goals.

In addition to these functions performed in the background, agents also offer the promise of a means of interaction with the analyst that emulates face-to-face conversation, and will ultimately allow information agents to collaborate as (near) peers with individuals and teams of human analysts. These *interactive agents* (or avatars) will track the analyst (or analytic team) activities and needs to conduct dialogue with the analysts—in terms of the semantic concepts familiar to the topic of interest—to contribute the following kinds of functions:

- *Agent conversationalists* that carry on dialogue to provide high-bandwidth interactions that include multimodal input from the analyst (e.g., spoken natural language, keyboard entries, and gestures and gaze) and multimodal replies (e.g., text, speech, and graphics). Such conversationalists will increase "discussions" about concepts, relevant data, and possible hypotheses [23].

- *Agent observers* that monitor analyst activity, attention, intention, and task progress to converse about suggested alternatives, potentials for denial and deception, or warnings that the analyst's actions imply cognitive shortcomings (discussed in Chapter 6) may be influencing the analysis process.

- *Agent contributors* that will enter into collaborative discussions to interject alternatives, suggestions, or relevant data.

The integration of collaborating information agents and information visualization technologies holds the promise of more efficient means of helping analysts find and focus on relevant information, but these technologies require greater maturity to manage uncertainty, dynamically adapt to the changing analytic context, and understand the analyst's intentions.

7.6   Summary

The analytic workflow requires a constant interaction between the cognitive and visual-perceptive processes in the analyst's mind and the explicit representations of knowledge in the intelligence enterprise. This chapter has reviewed the interactive tools that can aid the analyst in all phases of the knowledge-creation spiral and can also help the analyst manage the workflow process itself. Tools without a methodology become a nuisance; this chapter has emphasized the necessary understanding of how the wide variety of collaborative and analytic tools fit into an overall methodology to translate raw intelligence feeds into finished, digitally published intelligence. In the next chapter, we examine the automated services that capture and combine explicit knowledge to further aid the analyst in coping with massive volumes of arriving data.

Endnotes

[1]   Thompson, J. R., R. Hopf-weichel, and R. E. Geiselman, "The Cognitive Bases of Intelligence Analysis," Research Report 1362, U.S. Army Research Institute for the Behavioral Sciences, January 1984.

[2]   Information retrieval definition by the International Standards Organization, ISO/IEC 2382-1: 1993, Information Technology Vocabulary.

[3]   The table is based on the general classes presented by Moldovan, D., et al., "LASSO: A Tool for Surfing the Answer Net," *Proc. Of the Eighth Text Retrieval Conference* (TREC-8), NIST Special Pub. 500-246-2000, Table 5, p. 183.

[4]   *Strategic Investment Plan for Intelligence Community Analysis*, Washington D.C.: CIA, ADCI/AP200-01, p. 42.

[5]   *Intelligence Software Report 2002*, Fuld & Co., Cambridge, MA, 2002. This report, first issued in 2000 and updated in 2002, identified over 40 applicable packages, then reviewed and scored the applicability of over 10 dedicated CI packages in detail.

[6]   Druzdzel, M. J., and R. R. Flynn, "Decision Support Systems," in *Encyclopedia of Library and Information Science,* Kent, Allen (ed.), New York: Marcel Decker, 2000.

[7]   For an introduction to normative decision support tools, see Clemen, R. T.*, Making Hard Decisions: An Introduction to Decision Analysis,* Belmont CA: Duxbury, 1996.

[8]   See "Pathfinder: Pathfinder for the Web v4.6 and Pathfinder Portal Product Summary," Office of Naval Intelligence Pathfinder Support Office, Washington D.C., 2002; Mucks, J., "Web based Timeline Analysis System," WEBTAS, Air Force Research Laboratory, March 26, 2002.

[9]   Maybury, Mark T., "Analytic Tools for the New Millenium," in *Proc. of Annual JIVA Conference,* Albuquerque, NM, August 17, 2000.

[10] XML is a subset of the standard generalized markup language standard for creating markup languages; it is recommended by the W3C Web Consortium for creating markup languages for Web delivery.

[11] West, T. N., *Metadata and Metadata Markup and Homeland Security,* IC CIO, 21 August 2002.

[12] See http://www.xml.saic.com/icml/main/ic_core.html, accessed on-line on September 22, 2002.

[13]   A DTD is a specification of the permissible data elements for a class of documents, the parent-child relationships of the elements, and the order in which the elements may appear.

[14]   CINC-21Knowledge Management Objective Functional Architecture, U.S. Navy SPAWAR System Center, San Diego, SD-256, August 2000. See also Command–21 "Knowledge Web," U.S. Navy SPAWAR System Center, San Diego, SD-400, July 2001.

[15]   Portals are often categorized as: 1) *corporate portals*, which access the organization's intranet by employees (e.g., analysts and operations officers within an intelligence organization), 2) *customer portals*, which provide tailored services to the intelligence consumer outside the organization intranet, and 3) *vertical portals (vortals)*, which cross organization boundaries (e.g., collection, processing, analysis, and operations organizations).

[16]   "The Intel Gazette," Brochure, Office of the Intelligence Community CIO, Washington D.C., 2000.

[17]   Tufte, E. R., *Envisioning Information*, Cheshire, CT: Graphics Press, 1990, p.12. This book and Tufte's companions, *The Visual Display of Quantitative Information* and *Visual Explanations* provide excellent examples of methods to increase the number of dimensions that can be represented and density of data presented to the human perception on a flat plane (*escaping flatland*).

[18]   *P1000 Science and Technology Strategy for Information Visualization*, Washington D.C.: CIA DS&T, Version 2, December 16, 1996, p. 18.

[19]   Arnheim, R., *Visual Thinking*, Berkeley, CA: University California Press, 1969.

[20]   "Themescape: Award-Wining text Visualization," *AAT Delivers*, CIA Office of Advanced Analytic Tools (AAT), Issue 1, Feb. 2000.

[21]   This section presumes knowledge of agent technology. For an overview related to these applications, see Parunak, H. Van Dyke, "Go to the Ant: Engineering Principles from Natural Multi-Agent Systems," *Annals of Operations Research*, 1975, pp. 69–101; and Parunak, H. Van Dyke, "Agents in Overalls: Experiences and Issues in the Development and Employment of Industrial Agent-Based Systems," *International Journal of Cooperative Information Systems*, Vol. 9, No. 3, 2000 209–227. The proceedings of the annual Workshops on Collaborative Information Agents (CIA) are published by Springer-Verlag (Berlin) beginning with the first workshop (CIA-97) in 1997. The proceedings summarize research and development in critical areas necessary to make CIA technology viable: agent languages and specifications, uncertainty management, mobility and security, and human-agent conversation and dialogue.

[22]   See Chen, J. R., Mathe, N., and S. Wolfe, "Collaborative Information Agents on the World Wide Web," Report of the Computation Sciences Division, NASA Ames Research Center, May 6, 1998; see also Maes, Pattie, "Agents that Reduce Work and Information Overload," *Communications of the ACM*, Vol. 37, No. 7, July 1994.

[23]   Maybury, M. T., "Human Computer Interaction: State of the Art and Further Development in the International Context—North America," *Proc. of Human Computer Interaction International Status Conference,* Saarbruecken, Germany, October 26–27, 2001.

8

Explicit Knowledge Capture and Combination

In the last chapter, we introduced analytic tools that allow the intelligence analyst to interactively correlate, compare, and combine numerical data and text to discover clusters and relationships among events and entities within large databases. These interactive combination tools are considered to be *goal-driven* processes: the analyst is driven by a goal to seek solutions within the database, and the reasoning process is interactive with the analyst and machine in a common reasoning loop. This chapter focuses on the largely automated combination processes that tend to be *data driven*: as data continuously arrives from intelligence sources, the incoming data drives a largely automated process that continually detects, identifies, and tracks emerging events of interest to the user. These parallel goal-driven and data-driven processes were depicted as complementary combination processes in the last chapter (Figure 7.1). The automated processes described in this chapter are currently found in the processing phase of the intelligence cycle, especially the IMINT and SIGINT pipelines, where high-volume, disparate sources are correlated and combined. The data fusion and mining operations in these pipelines are performed on numerical data; the subsequent multi-INT fusion and mining operations in the analysis phase are more often performed on text reports generated from the IMINT, SIGINT, and other source analyses. In all cases, the combination processes help sources to cross-cue each other, locate and identify target events and entities, detect anomalies and changes, and track dynamic targets (refer to Figure 6.2).

8.1 Explicit Capture, Representation, and Automated Reasoning

The term *combination* introduced by Nonaka and Takeuchi in the knowledge-creation spiral is an abstraction to describe the many functions that are performed to create knowledge, such as correlation, association, reasoning, inference, and decision (judgment). This process requires the explicit representation of knowledge; in the intelligence application this includes knowledge about the world (e.g., incoming source information), knowledge of the intelligence domain (e.g., characteristics of specific weapons of mass destruction and their production and deployment processes), and the more general procedural knowledge about reasoning.

The process of capturing and representing knowledge has been a core challenge to the field of artificial intelligence. Progress has been made in moving from narrow domain-specific intelligence (e.g., expert systems of the early 1990s) to more robust reasoning systems that offer the potential for practical application in all-source intelligence applications. The DARPA Rapid Knowledge Formation (RKF) project and its predecessor, the High-Performance Knowledge Base project, represent ambitious research aimed at providing a robust explicit knowledge capture, representation, and combination (reasoning) capability targeted toward the intelligence analysis application [1]. The projects focused on developing the tools to create and manage shared, reusable knowledge bases on specific intelligence domains (e.g., biological weapons subjects); the goal is to enable creation of over one million axioms of knowledge per year by collaborating teams of domain experts. Such a knowledge base requires a *computational ontology*—an explicit specification that defines a shared conceptualization of reality that can be used across all processes. The abstract conceptualization relates phenomena in the real world to a machine-readable language, allowing reuse of the domain knowledge as well as automated analysis because the real-world context and meaning is provide by the structure (e.g., classes and hierarchies) of the ontology.

The challenge is to encode knowledge through the instantiation and assembly of generic knowledge components that can be readily entered and understood by domain experts (appropriate semantics) and provide sufficient coverage to encompass an expert-level of understanding of the domain. The knowledge base must have fundamental knowledge of entities (things that are), events (things that happen), states (descriptions of stable event characteristics), and roles (entities in the context of events). It must also describe knowledge of the relationships between (e.g. cause, object of, part of, purpose of, or result of) and properties (e.g., color, shape, capability, and speed) of each of these.

The knowledge base is organized in layers of abstractions (Table 8.1) representing knowledge in a variety of forms, shown in the table, including:

Table 8.1
Knowledge Representation Layers

Knowledge Abstraction Layer	Typical Knowledge Representations	Examples
Upper ontology	Man ⊃ Entity ⊃ Physical _Thing Attack⊃ Event ⊃ Temporal_Thing ∀x Man(x) ⟹ Mortal(x)	Fundamental reusable general knowledge of space, time, causality Common-sense knowledge
Core theories	(∀ *a*, *b*) *a* ∈ Event ∧ b ∈ Event causes(*a*, *b*) ⟹ precedes (*a*, *b*)	Fundamental and formal calculii of inference and mathematics, problem solving, planning
Domain-specific theories	Air_radar_23 ∧ Air_Missile_J5 ⟹ Mig_27 Missile_Range = F(alt,manueuver, K,p1, p2) ¬ (∃ Aircraft) [Fighter ∧ Commercial)	Domain-specific relationships, behaviors, models Domain expertise
Database of facts	Air_Attack ⊃ Act_of_War Mig-27 ∈ Fighter_Aircraft Kporanta ∈ NATION_STATE used-by (MIG-27, Kporanta) Date 10-24-02 Attack ∈ Air_Attack Used-in (Date 10-24-02 Attack, MIG-27)	Domain-specific instances of entities and events Relationships between instances of entities and events Extracted instances of current intelligence

- *Set representations* describe the relationships between concepts, such as the fact that an "air attack" is a subset of the concept "act of war" (Air_Attack ⊃ Act_of_War), and that "Kporanta" is an instance of a nation state (Kporanta ∈ Nation_State).

- *Logical operators* define the relationship between propositions (refer to Chapter 5 for an introduction to their use in basic reasoning) to enable prepositional logic to perform deduction. The examples in the table include the logical statement that there exists (∃) no aircraft(¬) that function as both fighter and commercial: ¬(∃ Aircraft) [Fighter ∧ Commercial), and all men are mortal :∀x Man(x) ⟹ Mortal(x).

- *Predicate operators* make statements about entities and events; they attribute properties, such as the fact that Kporanta employs Mig-27 aircraft: used-by (MIG-27, Kporanta) and allow manipulation by a predicate calculus.

- *Functional operators* allow normal mathematical operations to be performed on quantitative arguments; for example Missile_Range = F(alt,manueuver, K,p1, p2).

The structure of a high-performance knowledge base and associated inference engines to support the intelligence process is illustrated in Figure 8.1, which is adapted from the basic RKF architecture [2]. The figure shows the two stages of implementing and using such a knowledge base:

1. *Knowledge-base creation* is the initial (an ever ongoing) process of accumulating knowledge in the four layers of abstraction of knowledge in the knowledge base. Knowledge-base experts encode fundamental knowledge into the upper ontology and core theories layers. Domain-specific knowledge acquisition is performed by translating domain experts' natural language to formal expressions (described next), testing the inputs for consistency with prior knowledge, then diagramming the knowledge to allow reviewed, refined verification by the author or the collaborating team.

2. *Automated combination* is then performed as the populated knowledge base is coupled with inference engines to perform inductive, deductive, or abductive reasoning processes to combine current facts (e.g., current all-source intelligence feeds) with domain knowledge to infer combined intelligence. Entities and events are extracted from current all-source feeds and placed in the factual database layer, where they are used to derive new intelligence for delivery to the analyst. The figure illustrates two data-driven automatic inference processes that are

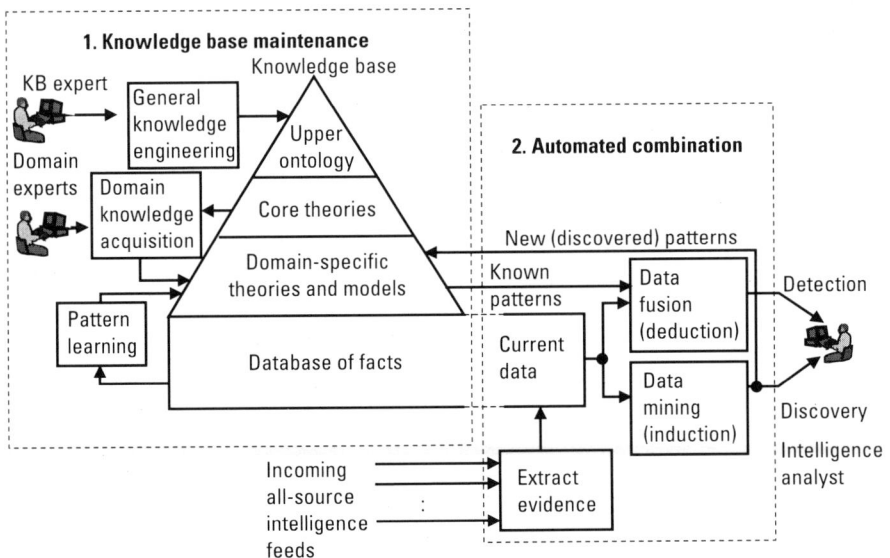

Figure 8.1 Explicit knowledge capture and combination.

described in subsequent sections in this chapter. A deductive *data-information fusion* process deduces (detects) and reports to the analyst the presence of entities, events, and situations of interest based on known domain-specific patterns within the knowledge base. An inductive *data-information mining* process induces (discovers) new, general patterns (relationships) in the incoming data and proposes these generalizations to the analyst for validation.

The data fusion and mining process are implemented in an inference engine, which applies the fundamental inferential axioms that are encoded in the core theory layer, guided by the problem-solving process in the core theory layer with a knowledge of the general world provided in the upper ontology. Of course, the system just described is a robust and ambitious, though not yet operational, approach to general-purpose reasoning applied to the intelligence domain. Even as this technology matures to provide powerful combination capabilities for the analyst, current data fusion and mining technologies are providing automated combination within narrow domains; these capabilities are described in the following section.

8.2 Automated Combination

Two primary categories of the combination processes can be distinguished, based on their approach to inference; each is essential to intelligence processing and analysis.

The inductive process of *data mining* discovers previously unrecognized patterns in data (new knowledge about *characteristics* of an unknown pattern class) by searching for patterns (relationships in data) that are in some sense "interesting." The discovered candidates are usually presented to human users for analysis and validation before being adopted as general cases [3].

The deductive process, *data fusion,* detects the presence of previously known patterns in many sources of data (new knowledge about the *existence* of a known pattern in the data). This is performed by searching for specific pattern templates in sensor data streams or databases to detect entities, events, and complex situations comprised of interconnected entities and events.

The characteristics of these two processes are contrasted in Table 8.2. The data sets used by these processes for knowledge creation are incomplete, dynamic, and contain data contaminated by noise. These factors make the following process characteristics apply:

- *Pattern descriptions.* Data mining seeks to induce general pattern descriptions (reference patterns, templates, or matched filters) to

Table 8.2
Comparison of Knowledge Detection and Discovery Methods

	Technology: Data Fusion	**Technology: Data Mining**
Knowledge Created	Detection of the presence of known entity or event types in time or space	Discovery of the existence of previously unknown entities or events in time or space
Reasoning Process	*Deduction:* detection of previously known patterns in data to infer the presence and identity of the entity or event represented by that pattern	*Induction:* Discovery of sufficient, correlated relationships in data to infer a general description (or rule set) that may be always or generally (to some quantified degree) true
Knowledge Patterns Used to Detect/ Discover Knowledge	Known: (specific) models are used as templates to detect similar patterns in data	Unknown: (general) model of interesting data properties is used as template to detect qualifying candidates for new knowledge in data
Detection/Discovery Process	Correlation of data with multiple specific models	Correlation of data with a simple general model (of interesting properties), followed by validation analysis
Object of Detection/ Discovery Process and Knowledge Gained	Detection of individual and related sets of entities and events Detection of the presence, type, and location of known types of entities or events in large volumes of data	Discovery of interesting general relationships and patterns of behavior, which may be validated as general models of relationships or behavior Discovery of new types of entities or events, by previously unidentified and unknown patterns, in large volumes of data
Applications	Testing known models of entities or events (templates) to detect those items: Target recognition Event detection Military network identification System status recognition	Learning new models of relationships or behavior to describe entities or events: Subtle behavior detection Machine learning (to provide specific models for data fusion) New statistical patterns in data sets

characterize data understood, while data fusion applies those descriptions to detect the presence of patterns in new data.

- *Uncertainty in inferred knowledge.* The data and reference patterns are uncertain, leading to uncertain beliefs or knowledge.

- *Dynamic state of inferred knowledge.* The process is sequential and inferred knowledge is dynamic, being refined as new data arrives.

- *Use of domain knowledge.* Knowledge about the domain (e.g., constraints, context) may be used in addition to collected raw intelligence data.

As noted in earlier chapters, these processes may be performed on numerical data or on text streams, or both, to detect or discover patterns in all-source content. Though beyond the current state of the art, the intelligence goal for such systems is to automatically correlate and combine, for example, the knowledge about a specific terrorist event contained in all foreign news reports (text and images), overhead imagery, relevant pre-event surveillance HUMINT reports, associated communication intercepts, and relevant MASINT measurements of terror cell movements. This list included a mixture of technical (quantitative) data and textual reports, both structured and unstructured.

The following sections provide a brief introduction to data fusion and mining approaches to automated deduction and induction, respectively. The introductory sections conclude with references to more in-depth texts that provide complete descriptions of the processes that implement each.

8.2.1　Data Fusion

Data fusion is an adaptive knowledge creation process in which diverse elements of similar or dissimilar observations (*data*) are aligned, correlated, and combined into organized and indexed sets (*information*), which are further assessed to model, understand, and explain (*knowledge*) the makeup and behavior of a domain under observation [4].

The process is performed cognitively by humans in daily life (e.g., combining sight, sound, and smells to detect a threat) and has been long applied for manual investigations in the military, intelligence, and law enforcement. In recent decades, the automation of this process has been the subject of intense research and development within the military, particularly to support intelligence and C2 [5]. As sensors and database sources of data become increasingly available, automated data fusion technologies are required to support humans to cope with the increasing data load.

The process is deductive in nature because it compares sensed data with previously learned (induced) templates or patterns to detect, identify, and model (or dynamically track behavior of) objects and groups of objects within the observed domain. Deduction is performed at the data, information, and knowledge levels.

The data-fusion process seeks to explain an adversary (or uncooperative) intelligence target by abstracting the target and its observable phenomena into a causal or relationship model, then applying all-source observation to detect entities and events to estimate the properties of the model. Consider the levels of representation in the simple target-observer processes in Figure 8.2 [6]. The adversary leadership holds to goals and values that create motives; these motives, combined with beliefs (created by perception of the current situation), lead to intentions. These intentions lead to plans and responses to the current situation; from alternative plans, decisions are made that lead to commands for action. In a hierarchical military, or a networked terrorist organization, these commands flow to activities (communication, logistics, surveillance, and movements). Using the three domains of reality terminology introduced in Chapter 5, the motive-to-decision events occur in the adversary's cognitive domain with no observable phenomena. The actions of explicitly recording and communicating these plans to operating units occur in the symbolic domain (communicated via human or machine language over communication systems) and the actions of those units appear in the physical domain. The data-fusion process uses observable evidence from both the symbolic and physical domains to infer the operations, communications, and even the intentions of the adversary.

The adversary's causal flow down is shown in the figure on the left, and the upward-flowing inference chain of data fusion is shown on the right. The deductive process is partitioned into levels to match the abstractions of the

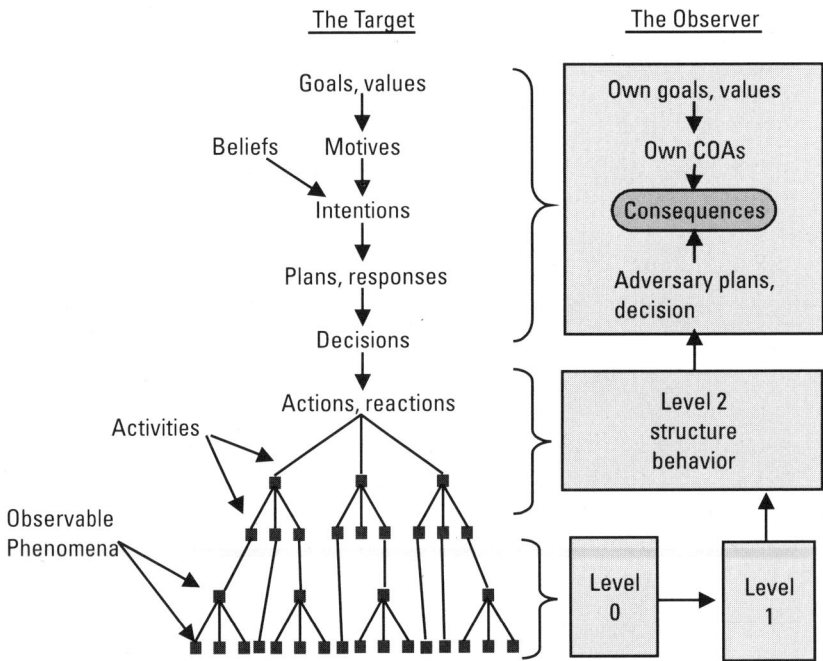

Figure 8.2 The target-observer model for data fusion.

adversary (target) behavior. Activities in the physical world are estimated at levels 0 and 1 by directly observing, detecting, identifying, and tracking entities over time. Level 2 estimates the composite situation of all entities and their relationships (e.g., aggregating entities into military units or terrorist cells and their command structures), and the temporal behavior of their activities. From this, the adversary's plans and intentions are estimated and compared with the observer's goals and alternative COAs to forecast consequences.

The emerging concept of effects-based military operations (EBO) requires intelligence products that provide planners with the ability to model the various effects influencing a target that make up a complex system. Planners and operators require intelligence products that integrate models of the adversary physical infrastructure, information networks, and leadership and decision making [7].

The U.S. DoD JDL has established a formal process model of data fusion that decomposes the process into five basic levels of information-refining processes (based upon the concept of levels of information abstraction) [8]:

- *Level 0: Data (or subobject) refinement.* This is the correlation across signals or data (e.g., pixels and pulses) to recognize components of an object and the correlation of those components to recognize an object.

- *Level 1: Object refinement.* This is the correlation of all *data* to refine individual objects within the domain of observation. (The JDL model uses the term object to refer to real-world entities, however, the subject of interest may be a transient event in time as well.)

- *Level 2: Situation refinement.* This is the correlation of all objects (*information*) within the domain to assess the current situation.

- *Level 3: Impact refinement.* This is the correlation of the current situation with environmental and other constraints to project the meaning of the situation (*knowledge*). The meaning of the situation refers to its implications to the user: threat, opportunity, change, or consequence.

- *Level 4: Process refinement.* This is the continual adaptation of the fusion process to optimize the delivery of knowledge against a defined mission objective.

A sequential flow of the data-fusion process, following our three-level information model, illustrates the five JDL levels (Figure 8.3). The process is characterized by the expected upward fusion flow from sources to data, then information, then knowledge, and also a downward feedback flow that controls the process and the sensors or sources acquiring data. The following paragraphs describe each functional level.

Level of abstraction	Representative information flow

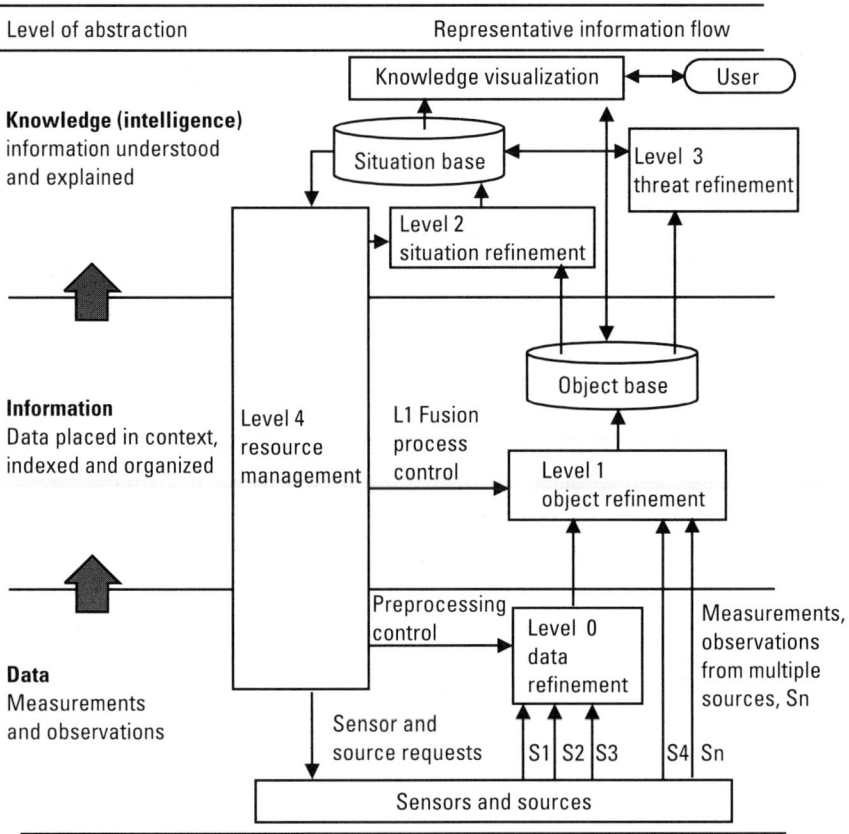

Figure 8.3 Data fusion is a process of deductive reasoning to correlate and combine multiple sources of data in order to understand a complex physical process.

8.2.1.1 Level 0: Data Refinement

Raw data from sensors may be calibrated, corrected for bias and gain errors, limited (thresholded), and filtered to remove systematic noise sources. Object detection may occur at this point—in individual sensors or across multiple sensors (so-called *predetection fusion*). The object-detection process forms observation *reports* that contain data elements such as observation identifier, time of measurement, measurement or decision data, decision, and uncertainty data.

8.2.1.2 Level 1: Object Refinement

Sensor and source reports are first aligned to a common spatial reference (e.g., a geographic coordinate system) and temporal reference (e.g., samples are propagated forward or backward to a common time.) These *alignment* transformations place the observations in a common time-space coordinate system to allow an *association* process to determine which observations from different sensors have their source in a common object. The association process uses a quantitative correlation metric to measure the relative similarity between observations. The typical correlation metric, C, takes on the following form:

$$c = \sum_{i1=1}^{n} w_i x_i$$

Where;

w_i = weighting coefficient for attribute x_i.

x_i = ith correlation attribute metric.

Values of x_i may include spatial distances (how close were the physical locations of the observations?), statistical distances (how similar were the measurements?), or spectral compatibility (how feasible was the measurement to occur from a common source?). The weighting coefficients, w_I, may be used to weight each contribution by relative importance or by absolute strength of contribution (e.g., inverse weighting by covariance statistics). The correlation metric may be used to make a hard decision (an *association*), choosing the most likely parings of observations, or a deferred decision, assigning more that one hypothetical paring and deferring a hard decision until more observations arrive. Once observations have been associated, two functions are performed on each associated set of measurements for common object:

1. *Tracking.* For dynamic targets (vehicles or aircraft), the current state of the object is correlated with previously known targets to determine if the observation can update a model of an existing model (*track*). If the newly associated observations are determined to be updates to an existing track, the state estimation model for the track (e.g., a Kalman filter) is updated; otherwise, a new track is initiated.

2. *Identification.* All associated observations are used to determine if the object identity can be classified to any one of several levels (e.g., friend/foe, vehicle class, vehicle type or model, or vehicle status or intent).

8.2.1.3　Level 2: Situation Refinement

All objects placed in space-time context in an information base are analyzed to detect relationships based on spatial or temporal characteristics. Aggregate sets of objects are detected by their coordinated behavior, dependencies, proximity, common point of origin, or other characteristics using correlation metrics with high-level attributes (e.g., spatial geometries or coordinated behavior). The synoptic understanding of all objects, in their space-time context, provides situation knowledge, or awareness.

8.2.1.4   Level 3: Impact (or Threat) Refinement

Situation knowledge is used to model and analyze feasible future behaviors of objects, groups, and environmental constraints to determine future possible outcomes. These outcomes, when compared with user objectives, provide an assessment of the implications of the current situation. Consider, for example, a simple counter-terrorism intelligence situation that is analyzed in the sequence in Figure 8.4.

8.2.1.5   Level 4: Process Refinement

This process provides feedback control of the collection and processing activities to achieve the intelligence requirements. At the top level, current knowledge (about the situation) is compared to the intelligence requirements required to achieve operational objectives to determine knowledge shortfalls. These shortfalls are parsed, downward, into information, then data needs, which direct the future acquisition of data (sensor management) and the control of internal processes. Processes may be refined, for example, to focus on certain areas of interest, object types, or groups. This forms the feedback loop of the data-fusion process.

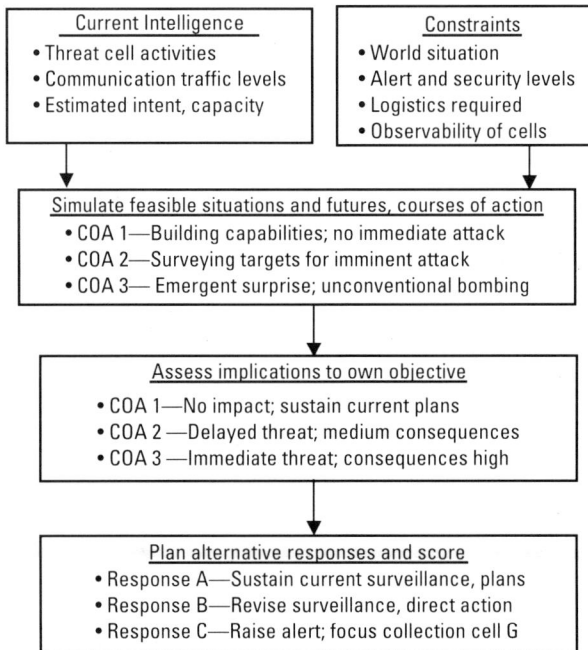

Figure 8.4 Typical situation analysis, prediction, and planning sequence.

General distinctions in the four correlation and combining levels (0, 1, 2, and 3) of the process are characterized in Table 8.3 to distinguish the difference in the resources, functions, and temporal focus at each level.

The theoretical foundations, systems architectures, and mathematical alternatives to implement data fusion are summarized in numerous texts that describe military and commercial applications (see [9–12]).

8.2.2 Data Mining

Data mining is the process by which large sets of *data* (or text in the specific case of text mining) are cleansed and transformed into organized and indexed sets (*information*), which are then analyzed to discover hidden and implicit, but previously undefined, patterns. These patterns are reviewed by domain experts to determine if they reveal new understandings of the general structure and relationships (*knowledge*) in the data of a domain under observation. The

Table 8.3
Distinctions Between the Data Fusion Processing Levels

	Level 0: Data Refinement	Level 1: Object Refinement	Level 2: Situation Refinement	Level 3: Impact Refinement
Level of Information Abstractions	Data (measurements and observations)	Objects (events and entities)	Situation (resulting from interacting objects)	Meaning (the implications of the situation)
Functions Performed	Signal estimation • Composite sensor detection	Object estimation • Detection • Association • Combination • Tracking • Classification	Group estimation • Group detection (aggregation) • Group association • Group combination • Group tracking • Group classify	Impact prediction • Model associations and behavior • Predict future behavior (courses of action) • Assess impact and implications to objective(s)
Temporal Focus	A single observation	A period: a small sequence of observations	Dynamic situation: a large number of observations	Implications to a future time
General Output Products	Object reports of entities or events	Object reports and behavior models	Group models and dynamic simulations	Predictions, alternatives, and implications (consequences)

data-mining process can also be applied to text, where the discovered patterns may include clusters of related articles, linked authors, or topics.

The object of discovery is a *pattern*, which is defined as a statement in some language, L, that describes relationships in subset F_s of a set of data, F, such that:

1. The statement holds with some certainty, c;

2. The statement is simpler (in some sense) than the enumeration of all facts in F_s [13].

This is the inductive generalization process described in Chapter 5. Mined knowledge, then, is formally defined as a pattern that is interesting, according to some user-defined criterion, and certain to a user-defined measure of degree. As an example, consider the following case:

Terrorist organization patterns:

1. *Interesting* criteria are *frequent* telecommunication between, or physical proximity or correlated statements by, different terrorist cells.

2. *Measures of degree* for these criteria are more than three messages within a week, travel to same city at the same time, or a statement opposed to common interests posted within one week [14].

In application, the mining process is extended from explanations of limited data sets to more general applications (induction). In this example, a relationship pattern between three terrorist cells may be discovered that includes intercommunication, periodic travel to common cities, and correlated statements posted on the Internet. This pattern may be more fully analyzed over many known terrorist cells and extended (by induction) to be a general pattern of behavior for detecting cells.

Data mining (also called *knowledge discovery*) is distinguished from data fusion by two key characteristics:

1. *Inference method.* Data fusion employs known patterns and deductive reasoning, while data mining searches for hidden patterns using inductive reasoning.

2. *Temporal perspective.* The focus of data fusion is *retrospective* (determining current state based on past data), while data mining is both retrospective and *prospective*—focused on locating hidden patterns that may reveal predictive knowledge.

The data mining literature has predominantly addressed business applications that seek to locate economic or buying patterns of warehouses of data, including point-of-sales data [15]. The increased availability of warehoused data and the potential economic benefits of improved knowledge of purchasing patterns have spurred significant research and development in the mining process. The term is used to refer to a range of processes, from manual analysis of data using visualization tools alone, to automated techniques that navigate and explore data searching for interesting patterns.

The Cross Industry Standard Process for Data Mining (CRISP-DM) is emerging as a standard reference model, as the JDL model is for data fusion [16]. The general functions of data mining can be structured (Figure 8.5) to

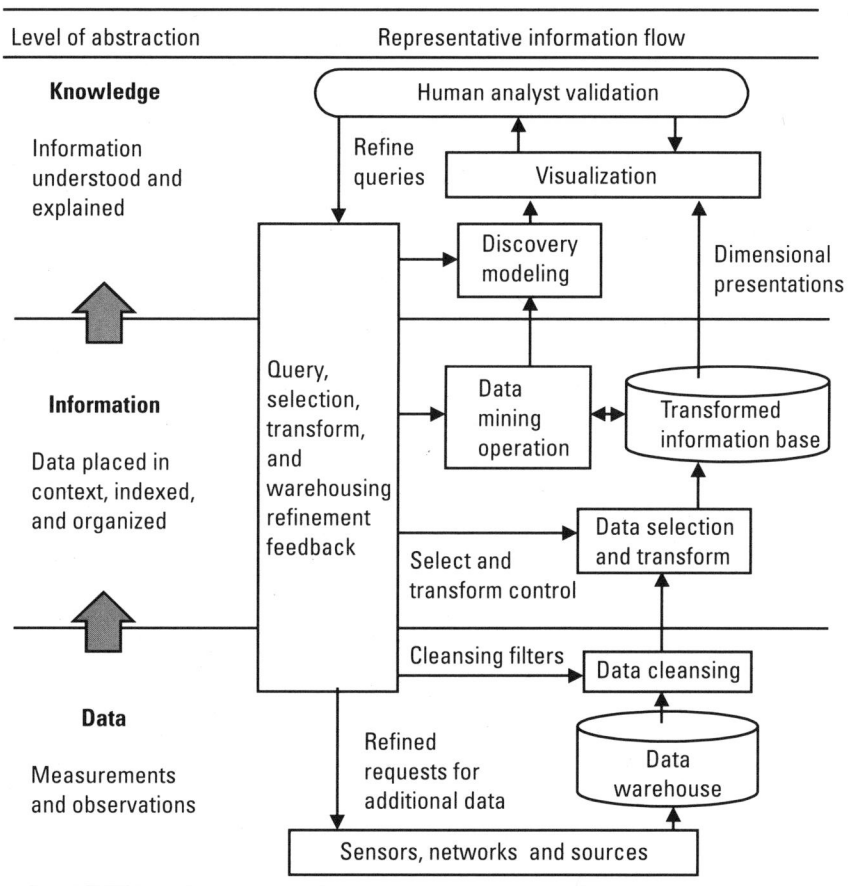

Figure 8.5 Data mining is an inductive process of evaluating data to locate patterns in the data that explain previously unknown general relationships in the underlying physical processes.

illustrate a similarity to the data-fusion process. (See [17–19] for introductions to the mining process.) Beginning with sensors and sources, the data warehouse is populated with data, and successive functions move the data toward learned knowledge at the top. The sources, queries, and mining processes may be refined, similar to data fusion. The functional stages in the figure are described next.

- *Data warehouse.* Data from many sources are collected and indexed in the warehouse, initially in the native format of the source. One of the chief issues facing many mining operations is the reconciliation of diverse database formats that have different formats (e.g., field and record sizes and parameter scales), incompatible data definitions, and other differences. The warehouse collection process (*flow in*) may mediate between these input sources to transform the data before storing in common form [20].

- *Data cleansing.* The warehoused data must be inspected and cleansed to identify and correct or remove conflicts, incomplete sets, and incompatibilities common to combined databases. Cleansing may include several categories of checks:

  1. *Uniformity checks* verify the ranges of data, determine if sets exceed limits, and verify that formats versions are compatible.

  2. *Completeness checks* evaluate the internal consistency of data sets to ensure, for example, that aggregate values are consistent with individual data components (e.g., "verify that total sales is equal to sum of all sales regions, and that data for all sales regions is present").

  3. *Conformity checks* exhaustively verify that each index and reference exists.

  4. *Genealogy checks* generate and check audit trails to primitive data to permit analysts to *drill down* from high-level information.

- *Data selection and transformation.* The types of data that will be used for mining are selected on the basis of relevance. For large operations, initial mining may be performed on a small set, then extended to larger sets to check for the validity of abducted patterns. The selected data may then be transformed to organize all data into common dimensions and to add derived dimensions as necessary for analysis.

- *Data mining operations.* Mining operations may be performed in a supervised manner in which the analyst presents the operator with a selected set of training data, in which the analyst has manually determined the existence of pattern classes. Alternatively, the operation may proceed without supervision, performing an automated search for patterns. A number of techniques are available (Table 8.4), depending upon the type of data and search objectives (interesting pattern types).

- *Discovery modeling.* Prediction or classification models are synthesized to fit the data patterns detected. This is the proscriptive aspect of mining: modeling the historical data in the database (the past) to provide a model to predict the future. The model attempts to abduct a generalized description that explains discovered patterns of interest and, using statistical inference from larger volumes of data, seeks to induct generally applicable models. Simple extrapolation, time-series trends, complex linked relationships, and causal mathematical models are examples of models created.

- *Visualization.* The analyst uses visualization tools that allow discovery of interesting patterns in the data. The automated mining operations *cue* the operator to discovered patterns of interest (candidates), and the analyst then visualizes the pattern and verifies if, indeed, it contains new and useful knowledge. OLAP refers to the manual visualization process

Table 8.4
Common Data Mining Operator Techniques

Mining Operator Methods	Description
Clustering	Segment the data into clusters (subsets of data) that share common properties; analyze the clusters for patterns that meet the interesting properties sought
Association or sequence discovery	Analyze the causal (sequence) or structural (association) relationships between sets of data to locate cause-effect relationships that meet interesting pattern properties
Statistical analysis	Determine the statistical (occurrence probabilities) characteristics of subsets of data and quantify the statistically significant (e.g., high occurrence) sets
Rule abduction	Analyze data to abduct IF-THEN-ELSE rules that describe the structure; test rules for validity in general and statistically characterize each
Link or tree abduction	Analyze the structural relationships between sets of data to locate links between data and tree structures that meet interesting connecting pattern properties
Deviation analysis	Locate deviations from statistically *normal* behavior and analyze for interest
Neural abduction	Train artificial neural networks to match data, then extract network coefficients (node weights) and network structure as abducted rules

in which a data manipulation engine allows the analyst to create data "views" from the human perspective and to perform the following categories of functions:

1. *Multidimensional analysis* of the data across dimensions, through relationships (e.g., command hierarchies and transaction networks) and in perspectives natural to the analyst (rather that inherent in the data);

2. *Transformation* of the viewing dimensions or *slicing* of the multidimensional array to view a subset of interest;

3. *Drill down* into the data from high levels of aggregation, downward into successively deeper levels of information;

4. *Reach through* from information levels to the underlying raw data, including reaching beyond the information base, back to raw data by the audit trail generated in genealogy checking;

5. *Modeling* of hypothetical explanations of the data, in terms of trend analysis, extrapolations.

- *Refinement feedback.* The analyst may refine the process, by adjusting the parameters that control the lower level processes, as well as requesting more or different data on which to focus the mining operations.

8.2.3   Integrated Data Fusion and Mining

In a practical intelligence application, the full reasoning process integrates the discovery processes of data mining with the detection processes of data fusion. This integration helps the analyst to coordinate learning about new signatures and patterns and apply that new knowledge, in the form of templates, to detect other cases of the situation. A general application of these integrated tools can support the search for *nonliteral* target signatures, the use of those learned and validated signatures to detect new targets [21]. (Nonliteral target signatures refer to those signatures that extend across many diverse observation domains and are not intuitive or apparent to analysts, but may be discovered only by deeper analysis of multidimensional data.) The integrated architecture (Figure 8.6) illustrates the complementary nature of the two processes. The mining component searches the accumulated database of sensor data, with discovery processes focused on relationships that may have relevance to the nonliteral target sets. Discovered models (templates) of target objects or processes are then tested, refined, and verified using the data-fusion process. Finally, the data-fusion process applies the models deductively for knowledge detection in incoming sensor data streams.

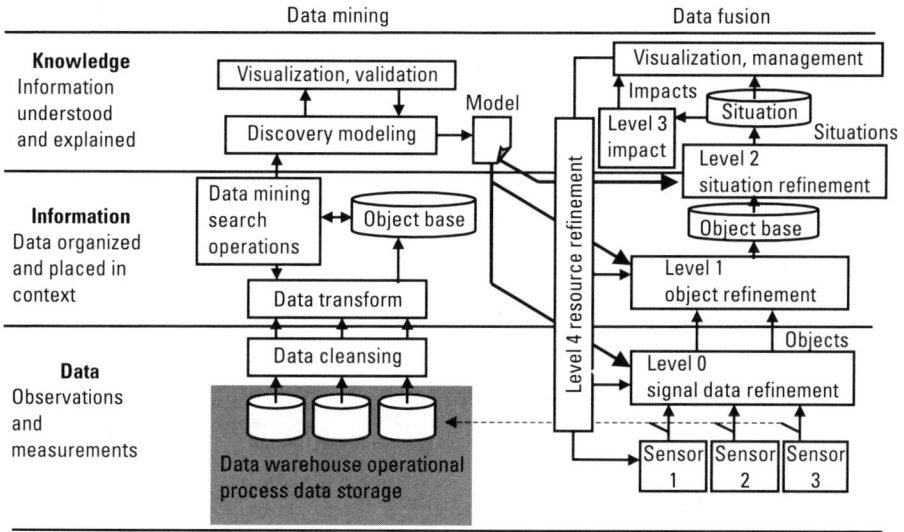

Figure 8.6 An integrated data mining and fusion architecture. (Source: [21]. © 1998 IEEE; used by permission.)

8.3 Intelligence Modeling and Simulation

In Chapter 5 (Section 5.4), we introduced analysis-synthesis as a model construction process, where the analyst synthesizes a representation of the subject of analysis (intelligence target) as a framework against which evidence is marshaled. Modeling activities take place in externalization (as explicit models are formed to describe mental models), combination (as evidence is combined and compared with the model), and in internalization (as the analyst ponders the matches, mismatches, and incongruities between evidence and model). Respected intelligence analyst Jack Davis has noted that the human mind is the "the most creative analytic tool in existence" but that it reaches limits in three areas: the volume of information it can store, the number of variables that can be brought to bear on a problem coherently, and the ability to track the consequences of the whole set of variables in one of the factors under considerations [22]. Modeling tools support the analyst in addressing these limitations for complex problems.

While we have used the general term *model* to describe any abstract representation, we now distinguish here between two implementations made by the modeling and simulation (M&S) community. *Models* refer to physical, mathematical, or otherwise logical representations of systems, entities, phenomena, or processes, while *simulations* refer to those methods to implement models over time (i.e., a simulation is a time-dynamic model) [23].

Models and simulations are inherently collaborative; their explicit representations (versus mental models) allow analytic teams to collectively assemble, and explore the accumulating knowledge that they represent. They support the analysis-synthesis process in multiple ways:

- *Evidence marshaling.* As described in Chapter 5, models and simulations provide the framework for which inference and evidence is assembled; they provide an audit trail of reasoning.
- *Exploration.* Models and simulations also provide a means for analysts to be immersed in the modeled situation, its structure, and dynamics. It is a tool for experimentation and exploration that provides deeper understanding to determine necessary confirming or falsifying evidence, to evaluate potential sensing measures, and to examine potential denial and deception effects.
- *Dynamic process tracking.* Simulations model the time-dynamic behavior of targets to forecast future behavior, compare with observations, and refine the behavior model over time. Dynamic models provide the potential for estimation, anticipation, forecasting, and even prediction (these words imply increasing accuracy and precision in their estimates of future behavior).
- *Explanation.* Finally, the models and simulations provide a tool for presenting alternative hypotheses, final judgments, and rationale.

The intelligence application of M&S is similar to the military operations analysis interest in tools (e.g., the U.S. Joint Simulation System) to gain insight into complex situations, but it does not focus on training, rehearsal, or weapons analysis aspects (though these may be of interest to intelligence operations officers) [24]. The focus of intelligence M&S is marshalling, exploring, tracking, and explaining. In the business text, *Serious Play: How the World's Best Companies Simulate to Innovate*, author Michael Schrage notes the value of simulation immersion by the business analyst. He observes, "the real value of a model or simulation may stem less from its ability to test a hypothesis than from its power to generate useful surprise. Louis Pasteur once remarked that 'chance favors the prepared mind.' It holds equally true that chance favors the prepared prototype: models and simulations can and should be media to create and capture surprise and serendipity" [25].

Earlier in Chapter 5 (Section 5.5) we introduced the concept of intelligence targets existing in three domains of reality; models and simulations likewise must represent these objects of analysis in three domains. Table 8.5 illustrates the representative intelligence models and simulations in each of the three fundamental categories of reality [26].

Table 8.5
Typical Intelligence Modeling and Simulation Applications

Domain	Models	Simulations
Cognitive	Leadership social networks	Agent-based simulations of decision making
	Military C2 doctrine models	Agent-based and -influenced net simulations of group social behaviors
	Models of adversary perceptions and beliefs	
Symbolic	Financial network flow models	Computer network simulations of network communication and computation
	Communication and network infrastructure models	Financial transaction and capital flow simulations
	Telephone call time chronologies and call nets	
Physical	CAD models of weapon platforms and systems	Simulations of manufacturing, logistics, transportation, or military operations processes
	CAD models of sites and facilities	Simulations of physical phenomena and sensors
	Geospatial layered models of area terrain, features, and lines of communication	

The table illustrates independent models and simulations in all three domains, however these domains can be coupled to create a robust model to explore how an adversary thinks (cognitive domain), transacts (e.g., finances, command, and intelligence flows), and acts (physical domain) [27]. Commercial models and simulations are currently used in each of these to conduct design and operations analyses. Geographic information systems, CAD packages, and simulations like the popular SimCity ™ are being used to evaluate urban design and development. Symbolic-domain financial models are used for business case analysis, and computer network simulations are used to evaluate network performance and plan operations. At the cognitive level, social network models and decision-making simulations are used to allow managers to explore complex business problems. The Modeling, Virtual Environments, and Simulation (MOVES) Institute of the Naval Postgraduate School, Monterey, California, is conducting research and development to apply M&S in all three domains suitable for intelligence analysis by simulating physical military targets and operations, network operations, and even adversarial decision making [28]. A recent study of the advanced methods required to support counter-terrorism analysis recommended the creation of scenarios using *top-down* synthesis (manual creation by domain experts and large-scale simulation) to create synthetic evidence for comparison with real evidence discovered by *bottom-up* data mining [29].

The following sections illustrate two intelligence applications of simulation tools.

8.3.1   M&S for I&W

The challenge of I&W demands predictive analysis, where "the analyst is look-ing at something entirely new, a discontinuous phenomenon, an outcome that he or she has never seen before. Furthermore, the analyst only sees this new pat-tern emerge in bits and pieces" [30]. The text, *Preventative Measures,* reports on a variety of M&S tools developed to provide warning of incipient crises (e.g., violent societal behavior, interstate conflict, or state failure) for the public poli-cymakers, the IC, and DoD [31]. The tools monitor world events to track the state and time-sequence of state transitions for comparison with indicators of stress. These analytic tools apply three methods to provide indicators to analysts:

1. *Structural indicator matching.* Previously identified crisis patterns (sta-tistical models) are matched to current conditions to seek indications in background conditions and long-term trends.

2. *Sequential tracking models.* Simulations track the dynamics of events to compare temporal behavior with statistical *conflict accelerators* in cur-rent situations that indicate imminent crises.

3. *Complex behavior analysis.* Simulations are used to support inductive exploration of the current situation, so the analyst can examine possi-ble future scenarios to locate potential *triggering events* that may cause instability (though not in prior indicator models).

A general I&W system architecture (Figure 8.7), organized following the JDL data-fusion structure, accepts incoming news feed text reports of current situations and encodes the events into a common format (by human or auto-mated coding). The event data is encoded into time-tagged actions (assault, kid-nap, flee, assassinate), proclamations (threaten, appeal, comment) and other pertinent events from relevant actors (governments, NGOs, terror groups). The level 1 fusion process correlates and combines similar reports to produce a single set of current events organized in time series for structural analysis of back-ground conditions and sequential analysis of behavioral trends by groups and interactions between groups. This statistical analysis is an automatic target-recognition process, comparing current state and trends with known clusters of unstable behaviors. The level 2 process correlates and aggregates individual events into larger patterns of behavior (situations). A dynamic simulation tracks the current situation (and is refined by the tracking loop shown) to enable the analyst to explore future excursions from the present condition. By analysis of the dynamics of the situation, the analyst can explore a wide range of feasible futures, including those that may reveal surprising behavior that is not intui-tive—increasing the analyst's awareness of unstable regions of behavior or the potential of subtle but potent triggering events.

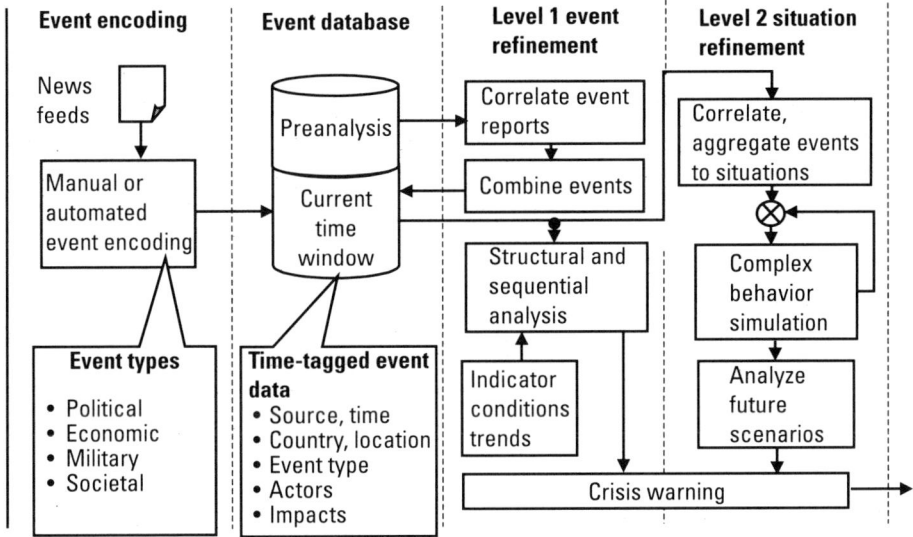

Figure 8.7 Data flow within a crisis warning system.

8.3.2 Modeling Complex Situations and Human Behavior

The complex behavior noted in the prior example may result from random events, human free will, or the nonlinearity introduced by the interactions of many actors. The most advanced applications of M&S are those that seek to model environments (introduced in Section 4.4.2) that exhibit complex behaviors—emergent behaviors (surprises) that are not predictable from the individual contributing actors within the system. Complexity is the property of a system that prohibits the description of its overall behavior even when all of the components are described completely. Complex environments include social behaviors of significant interest to intelligence organizations: populations of nation states, terrorist organizations, military commands, and foreign leaders [32]. Perhaps the grand challenge of intelligence analysis is to understand an adversary's cognitive behavior to provide both warning and insight into the effects of alternative preemptive actions that may avert threats. The U.S. DCI has articulated this challenge:

> To this day, Intelligence is always much better at counting heads than diving what is going on inside them. That is, we are very good at gauging the size and location of militaries and weaponry. But for obvious reasons, we can never be as good at figuring out what leaders will do with them. Leadership analysis remains perhaps the most difficult of analytic specialties [33].

Nonlinear mathematical solutions are intractable for most practical problems, and the research community has applied dynamic systems modeling and agent-based simulation (ABS) to represent systems that exhibit complex behavior [34]. ABS research is being applied to the simulation of a wide range of organizations to assess intent, decision making and planning (cognitive), command and finances (symbolic), and actions (physical). The applications of these simulations include national policies [35], military C2 [36], and terrorist organizations [37]. As these technologies mature, such tools will increasingly aid analysts in the study of the cognitive domain.

8.4 Summary

This chapter has introduced the potential contribution of automated reasoning systems that can capture and apply explicit knowledge to locate critical information, focus attention to key issues, detect the presence of known patterns, or discover new patterns in massive volumes of incoming data. These tools will also support the creation of complex models and simulations of physical, symbolic, and cognitive systems of interest to the analyst, enabling the analyst to explore and experience these targets to gain deeper understanding of their structure and behavior. A recent *Harvard Business Review* article chronicled the promise of "Predicting the Unpredictable" by agent-based simulation; this technology may indeed enable just a degree of this capability for the analyst, too [38]. But these intelligence models and simulations must always be subject to an evaluation of appropriateness, validity, and usefulness; they can never be used blindly or trusted implicitly. These combination tools must never be viewed as *black boxes* that provide answers without explanation; rather, they must viewed as *transparent boxes* that allow the analyst to enter into a problem deeply to gain insight not available by a more cursory examination [39].

Endnotes

[1]   Cohen, P. et al., "The DARPA High-Performance Knowledge Bases Project," *AI Magazine,* Winter 1998, pp. 25–49; and Burke, M., "Rapid Knowledge Formation," briefing, DARPA Information Technology Office, 2000.

[2]   This figure is adapted from Burke, M., "Rapid Knowledge Formation," briefing, DARPA Information Technology Office, 2000.

[3]   Section 8.2 and subsections are adapted from material in the author's text, *Information Warfare Principles and Operations*, Norwood, MA: Artech House, 1998.

[4]   Definition from the article written by Buede, D., and E. Waltz, "Data Fusion," in *McGraw Hill Encyclopedia of Science and Technology*, New York: McGraw Hill, 1998. An

expanded definition provided by the U.S. DoD JDL is: "A process dealing with the association, correlation, and combination of data and information from single and multiple sources to achieve refined position and identity estimates, and complete and timely assessments of situations and threats, and their significance. The process is characterized by continuous refinements of its estimates, and by evaluation of the need for additional sources, or modification of the process itself to achieve improved results."

[5] Waltz, E. L., and D. M. Buede, "Data Fusion and Decision Support for Command and Control," *IEEE Trans. on Systems, Man and Cybernetics,* Vol. SMC-16, No. 6, November–December 1986.

[6] It is important to note that this basic model is used for illustrative purposes; real-world adversarial systems are complex and not so easily represented by rigid doctrinal and hierarchical models. The reader is referred to Section 8.3, where simulation approaches are applied to complex situations.

[7] For an introduction to EBO and intelligence implications, see Davis, P. K., *Effects Based Operations: A Grand Challenge for the Analytical Community,* Santa Monica, CA: RAND, 2001, accessed on-line on October 30, 2002 at http://www.rand.org/publications/MR/MR1477. See also Christian, M. C., and J. E. Dillard, "Why We Need a National Joint Targeting Center," *Air and Space Power Chronicles,* Jan 2000, accessed on-line on October 30, 2002 at http://www.airpower.maxwell.af.mil/airchronicles/cc/Dillard.html. In the wake of Operation ALLIED FORCE, the United Kingdom has also analyzed the need for integrated targeting and has established a Directorate of Targeting and Information Operations. See *Defence Second Report on the Lessons of Kosovo* (HC 347-I), U.K. House of Commons, Select Committee on Defence, January 24, 2001, ANNEX.

[8] The evolution of the JDL model is articulated in a series of documents: White Jr., F. E., *Data Fusion Lexicon,* Joint Directors of Laboratories, Technical Panel for C3, Data Fusion SubPanel, Naval Ocean System Center, San Diego, CA, 1987; White Jr., F. E., "A Model for Data Fusion," *Proc. First National Symp. on Sensor Fusion,* Vol. 2, 1988; and Steinberg, A. N., C. L. Bowman, and White Jr., F. E., "Revisions to the JDL Data Fusion Model," *Proc. of Third NATO/IRIS Conf.,* Quebec City, Canada, 1998.

[9] Waltz, E. L., and J. Llinas, *Multisensor Data Fusion,* Norwood, MA: Artech House, 1990.

[10] Hall, D., *Mathematical Techniques in Data Fusion,* Norwood, MA: Artech House, 1992.

[11] Antony, R., *Principles of Data Fusion Automation,* Norwood MA: Artech House, 1995.

[12] Hall, D. L., and J. Llinas, *Handbook of Multisensor Data Fusion,* Boca Raton: CRC Press, 2001.

[13] Piatetsky-Shapiro, G., and W. J. Frawley (eds.), *Knowledge Discovery in Databases,* Menlo Park, CA: AAAI Press/MIT Press, 1991, p. 3.

[14] This example is very specific for purposes of illustration, however, mining criteria can be much more general in nature (yielding more possible relationships.)

[15] *An Overview of Data Mining at Dun & Bradstreet,* DIG white paper 95/01, Pilot Software, Cambridge, MA, September 1995.

[16] Chapman, P., et al., "CRISP-DM 1.0: Step-by-Step Data Mining Guide," CRISP-DM Consortium, 2000, accessed on-line on October 30, 2002 at http://www.spss.com/CRISPDM.

[17]   Mattison, R., and R. M. Mattison, *Data Warehousing and Data Mining for Telecommunications*, Norwood, MA: Artech House, 1997.

[18]   Gardner, C., *IBM Data Mining Technology*, IBM Corporation, April 11, 1996.

[19]   Fayyad, U. M., et al. (eds.), *Advances in Knowledge Discovery and Data Mining*, Cambridge, MA: MIT Press, 1996.

[20]   Wiederhold, G., "Mediators in the Architecture of Future Information Systems," *IEEE Computer*, March 1992, pp. 38–49.

[21]   Waltz, E. L., "Information Understanding: Integrating Data Fusion and Data Mining Processes," *Proc. of IEEE International Symposium on Circuits and Systems*, Monterey CA, May 31–June 4, 1997.

[22]   Davis, J., "Combatting Mind Set," *Studies in Intelligence*, Vol. 36, No. 5, 1992, p. 35.

[23]   Definitions from the Defense Modeling and Simulation Office Modeling & Simulation Glossary.

[24]   The Military Operations Research Society has studied the specific role of modeling and simulation for military operations analysis. See Clements, D., and S. Iwanski, *Evolving Principles of Operations Analysis in DoD Workshop*, February 29–March 2, 2000, Naval Postgraduate School, Monterey, CA.

[25]   Schrage, M., *Serious Play: How the World's Best Companies Simulate to Innovate*, Boston: HBR Press, 2001, p. 117.

[26]   In Chapter 5 (Section 5.5), we introduced these three domains as the fundamental three representations of reality, based on historical philosophical thought. Some have proposed intelligence models with more than four domains (e.g., physical, information, political, cultural, financial, and legal), though all can be organized under the fundamental three categories. See Smith, R., "Counter Terrorism Simulation: A New Breed of Federation," *Proc. Spring 2002 Simulation Interoperability Workshop*, March 2002.

[27]   See a description of a coupled three-domain architecture in the author's paper: Waltz, E., "Data Fusion in Offensive and Defensive Information Operations," *Proc. of National Symposium of Sensor and Data Fusion*, San Antonio TX, June 2000, pp. 219–232.

[28]   The MOVES Institute Brochure, Naval Postgraduate School, 2002, accessed on-line on September 27, 2002, at http://www.movesinstitute.org/MOVESbrochure8-02.pdf.

[29]   "Protecting America's Freedom in the Information Age: A Report of the Markle Foundation Task Force," report of the Working Group on Analytic Methods, The Markle Foundation, New York, October 2002, p. 144.

[30]   McCarthy, M., "The National Warning System: Striving for an Elusive Goal," *Defense Intelligence Journal*, Vol. 3, 1994, p. 9.

[31]   Davies, J., and T. Gurr (eds.), *Preventative Measures: Building Risk Assessment and Crisis early Warning Systems*, Lanham, MD: Rowman & Littlefield, 1998.

[32]   See Bremer, S., *Simulated Worlds: A Computer Model of National Decision Making*, Princeton, NJ: Princeton University Press, 1977; Czerwinski, T. (ed.), *Coping With the Bounds: Speculations on Nonlinearity in Military Affairs*, Washington D.C.: NDU Press, 1997;

Alberts, D. S., and T. Czerwinski (eds.), *Complexity, Global Politics and National Security,* Washington D.C.: NDU Press, 1998.

[33] Remarks of DCI George J. Tenet, "Opening Remarks," The Conference on CIA's Analysis of the Soviet Union, 1947–1991, Princeton University, March 8, 2001.

[34] For an introduction to these topics, see Robinson, C., *Dynamical Systems,* Boca Raton: CRC Press, 1995; and Axelrod, R., *The Complexity of Cooperation,* Princeton, NJ: Princeton University Press, 1997.

[35] Axelrod, R., and M. Cohen, *A Complex Adaptive Systems Approach to Information Policy,* Report Sponsored by OASD for C3I, June 8, 1997.

[36] Hunt, C. L. T. C., and I. Sais, "Complexity-Based Modeling and Simulation: Modeling Innovation at the Edge of Chaos," *Proc. of 1998 Command and Control Research and Technology Symposium,* CCRP, Monterey CA, June 29–July 1, 1998.

[37] The MOVES Institute Brochure, p. 11.

[38] Bonabeau, E., "Predicting the Unpredictable," *Harvard Business Review,* March 2002, pp. 109–116.

[39] For an excellent cautionary article on the use of models and simulations, see the classic article by Sterman, J. D., "A Skeptic's Guide to Computer Models," MIT Sloan School of Management, accessed on-line on September 27, 2002 at http://web.mit.edu/jsterman/www/Skeptic's_Guide.html.

9

The Intelligence Enterprise Architecture

The processing, analysis, and production components of intelligence operations are implemented by enterprises—complex networks of people and their business processes, integrated information and communication systems and technology components organized around the intelligence mission. As we have emphasized throughout this text, an effective intelligence enterprise requires more than just these components; the people require a collaborative culture, integrated electronic networks require content and contextual compatibility, and the implementing components must constantly adapt to technology trends to remain competitive. The effective implementation of KM in such enterprises requires a comprehensive requirements analysis and enterprise design (synthesis) approach to translate high-level mission statements into detailed business processes, networked systems, and technology implementations.

The central conceptual property of an enterprise is called its *architecture.* The means of representing the architecture is by architectural *descriptions—* the explicit representations of the enterprise's organizational structure of components and their relationships [1]. Because the enterprise includes people, systems, and technology components, its architecture descriptions must analyze and describe the many complicated facets and interaction of these human, nformation, and physically implemented technology components.

In this chapter, we introduce the structured methodologies to perform the translation from mission objectives to the enterprise architecture descriptions that enable the implementation of KM in real enterprises.

9.1   Intelligence Enterprise Operations

As noted in earlier chapters, the U.S. IC is transitioning to unify the agencies within the community into a single enterprise to enhance collaborative analysis and problem solving. In the early 1990s the community implemented Intelink, a communitywide network to allow the exchange of intelligence between agencies that maintained internal compartmented networks [2]. The DCI vision for "a unified IC optimized to provide a decisive information advantage..." in the mid-1990s led to the IC CIO to establish an IC Operational Network (ICON) office to perform enterprise architecture analysis and engineering to define the system and communication architectures in order to integrate the many agency networks within the IC [3]. This architecture is required to provide the ability to collaborate securely and synchronously from the users' desktops across the IC and with customers (e.g., federal government intelligence consumers), partners (component agencies of the IC), and suppliers (intelligence data providers within and external to the IC). The undertaking illustrates the challenge of implementing a mammoth intelligence enterprise that is comprised of four components:

1. *Policies.* These are the strategic vision and derivative policies that explicitly define objectives and the approaches to achieve the vision.

2. *Operational processes.* These are collaborative and operationally secure processes to enable people to *share* knowledge and assets *securely* and *freely* across large, diverse, and in some cases necessarily compartmented organizations. This requires processes for dynamic modification of security controls, public key infrastructure, standardized intelligence product markup, the availability of common services, and enterprisewide search, collaboration, and application sharing.

3. System (network). This is an IC system for information sharing (ICSIS) that includes an agreed set of databases and applications hosted within shared virtual spaces within agencies and across the IC. The system architecture (Figure 9.1) defines three virtual collaboration spaces, one internal to each organization and a second that is accessible across the community (an intranet and extranet, respectively). The internal space provides collaboration at the Special Compartmented Intelligence (SCI) level within the organization; owners tightly control their data holdings (that are organizationally sensitive). The community space enables IC-wide collaboration at the SCI level; resource protection and control is provided by a central security policy. A separate collateral community space provides a space for data shared with DoD and other federal agencies.

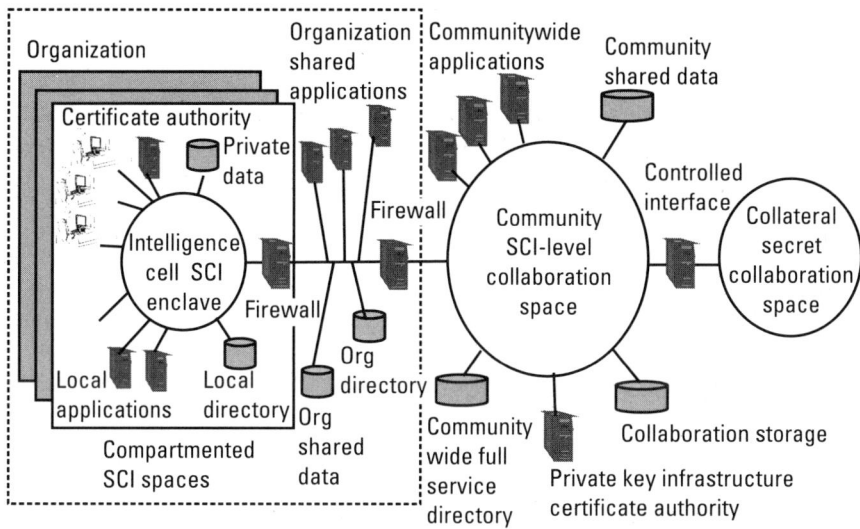

Figure 9.1 ICSIS system architecture.

4. Technologies. The enterprise requires the integration of large installed bases of legacy components and systems with new technologies. The integration requires definition of standards (e.g., metadata, markup languages, protocols, and data schemas) and the plans for incremental technology transitions.

The scope of such an undertaking—typical of many KM enterprise projects—illustrates the need for an integrated analysis of the entire enterprise from people and processes to hardware and software. The undertaking also requires a methodology to describe the architecture for simulation, evaluation, and implementation. The ICON effort preceded the call for a more collaborative and data-sharing culture and smoother collaboration capabilities following terrorist attacks on the United States in 2001 [4]. The scope and scale of the ICON effort will take years to fully implement in a series of phases; while this effort is a great challenge, even modest intelligence architectures pose challenges to the architect to define and describe its many facets. In addition to this community-wide effort, agencies within the community have ongoing efforts to enhance collaboration within their own networks, such as the Joint Intelligence Virtual Architecture at the Defense Intelligence Agency [5].

In the following sections, we introduce the methodology to describe and develop intelligence enterprise architectures, and we provide an example to illustrate the process.

9.2   Describing the Enterprise Architecture

Two major approaches to architecture design that are immediately applicable to the intelligence enterprise have been applied by the U.S. DoD and IC for intelligence and related applications. Both approaches provide an organizing methodology to assure that all aspects of the enterprise are explicitly defined, analyzed, and described to assure compatibility, completeness, and traceability back to the mission objectives. The approaches provide guidance to develop a comprehensive abstract *model* to describe the enterprise; the model may be understood from different *views* in which the model is observed from a particular perspective (i.e., the perspectives of the user or developer) and described by specific products that makeup the *viewpoint.*

The first methodology is the Zachman Architecture Framework™, developed by John Zachman in the late1980s while at IBM. Zachman pioneered a concept of multiple perspectives (views) and descriptions (viewpoints) to completely define the information architecture [6]. This framework is organized as a matrix of 30 perspective products, defined by the cross product of two dimensions:

1. Rows of the matrix represent the viewpoints of architecture stakeholders: the owner, planner, designer, builder (e.g., prime contractor), and subcontractor. The rows progress from higher level (greater degree of abstraction) descriptions by the *owner* toward lower level (details of implementation) by the subcontractor.

2. Columns represent the descriptive aspects of the system across the dimensions of data handled, functions performed, network, people involved, time sequence of operations, and motivation of each stakeholder.

Each cell in the framework matrix represents a descriptive *product* required to describe an aspect of the architecture. The intelligence planner's view of functions in an intelligence enterprise, for example, includes the CIO's list of the business processes to be performed; the owner's view of functions includes the intelligence unit's detailed business model describing activities, flows, and specific responsibilities. This framework identifies a single descriptive product per view, but permits a wide range of specific descriptive approaches to implement the products in each cell of the framework:

- Mission needs statements, value propositions, balanced scorecard, and organizational model methods are suitable to structure and define the owner's high-level view.

- Business process modeling, the object-oriented Unified Modeling Language (UML), or functional decomposition using Integrated Definition Models (IDEF) explicitly describe entities and attributes, data, functions, and relationships. These methods also support enterprise functional simulation at the owner and designer level to permit evaluation of expected enterprise performance.

- Detailed functional standards (e.g., IEEE and DoD standards specification guidelines) provide guidance to structure detailed builder- and subcontractor-level descriptions that define component designs.

The second descriptive methodology is the U.S. DoD Architecture Framework (formally the C4ISR Architecture Framework), which defines three interrelated perspectives or architectural views, each with a number of defined products [7]. The three interrelated views (Figure 9.2) are as follows:

1. *Operational architecture* is a description (often graphical) of the operational elements, intelligence business processes, assigned tasks, workflows, and information flows required to accomplish or support the intelligence function. It defines the type of information, the frequency of exchange, and what tasks are supported by these information exchanges.

2. *Systems architecture* is a description, including graphics, of the systems and interconnections providing for or supporting intelligence functions. The system architecture defines the physical connection, location, and identification of the key nodes, circuits, networks, and users and specifies system and component performance parameters. It is constructed to satisfy operational architecture requirements per standards defined in the technical architecture. This architecture view shows how multiple systems within a subject area link and interoperate and may describe the internal construction or operations of particular systems within the architecture.

3. *Technical architecture* is a minimal set of rules governing the arrangement, interaction, and interdependence of the parts or elements whose purpose is to ensure that a conformant system satisfies a specified set of requirements. The technical architecture identifies the services, interfaces, standards, and their relationships. It provides the technical guidelines for implementation of systems upon which engineering specifications are based, common building blocks are built, and product lines are developed.

The three views are fully interrelated, although they may also be viewed in *layers* to distinguish the elements, models, and metrics that comprise enterprise (Figure 9.2). Specific DoD architecture *products* are defined for each of the three views (Table 9.1) and for a comprehensive (all-view) top-level description that is similar to the owner row of the Zachman framework. It is important to note that the CIOs of U.S. federal organizations have also defined The *Federal Enterprise Architecture Framework* suitable for non-DoD organizations to organize and manage the development and maintenance of architecture descriptions [8].

The DoD framework products have been mapped into the Zachman framework by Sewell to illustrate the similarity and compatibility between the two approaches [9]. Both approaches provide a framework to decompose the enterprise into a comprehensive set of perspectives that must be defined before building; following either approach introduces the necessary discipline to structure the enterprise architecture design process.

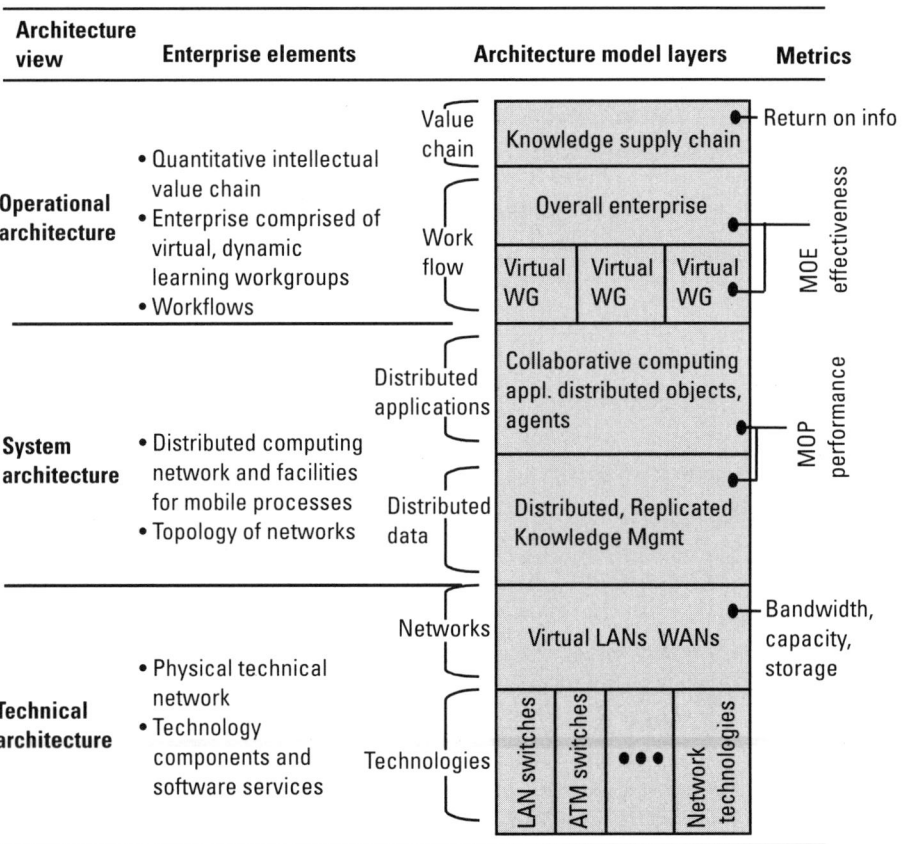

Architecture view	Enterprise elements	Architecture model layers		Metrics
Operational architecture	• Quantitative intellectual value chain • Enterprise comprised of virtual, dynamic learning workgroups • Workflows	Value chain Work flow	Knowledge supply chain Overall enterprise Virtual WG / Virtual WG / Virtual WG	Return on info MOE effectiveness
System architecture	• Distributed computing network and facilities for mobile processes • Topology of networks	Distributed applications Distributed data	Collaborative computing appl. distributed objects, agents Distributed, Replicated Knowledge Mgmt	MOP performance
Technical architecture	• Physical technical network • Technology components and software services	Networks Technologies	Virtual LANs WANs LAN switches / ATM switches / ••• / Network technologies	Bandwidth, capacity, storage

Figure 9.2 Three DoD views of an intelligence enterprise.

Table 9.1
Major DoD Architecture Framework Products

View	Ref	Product	Description
All	AV-1	Overview and summary information	Scope, purpose, intended users, environment depicted, analytical findings if applicable
	AV-2	Integrated dictionary	Definition of all terms used in all products
Operational	OV-1	High-level operational concept graphic	High-level graphical description of operational concept (high-level organizations, missions, geographic configuration, and connectivity)
	OV-2	Operational node connectivity description	Operational nodes, activities performed at each node, connecting paths and information flows (data and control) between nodes
	OV-3	Operational information exchange matrix	Information exchanged between nodes and the relevant attributes of that exchange
	OV-6a	Operation rules model	Operational activity sequence and timing of business rules
	OV-6b	Operation state transition model	
	OV-6c		Responses of business processes to events
	OV-7	Operation event/trace model	Critical activity sequence tracing in scenarios
		Logical data model	Data requirements and business process rules of the operational view
System	SV-1	System interface description	Identification of systems and system components and their interfaces within and between nodes
	SV-2	Communication description	Physical nodes and communication lay downs
	SV-3	Systems matrix	System-to-system relationships
	SV-4	Functional description	Functions of each system; info flows
	SV-10a	Rules model	Timing rule constraints
	SV-10b	State transition model	States and responses of system to events
	SV-10c	Event/trace model	Critical sequences described in ops view
	SV-11	Physical data model	Formats and file structures of logical data model
Technical	TV-1	Technical architecture profile	Delineation of all standards that apply to the architecture
	TV-2	Standards technology forecast	Projection and time-phasing of transition to emerging technical standards

From: [10].

The emerging foundation for enterprise architecting using framework models is distinguished from the traditional systems engineering approach, which focuses on optimization, completeness, and a build-from-scratch originality [11]. Enterprise (or system) architecting recognizes that most enterprises will be constructed from a combination of existing and new integrating components:

- Policies, based on the enterprise strategic vision;
- People, including current cultures that must change to adopt new and changing value propositions and business processes;
- Systems, including legacy data structures and processes that must work with new structures and processes until retirement;
- IT, including legacy hardware and software that must be integrated with new technology and scheduled for planned retirement.

The adoption of the architecture framework models and system architecting methodologies are developed in greater detail in a number of foundational papers and texts [12].

9.3   Architecture Design Case Study: A Small CI Enterprise

The enterprise architecture design principles can be best illustrated by developing the architecture description for a fictional small-scale intelligence enterprise: a typical CI unit for a *Fortune* 500 business. This simple example defines the introduction of a new CI unit, deliberately avoiding the challenges of introducing significant culture change across an existing organization and integrating numerous legacy systems. Though small in scale, the CI unit is an enterprise comprised of people on the CI analytic unit, CI business processes, and integrated ITs. The CI unit provides legal and ethical development of descriptive and inferential intelligence products for top management to assess the state of competitors' businesses and estimate their future actions within the current marketplace. The unit is not the traditional *marketing* function (which addresses the marketplace of customers) but focuses specifically on the competitive environment, especially competitors' operations, their business options, and likely decision-making actions.

- *The situation:* FaxTech recognizes that its operations are threatened by the rapidly changing competitive landscape in its high-tech news business [13]. The emergence of new competitive threats from cable and traditional news networks that broadcast business and technology segments are capturing increasing shares of FaxTech's market share. The FaxTech board has issued an urgent requirement for the rapid implementation of a CI unit. The enterprise architect recognizes the assignment as a corporate KM project that should be evaluated against O'Dell and Grayson's four-question checklist for KM projects [14]:

1. *Select projects to advance your business performance.* This project will enhance competitiveness and allow FaxTech to position and adapt its product and services (e.g., reduce cycle time and enhance product development to remain competitive).

2. *Select projects that have a high success probability.* This project is small, does not confront integration with legacy systems, and has a high probability of technical success. The contribution of KM can be articulated (to deliver competitive intelligence for executive decision making), there is a champion on the board (the CIO), and the business case (to deliver decisive competitor knowledge) is strong. The small CI unit implementation does not require culture change in the larger Fax-Tech organization—and it may set an example of the benefits of collaborative knowledge creation to set the stage for a larger organizationwide transformation.

3. *Select projects appropriate for exploring emerging technologies.* The project is an ideal opportunity to implement a small KM enterprise in FaxTech that can demonstrate intelligence product delivery to top management and can support critical decision making.

4. *Select projects with significant potential to build KM culture and discipline within the organization.* The CI enterprise will develop reusable processes and tools that can be scaled up to support the larger organization; the lessons learned in implementation will be invaluable in planning for an organizationwide KM enterprise.

- *The architecture framework.* Meeting the criteria on the checklist, the CI enterprise architect considers the multiple views of the Zachman architecture framework and creates a custom matrix of views appropriate for the small CI enterprise (Figure 9.3). The appropriate (generally not all) products to describe the enterprise are identified (unnecessary products are blank in the figure). Each cell of the matrix contains the title of the specific product(s) to be produced for the view that addresses a focus (column) and stakeholder (row). The matrix identifies 10 numbered cells that are provided (at the top level only) in the following subsections to illustrate the design approach. The arrows in the matrix also illustrate the top-down design flow from the high-level planner's viewpoint (FaxTech's top management) to the owner's viewpoint (the director of the CI unit), and on to the designer, builder, and subcontractor viewpoints.

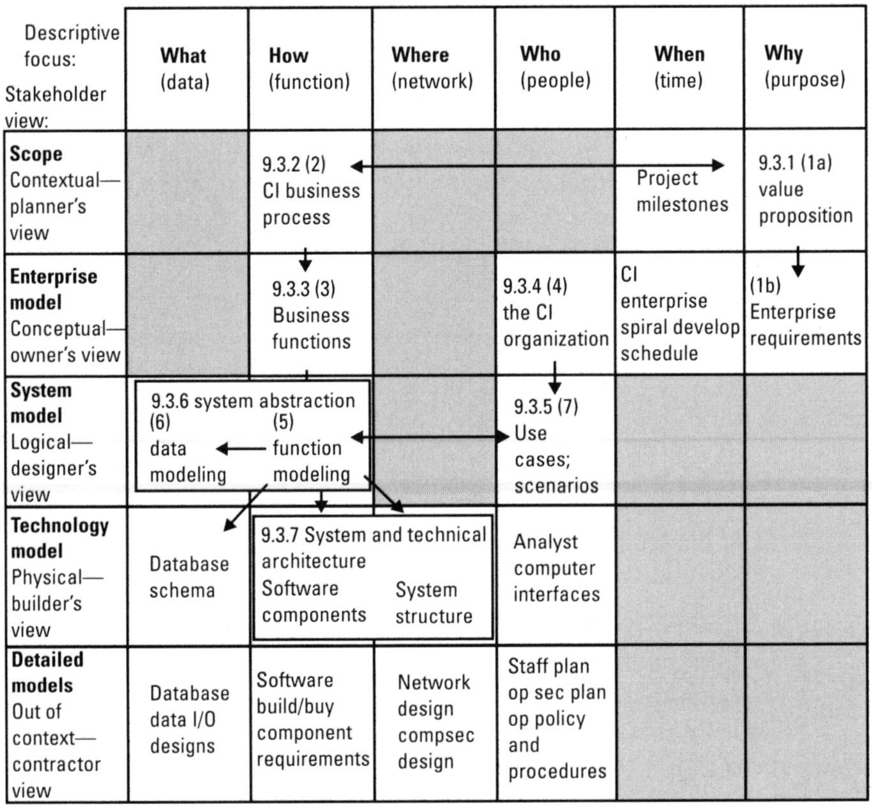

Descriptive focus: Stakeholder view:	What (data)	How (function)	Where (network)	Who (people)	When (time)	Why (purpose)
Scope Contextual— planner's view		9.3.2 (2) CI business process			Project milestones	9.3.1 (1a) value proposition
Enterprise model Conceptual— owner's view		9.3.3 (3) Business functions		9.3.4 (4) the CI organization	CI enterprise spiral develop schedule	(1b) Enterprise requirements
System model Logical— designer's view	9.3.6 system abstraction (6) data modeling	(5) function modeling		9.3.5 (7) Use cases; scenarios		
Technology model Physical— builder's view	Database schema	9.3.7 System and technical architecture Software System components structure		Analyst computer interfaces		
Detailed models Out of context— contractor view	Database data I/O designs	Software build/buy component requirements	Network design compsec design	Staff plan op sec plan op policy and procedures		

Figure 9.3   CI enterprise architecture descriptive products.

A complete design will certainly include more products and products in greater detail than the following illustrative subsections. The following 10 sub-section numbers follow the numbering of the 10 major products identified in the matrix.

9.3.1   The Value Proposition

The CI value proposition must define the value of competitive intelligence, how it will impact FaxTech, who will use the intelligence products, and how the intelligence will be valued. The mission statement succinctly defines the CI mission:

> The competitive intelligence unit provides FaxTech executive management timely and accurate assessments of competitor's activities and market posi-tions, using ethical means, to permit an assessment of FaxTech's relative

competitive position and to enable sound business judgments based on a comprehensive understanding of the competitive landscape.

The value added by the CI unit must be explicitly defined and top-level performance goals identified for management by further decomposing the mission statement into the elements of value using the balanced scorecard. The scorecard method, introduced in Chapter 3, Section 3.5, identifies specific measures in each of the areas of the scorecard. This value proposition for CI must include three components: the delivery of reliable and relevant information, interpretation of findings to impact strategy, and creation of special insight to support critical decisions [15]. Competitive intelligence measures include both hard quantitative metrics of intelligence timeliness and accuracy and soft subjective measures of the unit's organizational performance [16].

The quantitative measures may be difficult to define; the financial return on CI investment measure, for example, requires a careful consideration of how the derived intelligence couples with strategy and impacts revenue gains. Kilmetz and Bridge define a top-level measure of CI return on investment (ROI) metric that considers the time frame of the payback period (t, usually updated quarterly and accumulated to measure the long-term return on strategic decisions) and applies the traditional ROI formula, which subtracts the cost of the CI investment (C_{CI+I}, the initial implementation cost, plus accumulating quarterly operations costs using net present values) from the revenue gain [17]:

$$ROI_{CI} = \sum \left[(P \times Q) - C_{CI+I} \right]_t$$

The expected revenue gain is estimated by the increase in sales (units sold, Q, multiplied by price, P, in this case) that are attributable to CI-induced decisions. Of course, the difficulty in defining such quantities is the issue of assuring that the gains are uniquely attributable to decisions possible only by CI information [18].

In building the scorecard, the enterprise architect should seek the lessons learned from others, using sources such as the Society for Competitive Intelligence Professionals or the American Productivity and Quality Center [19]. These sources can also provide peers by which FaxTech may benchmark its processes and expected values. The scorecard (Table 9.2) provides the highest level definition of expected values and measures for the CI unit to guide its operations and measure its value-added contribution to FaxTech management.

9.3.2 The CI Business Process

The Society of Competitive Intelligence Professionals has defined a CI business cycle that corresponds to the intelligence cycle introduced in Chapter 2; the

Table 9.2
CI Unit Balanced Scorecard

Balanced Scorecard Measures—The Faxtech CI Enterprise		
Vision and Strategy Goals		**Vision and Strategy Measures**
1. Enhance competitiveness and responsiveness of FaxTech strategy; provide continuous assessment of competitors' relative performance and market positions		1. Identify top 80% of competitor's operations and likely strategies; 75% accuracy of financial performance compared to 6 month trailing data
2. Support competitive decision making; create operations and pricing comparison data, strengths, weaknesses, opportunities, and threats (SWOT)		2. Estimate to 70% accuracy, cost of operations of major competitors; maintain SWOT on all competitors
3. Provide predictive analyses of competitors' research and development and product development activities, production changes, and new product launches		3. Predict competitor actions within two months (plus or minus); produce research and development timelines
4. Reduce risk of surprise of mergers and acquisitions within the marketplace		4. Identify, assess, and rank top three strategic alliance candidates, estimate likely actions
Scorecard Area	**System Measures**	**Outcome Measures**
Shareholder (Financial)	Estimated return on the investment	ROI_{CI}
Customer (Management)	Timeliness of alerts	Alert lead time
	Accuracy and coverage of term CI reports	Reporting accuracy
		Decision impact; assessed by management
	Impact on decision making	surveys
Internal	Intelligence requirement fulfillment	Percentage of requirements completed
	Task completion rate	Time to task completion
	Competitor data coverage	Percentage of data profiles updated
	Competitor data depth	Percentage of strategy profiles validated
Learning and Growth	CI team collaboration productivity	Team cohesion measures
	Subject matter insight	Percentage of predictive accuracy
	Intelligence workflow efficiency	Response time
	Intelligence product improvements	Product completeness

cycle differs by distinguishing primary and published source information, while eliminating the automated processing of technical intelligence sources. The five stages, or business processes, of this high-level business model include:

1. *Planning and direction.* The cycle begins with the specific identification of management needs for competitive intelligence. Management defines the specific categories of competitors (companies, alliances) and threats (new products or services, mergers, market shifts, technology discontinuities) for focus and the specific issues to be addressed. The priorities of intelligence needed, routine reporting expectations, and schedules for team reporting enables the CI unit manager to plan specific tasks for analysts, establish collection and reporting schedules, and direct day-to-day operations.

2. *Published source collection.* The collection of articles, reports, and financial data from open sources (Internet, news feeds, clipping services, commercial content providers) includes both manual searches by analysts and active, automated searches by software agents that explore (*crawl*) the networks and cue analysts to rank-ordered findings. This collection provides broad, background knowledge of CI targets; the results of these searches provide cues to support deeper, more focused primary source collection.

3. *Primary source collection.* The primary sources of deep competitor information are humans with expert knowledge; ethical collection process includes the identification, contact, and interview of these individuals. Such collections range from phone interviews, formal meetings, and consulting assignments to brief discussions with competitor sales representatives at trade shows. The results of all primary collections are recorded on standard format reports (date, source, qualifications, response to task requirement, results, further sources suggested, references learned) for subsequent analysis.

4. *Analysis and production.* Once indexed and organized, the corpus of data is analyzed to answer the questions posed by the initial tasks. Collected information is placed in a framework that includes organizational, financial, and product-service models that allow analysts to estimate the performance and operations of the competitor and predict likely strategies and planned activities. This process relies on a synoptic view of the organized information, experience, and judgment. SMEs may be called in from within FaxTech or from the outside (consultants) to support the analysis of data and synthesis of models.

5. *Reporting.* Once approved by the CI unit manager, these quantitative models and more qualitative estimative judgments of competitor strategies are published for presentation in a secure portal or for formal presentation to management. As result of this reporting, management provides further refining direction and the cycle repeats.

This high-level description can be further detailed in a high-level operational concept graphic (the DoD OV-1 product), which illustrates the component parts that accomplish the CI cell operations (Figure 9.4). This simple model distinguishes the interfaces to management (at the top), the delivery of intelligence products vial a portal (on the right), and the sources of data (on the left) centered around a knowledge base in the center. The knowledge base is partitioned into three sections. The requirements section holds specific needs and the status of accumulating knowledge against those needs, the holdings section accumulates collected materials, and the production section maintains produced models and completed reports.

The annotated flows in the graphic show the major flow paths that lead from planning and direction to collection, analysis, and reporting. This high-level operational graphic (DoD OV-2 format) provides a general description of the flows of data without distinguishing the controlling activities that translate the intelligence cycle into an efficient workflow.

9.3.3   The CI Business Process Functional Flow

The owner's view of the business process describes the CI process in greater detail, distinguishing the flow of data between the basic functions of the cycle, controls to those functions, and the mechanisms (means) that provide resources. Of course, this cyclic model is the most abstract representation of the overall CI process; in reality, the CI process is an intelligence continuum of these processes,

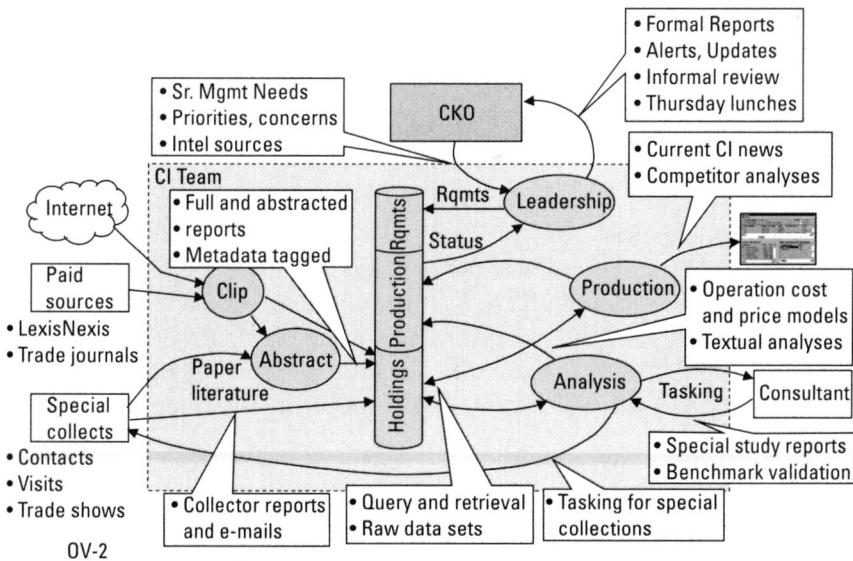

Figure 9.4 CI business operational concept.

which occur concurrently, rather than in a pure cycle. On a day-to-day level, many analyses are being conducted concurrently: weekly competitor summaries, crisis analyses of a competitor's new product launch, and long-term analyses of potential strategic alliance partners. A more detailed representation of these concurrent activities can be illustrated in a functional flow diagram using the integrated definition for function modeling (IDEF0) format [20]—Figure 9.5(a)—to show the five processes with all major interconnecting flows of data (inputs on the left, outputs on the right), controls (e.g., policies and procedures on the top), and mechanisms (e.g., resources, on the bottom).

The model provides greater detail, depicting the simple CI cycle as a *waterfall flow*, but notice that this format highlights numerous other details of interaction within the five major processes:

1. Feedback to planning and direction processes from analysis provides continual status of progress toward fulfillment of current requirements.

2. Critical cross-cueing between published and primary source collection activities is identified.

3. Feed forward of requirements and schedules from planning and direction to all other functions is identified.

4. Identification is made of governing policies, procedures, and templates used in each process.

5. Status and intermediate product outputs from each process are posted to the dissemination portal.

The IDEF0 format is defined to allow structured decomposition of each of the processes to lower levels of functional and interface detail, as illustrated by further decomposing the analyze and produce process in Figure 9.5(b). The same I/O interfaces, controls, and mechanisms are retained, while revealing inner detail in four subprocesses. This level of detail shows the receipt, indexing, and filtering of incoming reports (subprocess 4.1) before entry into the knowledge base (4.2). Data-driven data fusion also matches incoming data to predefined warning cues (for example, a critical competitor action, such as detectable merger indicators in financial data or comments in business reports) in subprocess 4.1. The analysis and synthesis process (4.3) performed manually by analysts with supporting modeling and simulation tools (4.4) is constrained by budget and schedule, guided to meet CI requirements that have been translated to reporting templates, and guided by collaborative analytic procedures that define the organization's team decision-making process. These subprocesses can be further decomposed, with corresponding supporting detailed definitions of the data exchanged between processes [21].

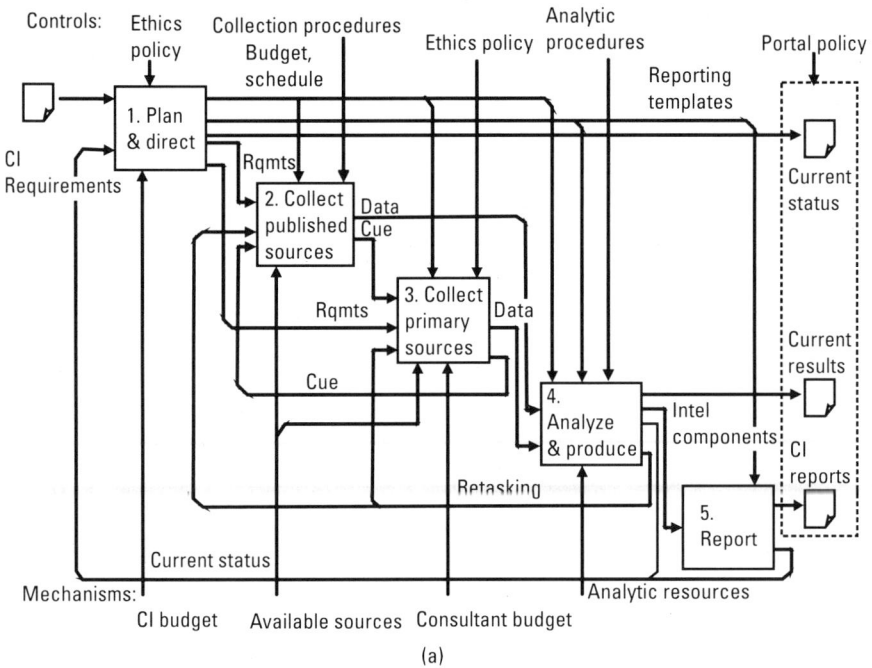

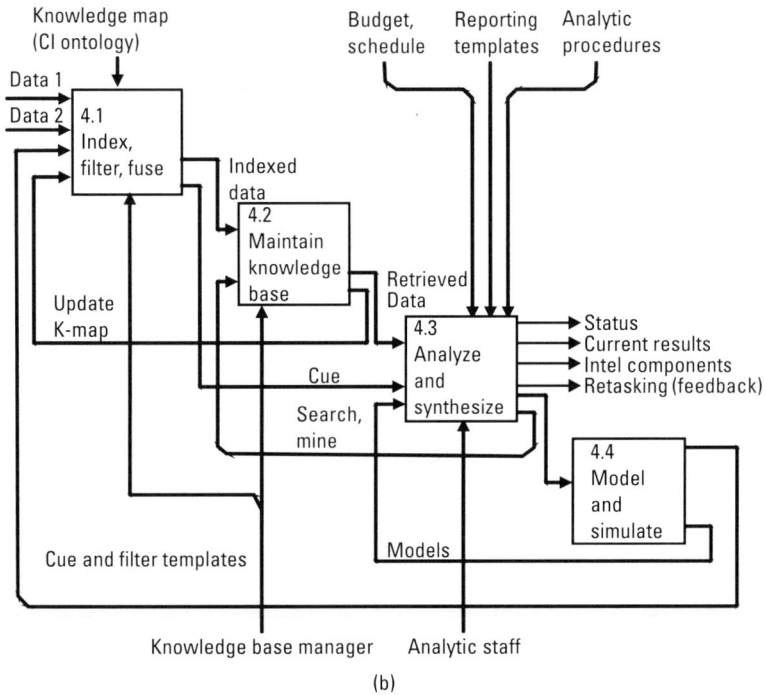

Figure 9.5 (a) CI business process functional flow, and (b) analyze and produce process functional flow.

9.3.4   The CI Unit Organizational Structure and Relationships

The FaxTech CI unit must monitor a one-billion-dollar market with five major competitors and over a dozen related market players that can have a significant influence on the marketplace. This scale of study influences the intelligence breadth and volume to be studied, the size, structure, and composition of the CI organization (operational architecture), and the supporting system architecture. Of course, the complexity and dynamics of FaxTech's rapidly changing business also influence the rate and volume of raw data to be analyzed and the depth of analysis required (e.g., financial simulations). In turn, these factors define the required budget to support the enterprise installation and annual support.

The FaxTech CI unit (Figure 9.6) includes a team of five full-time individuals led by a senior manager with intelligence analysis experience. This manager accepts tasking from executive management, issues detailed tasks to the analytic team, and then reviews and approves results before release to management. The manager also manages the budget, secures consultants for collection or analysis support, manages special collections, and coordinates team training and special briefings by SMEs. The deputy and two competitor analysts perform the day-to-day term analysis of the competitor set and special analyses either

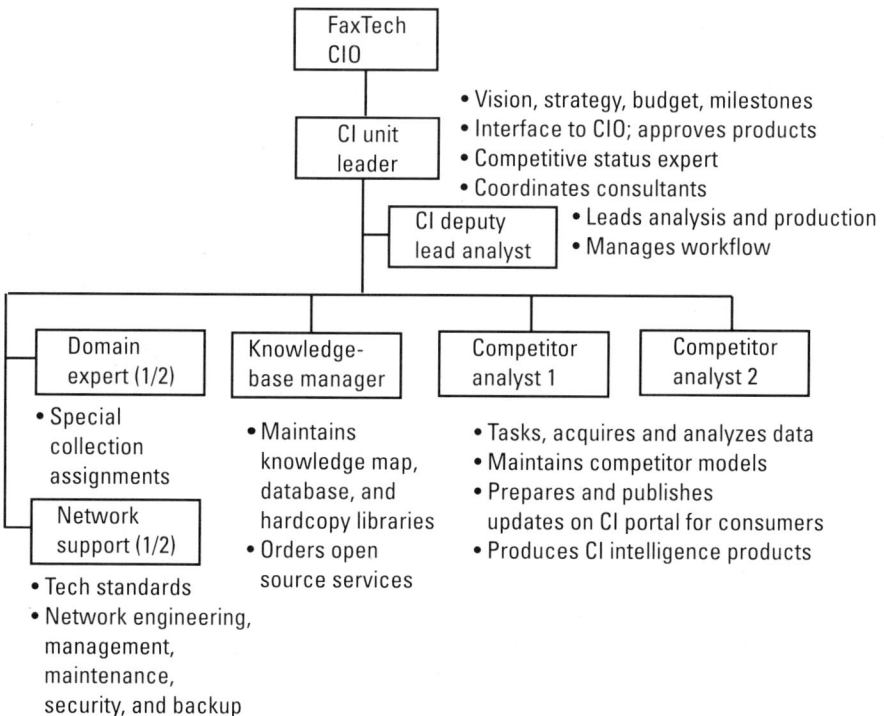

Figure 9.6 CI unit organization and relationships.

requested by management or triggered by events in the marketplace (crisis analysis). A single knowledge-base manager maintains the CI knowledge map, creates the database structures defined in the data model, and oversees the collection of data from commercial sources (e.g., clipping services by subscription and purchased data sets) and the Internet. In support of this team is a half-time network support administrator (shared with FaxTech's IT unit) and one half-time equivalent support from special subject experts to support special collections (e.g., at trade shows) and special analyses (e.g., product teardowns). This expertise may be provided by shared time of experts within FaxTech's organization or by consultants operating under nondisclosure agreements.

The analytic team must address the need for access across numerous foreign languages to understand foreign media reports, industry documents, and market information. The small analytic staff may have skills in several languages, but will require translation tools and access to contracted translation support (consultants) for special documents.

9.3.5   A Typical Operational Scenario

For each of the five processes, a number of use cases may be developed to describe specific actions that actors (CI team members or system components) perform to complete the process. In object-oriented design processes, the development of such use cases drives the design process by first describing the many ways in which actors interact to perform the business process [22]. A scenario or process thread provides a view of one completed sequence through a single or numerous use case(s) to complete an enterprise task. A typical crisis response scenario is summarized in Table 9.3 to illustrate the sequence of interactions between the actors (management, CI manager, deputy, knowledge-base manager and analysts, system, portal, and sources) to complete a quick response thread. The scenario can be further modeled by an activity diagram [23] that models the behavior between objects.

The development of the operational scenario also raises nonfunctional performance issues that are identified and defined, generally in parametric terms, for example:

- Rate and volume of data ingested daily;
- Total storage capacity of the on-line and offline archived holdings;
- Access time for on-line and off-line holdings;
- Number of concurrent analysts, searches, and portal users;
- Information assurance requirements (access, confidentiality, and attack rejection).

Table 9.3
A Crisis Response Scenario

1. Scenario: Crisis response

2. Summary: The CI team responds to an urgent request from management for a 72-hour turnaround to analyze the potential for a merger between two competitors.

3. Preconditions	Postconditions (results)
• Competitors and consultants are identified in the current knowledge base	• Status reported daily • Final report delivered at 72 hours

4. Basic flow

1. *Management* issues a CI crisis response memo to the *CI manager*, identifying specific intelligence needs, priorities, and response time.

2. *CI manager* enters the requirement template and translates the memo to specific direction:

2.1 CI manager identifies the target competitors of interest.

2.2 CI manager checks specific key information needs on requirements template.

2.3 CI manager prioritizes the needs.

2.4 CI manager establishes security requirements, schedule, and reporting-dissemination requirements.

2.5 CI manager approves requirement.

3. *CI deputy* creates tasking plan to accomplish the requirement direction.

3.1 *Deputy* opens tasking template, plans activities, and allocates task assignments to team members.

3.2 *System* issues tasking assignments to the collection and analysis team.

3.3 *Deputy* creates special collect request form; *system* issues request for immediate consultant services.

4. *Knowledge-base manager* sets published data collection parameters.

4.1 *Knowledge-base manager* translates CI requirements to sources, search, and reply format parameters.

4.2 *Knowledge-base manager* sets source, search, and format parameters and issues search orders to commercial content providers.

4.3 *Knowledge-base manager* creates crisis holdings partition in knowledge base.

4.4 *System* searches *sources* and identifies existing relevant holdings and creates abstracts and links within partition.

4.5 *System* populates partition with accumulating data, creates abstracts, and ranks data against requirements.

4.6 *System* maintains crisis holdings summary metrics.

5. *Consultant* issues primary source reports.

5.1 *Consultant* reviews special collect requirements from the field via the *portal*.

Table 9.3 (continued)

5.2 *Consultant* reviews updated cues from ongoing analysis via the *portal*.

5.3 *Consultant* prepares and issues reports by completing the report template on the *portal*.

5.4 *System* alerts analyst to special collect report availability.

6. *Analysts* conduct analysis.

6.1 *Analyst* reviews task assignment and issues manual searches concurrent with auto searches.

6.2 *Analyst* creates report templates to provide status, intermediate, and final results for *portal*.

6.3 *System* automatically organizes manual and automated search holdings by template requirements.

6.4 *System* generates titles, text summaries of accumulated holdings, and rank orders linked listing

6.5 *System* links published and special collection holdings to requirements and analysis templates.

6.6 *Analyst* authors text analyses in report format templates.

6.7 *Analyst* completes custom merger-SWOT template, linking data sources to conclusions.

6.8 *Analyst* runs financial analysis simulation to compute merger performance predictor metrics.

6.9 *System* compares template completion with requirement to produce status summary.

6.10 *Analyst* issues task refinements, identifying new needs from published and special sources.

7. *CI deputy* creates and distributes intermediate status reports.

7.1 *Deputy* reviews intermediate analytic products.

7.2 *Deputy* enters status summary, abstract of current results (if any), and links to intermediate products.

7.3 *Manager* approves and releases status for distribution to *portal*.

8. *CI manager* publishes final report.

8.1 *Manager* approves final report.

8.2 *Manager* publishes final report for distribution to *portal*.

8.3 *System* maintains configuration management of final report.

9. The scenario ends

5. Alternative flows: the crisis response flow may be extended in time and converted to a standing watch requirement	**6. Revision:** 3.2 dated 9-3-02

9.3.6   CI System Abstraction

The purpose of use cases and narrative scenarios is to capture enterprise behavior and then to identify the classes of object-oriented design. The italicized text in the scenario identifies the actors, and the remaining nouns are candidates for objects (instantiated software classes). From these use cases, software designers can identify the objects of design, their attributes, and interactions. Based upon the use cases, object-oriented design proceeds to develop sequence diagrams that model messages passing between objects, state diagrams that model the dynamic behavior within each object, and object diagrams that model the static description of objects. The object encapsulates state attributes and provides services to manipulate the internal attributes.

Based on the scenario of the last section, the enterprise designer defines the class diagram (Figure 9.7) that relates objects that accept the input CI requirements through the entire CI process to a summary of finished intelligence. This diagram does not include all objects; the objects presented illustrate those that acquire data related to specific competitors, and these objects are only a subset of the classes required to meet the full enterprise requirements defined earlier. (The objects in this are included in the analysis package described in the next section.) The requirement object accepts new CI requirements for a defined competitor; requirements are specified in terms of essential elements of information (EEI), financial data, SWOT characteristics, and organization structure. In this object, key intelligence topics may be selected from predefined templates to specify specific intelligence requirements for a competitor or for a marketplace event [24]. The analyst translates the requirements to tasks in the task object; the task object generates search and collect objects that specify the terms for automated search and human collection from primary sources, respectively. The results of these activities generate data objects that organize and present accumulated evidence that is related to the corresponding search and collect objects.

The analyst reviews the acquired data, creating text reports and completing analysis templates (SWOT, EEI, financial) in the analysis object. Analysis entries are linked to the appropriate competitor in the competitor list and to the supporting evidence in data objects. As results are accumulated in the templates, the status (e.g., percentage of required information in template completed) is computed and reported by the status object. Summary of current intelligence and status are rolled up in the summary object, which may be used to drive the CI portal.

9.3.7 System and Technical Architecture Descriptions

The abstractions that describe functions and data form the basis for partitioning packages of software services and the system hardware configuration. The system architecture description includes a network hardware view (Figure 9.8, top) and a comparable view of the packaged software objects (Figure 9.8, bottom). The figure illustrates the allocation of packages to servers on the network.

The network view is organized in an n-tier architecture (n = 4), partitioning security, data storage, CI business logic, and client workstation functions. The CI unit operates as a protected enclave within the larger FaxTech organization; the security zone includes the interface to the Internet, dedicated lines, and

Status

Schedule
Pct_Complete
Est_Complete
Analysis_Link
Issues_Text

Requirement

Competitor_ID
Rqmt_Text
Rqmt_Priority
EEI_Rqmts
Org_Rqmts
Financial_Rqmts
SWOT_Rqmts
Other_Rqmts

CompanyList

Competitor_ID
Competitor_Name
Competiror_Text
Competitor_Data
SEC_Code

Generates

Involves

Task

Task_ID
Competitor_ID
Task_Text
Due_Date
Est_Budget
Source_List
Task_Priority
Assign_Team

Generates

Generates

Search

Request_ID
Competitor_ID
Topic_Category
Topic_Context
Topic_Keywords
Topic_Text
Search_Param1
Search_Param2

Collect

Request_ID
Collector_ID
Request_Text
Collect_Range
Collect_Class

Data

Data_Item_ID
Data_Source
Data_Date
Data_Pedigree
Competitor_ID
Data_Abstract
Data_Text
Request_ID

Generates

Generates

Analytic process

Determines

Is linked to

Analysis

Competitor_ID
Timeline_Analysis
SWOT_Analysis
Fin_Analysis
Other_Analysis

Is linked to

Is reported in

CI summary

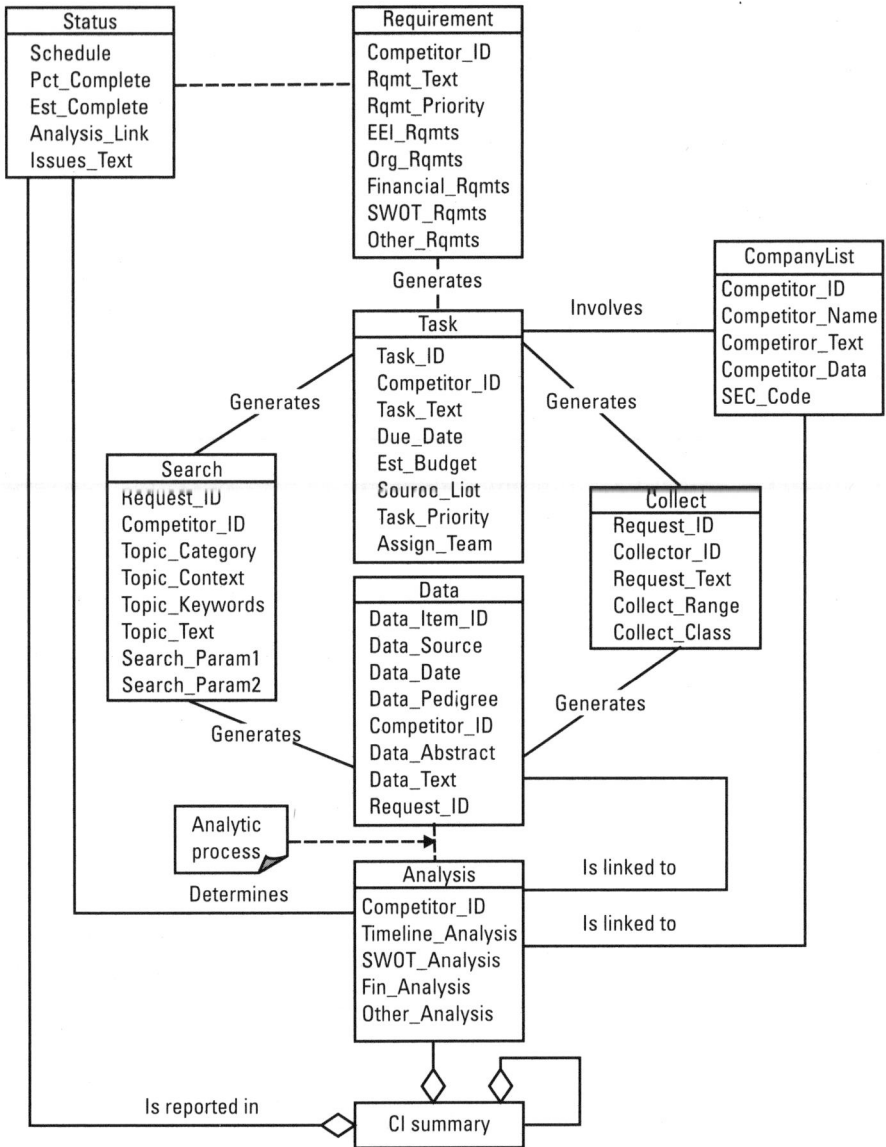

Figure 9.7 CI system class diagram relating requirements to analytic results.

the FaxTech Intranet. A firewall and separate intrusion detection unit separates the three tiers of the enclave from the Internet and the local intranet, providing enclave boundary control. Not shown are other security services (e.g., public key infrastructure and encryption services) to provide:

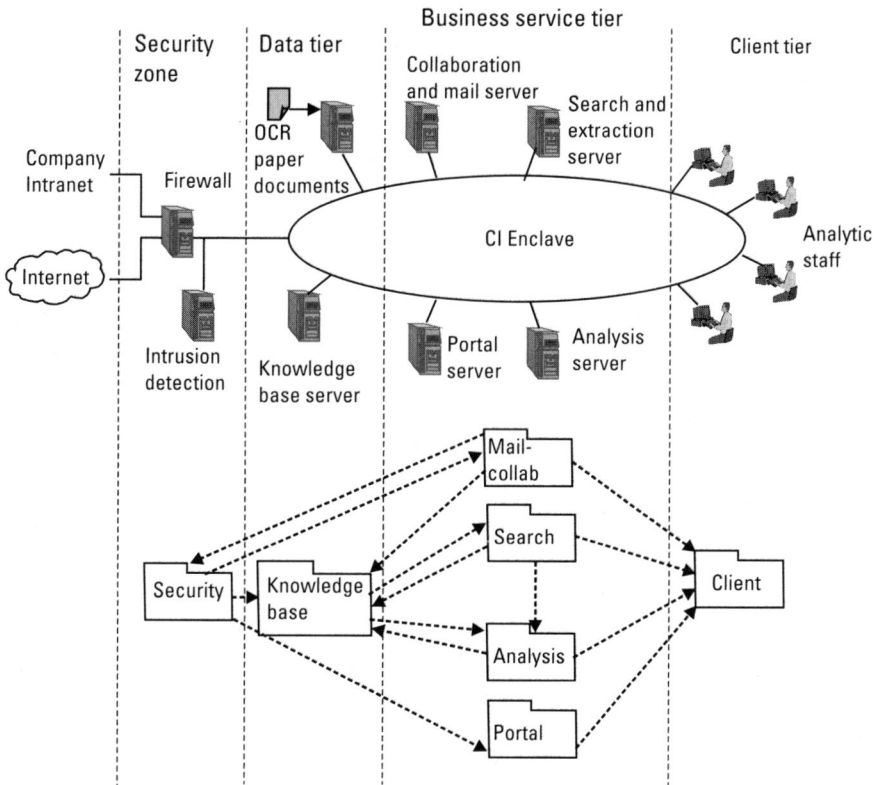

Figure 9.8 System network architecture (top) and package interfaces (bottom).

- *Authorized access.* Multiple levels of security and individual access control are required to label and protect sensitive information.

- *Resistance to attacks and malicious code.* Detection, resistance, and response to external and internal attempts to access data or insert unauthorized processes is required.

- *Disaster recovery.* The enterprise must maintain regular backups of data and can also provide colocated or offsite mirrored servers and data storage to enable rapid recovery from physical or electronic disasters

The data tier includes the scalable relational database (knowledge base) and a dedicated hardcopy data entry machine with a scanner and optical character recognition capability. The business tier includes the software applications that perform the CI business functions identified earlier in the functional analysis (Section 9.3.3). This tier requires careful definition of the objects (as in class description in the prior section) to assure compatibility between application packages, the knowledge base, and client objects. These middle-tier business objects are implemented by a combination of commercial applications and custom software components to provide portal, mail and collaboration, search, and special analytic services. The client tier includes the package of common objects (e.g., browser, mail, collaboration, and office suite) at each analyst's workstation. The package view (Figure 9.8, bottom) describes the partitioning of the objects into packages and their implementation on hardware servers.

The enterprise technical architecture is described by the standards for commercial and custom software packages (e.g., the commercial and developed software components with versions, as illustrated in Table 9.4) to meet the requirements developed in system model row of the matrix. Fuld & Company has published periodic reviews of software tools to support the CI process; these reviews provide a helpful evaluation of available commercial packages to support the CI enterprise [25]. The technical architecture is also described by the standards imposed on the implementing components—both software and hardware. These standards include general implementation standards [e.g., American National Standards Institute (ANSI), International Standards Organization (ISO), and Institute of Electrical and Electronics Engineers (IEEE)] and federal standards regulating workplace environments and protocols. The applicable standards are listed to identify applicability to various functions within the enterprise.

A technology roadmap should also be developed to project future transitions as new components are scheduled to be integrated and old components are retired. It is particularly important to plan for integration of new software releases and products to assure sustained functionality and compatibility across the enterprise.

9.4 Summary

Architecture frameworks provide a structured modeling methodology to guide the design of a complete enterprise—taking into account the operations, systems, and technologies necessary to meet the enterprise mission and deliver on the value proposition. The simple example in this chapter has illustrated the approach to develop a comprehensive set of descriptive products for even a small competitive intelligence enterprise. Larger scale enterprises and those with existing legacy systems (like the U.S. IC and large business KM enterprises) require significant rigor in applying these methodologies to comprehensively model their enterprises to achieve increased intelligence value—but the principles illustrated in this example cell are applicable across the larger enterprise.

The extension of KM principles from the CI project across the entire Fax-Tech organization requires one last component—a means to communicate

Table 9.4
Representative Technical Architecture Components

Software Component	Package Functions Performed
Security	Firewall; intrusion detection system
	Private key infrastructure
Knowledge base	Relational database
	Define CI process, audit workflow, and produce status reports
	XML template creation, configuration management
	Metadata tagging—manual summarization
Mail and collaboration	Electronic messaging environment
	Collaboration environment
	Expert directory and finder
Search and retrieval	Web crawler, search agents, text search by key word and context, text summarization
	Image search, video search, audio search
	Multilingual search and text translation
Analysis	Automated source data linking
	Text analysis
	Data structuring, visualization tools
	Financial simulation
Portal	Digital product version publication and maintenance
	Team collaboration portal
	Hyperlinked document production
Client	Browser
	Mail and collaboration clients
	Office tool suite
	Analytic tool clients

tacitly the value of organizational transformation. As we introduced in Chapter 3, this is best conveyed by the narrative story rather than by the preceding engineering architecture diagrams. From the CI project, FaxTech management chooses one story that conveys that tacit experience to the workforce: "I joined the new competitive intelligence unit here at FaxTech in January without previous experience in news or in research and development. (I was an MI analyst with experience in foreign language and weapons systems.) I was given one week of training and began collaborating with our product specialists around the globe to find out why our competitors 'owned' one particular market segment.

Everyone was very open with me and shared their frustrations; they also shared their lists of customers and showed me our competitor's products. I organized that data in our warehouse and created a simple portal to share the data with our marketing and sales teams. Within 2 months, I was surprised to see 65 employees conversing on the portal about this forgotten market and our competitor's products. I organized their suggestions in a lessons-learned base. The sales force logged into the portal and gained new insight and the marketing team developed a new 'TechAlert' news product tailored to research and development labs. Now, a year later, we have over 100 research and development lab accounts. The sales team even included me on their team to receive the industry new product award—and I didn't even know what an research and development lab was when I started here a year ago."—*Jenny Crenshaw, Competitive Analyst, Corporate CI Unit*

Endnotes

[1]   The definition of architecture by the International Council on Systems Engineering System Architecture Working Group is: "The fundamental and unifying system structure defined in terms of system elements, interfaces, processes, constraints and behaviors." The more compact IEEE 610.2 definition is, "The organizational structure of a system or component."

[2]   Martin, F. T., *Top Secret Intranet: How U.S. Intelligence Built Intelink—The World's Largest, Most Secure Network*, New York: Prentice Hall, 1998.

[3]   Swindle, M., "Intelligence Community System for Information Sharing (ICSIS)," presented at DoD Secure E-Business Summit, May 7, 2001, accessed on-line on June 20, 2002 at http://www.secure-biz.net/OSD2001/presentations/Presentations/T06_Swindle.ppt. See also Onley, D. S., "Intelligence Community Welcomes Data Sharing," *Government Computer News*, May 8, 2001, accessed on-line on June 28, 2002 at http://www.gcn.com/vol1_no1/ daily-updates/4217-1.html.

[4]   After the Sept. 11, 2001, terrorist attacks in the United States, numerous media reports focused on the capabilities of the IC to collaborate and share data. See, for example, Myers, L., "Report Rips Intelligence Priorities," NBC and MSNBC News Service, July 17, 2002, accessed on-line on August 31, 2001, at http://www.msnbc.com/news/781308.asp; and Waller, D., "Is the Intelligence Community Ready for this War?" *Time*, October 9, 2001, accessed on-line on September 30, 2002, at http:// www.time.com/time/columnist/waller/article/0,9565,178770,00.html.

[5]   Verton, D., "DIA Tackles Flow of Intelligence," *Federal Computer Week*, October 18, 1999.

[6]   See, for example, the following classic articles: Zachman, J. A., "A Framework for Information Systems Architecture," *IBM Systems Journal*, Vol. 26, No. 3, 1987, and Sowa, J. F., and J. A. Zachman, "Extending and Formalizing the Framework for Information Systems Architecture," *IBM Systems Journal*, Vol. 31, No. 3, 1992.

[7] *C4ISR Architecture Framework Version 2.0.* U.S. DoD Office of the Assistant Secretary of Defense for Command, Control, Communications, and Intelligence, Washington, DC, November 1997. (The DoD is reviewing a version 2.1 at the time of this manuscript and is considering revising the name to DoD Architecture Framework, or DoDAF.)

[8] Federal Enterprise Architecture Framework, "The Chief Information Officers Council," Version 1.1, September 1999.

[9] Sowell, P. K., "The C4ISR Architecture Framework: History, Status, and Plans for Evolution," *Proc of 5th International Command and Control Research and Technology Symposium,* June 2000; see also Sowell, P. K., "Mapping the Zachman Framework to the C4ISR Architecture Framework," *MITRE,* September 3, 1999.

[10] C4ISR Framework Ver 2.0, Table of Essential and Supporting Framework Products.

[11] See the author's "Requirements Derivation for Data Fusion Systems," in *Handbook of Multisensor Data Fusion,* Hall, D. L., and J. Llinas, Boca Raton: CRC Press, 2001, pp. 15-1–15-8.

[12] See, for example, Rechtin, E., *System Architecting: Creating and Building Complex Systems,* Engelwood Cliffs, NJ: Prentice Hall, 1991; Rechtin, E., and M. Maier, *The Art of System Architecting,* Boca Raton: CRC Press, 1997; Maier, M. W., "Architecting Principles for Systems-of-Systems," *Systems Engineering,* Vol. 2, No. 1, 1999, pp. 1–18; and Rechtin, E., "The Art of Systems Architecting," *IEEE Spectrum,* October, 1992, pp.66–69.

[13] The name FaxTech has no relationship to any actual business; it has been created solely for this example. The name has been checked against the U.S. patent and trademark base to ensure that it is not a registered trademark.

[14] Adapted from the KM project checklist in: O'Dell, C. and C. J. Grayson, *If Only We Knew What We Know,* NY: Free Press, 1998, pp.194–195.

[15] Langabeer II, J. R., "Exploring the CI Value Equation," *Competitive Intelligence Review,* Vol. 10, No. 3,1999, pp. 27–32.

[16] Simon, N. J., "Determining Measures of Success," *Competitive Intelligence,* Vol. 1, No. 2, July–September 1998, pp. 45–47.

[17] Kilmetz, S. D., and R. S. Bridge, "Gauging the Returns on Investments in Competitive Intelligence: A Three-Step Analysis for Executive Decision Makers," *Competitive Intelligence Review,* Vol. 10, No. 1, 1999, pp. 4–11.

[18] Davison, L., "Measuring Competitive Intelligence Effectiveness: Insights from the Advertising Industry," *Competitive Intelligence Review,* Vol. 12, No. 4, 2001, pp.25–38.

[19] The Society of Competitive Intelligence Professionals Web site is http://www.scip.org. The American Productivity and Quality Center Web site is http://www.apqc.org.

[20] Integrated Definition for Function Modeling (IDEF0), Federal Information Processing Standards Publication 183, NIST, December 21, 1993. This is but one of a number of modeling standards that may be adopted.

[21] The IDEF1X companion standard to IDEF0 provides a data modeling standard for describing relational data: Integrated Definition for Information Modeling (IDEF1X), Federal Information Processing Standards Publication 184, NIST, December 21, 1993.

[22] Rosenberg, D., *Use Case Driven Object Modeling with UML*, Boston: Addison-Wesley, 1999.

[23] The diagrams referenced in this section are standard unified modeling language constructs. See the Object Management Group for latest UML standards, http://www. omg.org.

[24] See an enumeration of typical key intelligence topics in Herring, J. P., "Key Intelligence Topics: A Process to Identify and Define Intelligence Needs," *Competitive Intelligence Review*, Vol. 10, No. 2, 1999, pp. 4–14.

[25] *Intelligence Software Report 2002*, Fuld & Co., Cambridge, MA, 2002. This report, first issued in 2000 and updated in 2002, identified over 40 applicable packages, then reviewed and scored the applicability of over 10 dedicated CI packages in detail.

10

Knowledge Management Technologies

IT has enabled the growth of organizational KM in business and government; it will continue to be the predominant influence on the progress in creating knowledge and foreknowledge within intelligence organizations. As we have pointed out throughout this book, the successful application of new IT to improve organizational effectiveness is entirely dependent upon the success of process and cultural transitions necessary to reap the benefits. Envisioning the future of KM requires an understanding of the trends in IT, particularly in those technologies that we distinguish here as KM technologies, which will continue to expand human thinking, collaboration, and problem solving. As we noted in earlier chapters, the future intelligence competitions will be for rapid knowledge discovery and delivery. When sensors and communications become global commodities, the critical technologies for the intelligence organization will be those that support deep and rapid analysis-synthesis. In this chapter, the key enabling and emerging KM technologies that will enable future generations of intelligence enterprises are introduced.

10.1 Role of IT in KM

When we refer to *technology,* the application of science by the use of engineering principles to solve a practical problem, it is essential that we distinguish the difference between three categories of technologies that all contribute to our ability to create and disseminate knowledge (Table 10.1). We may view these as three technology layers, with the basic computing materials sciences providing the foundation technology applications for increasing complexity and scale of communications and computing. These materials technologies produce computing

Table 10.1
Technology Contributors to KM

Technology Category	Description	Technology Examples
KM Technologies	The integration and application of cognitive and organizational sciences to implement enterprises comprised of humans and computing systems to achieve operational goals	Collaboration messaging and knowledge exchange tools Cognitive support: data fusion, mining, analysis, and visualization tools Intelligent agents, artificial intelligence
Information Technologies	The integration and application of computer science (software) and computing devices (hardware) to implement applied computing networks that transmit, store, or manipulate information by electronic, optical, or other physical phenomena	Object-relational databases Network computing Quantum and biological computing algorithms Multiple level security Cryptography Computational mathematics Digital computing-communication convergence
Computing Materials Technologies	The application of materials, biological, and other physical sciences to implement computing, storage, and communication components	Silicon microelectronics, optoelectronics, photonics Biotechnology Nanotechnology Optical and magnetic storage Quantum scale structures

devices (hardware) that are integrated into computing network structures to provide the platform to process data. The computer sciences produce the next layer of abstract structures and computational methods that are the basis for ITs (software), which are applied to organize and combine data regarding real-world applications. These technologies are further integrated into enterprises, where KM technologies apply the cognitive and organizational sciences to orchestrate networks of people and computing machines to create and apply knowledge to achieve business goals. Notice that the first layer is physical (materials), the second integrates the physical and the abstract (devices and information), while the KM technology layer focuses almost entirely in abstractions (knowledge).

KM technology is, of course, dependent upon the capabilities provided by the layers below, and revolutionary improvements in the underlying layers can be expected to enable increased capabilities in KM technology growth. Because this can have a significant effect on intelligence, the U.S. IC commissioned a specific study of the potential synergistic effects of materials technologies on IT [1]. That study noted that the future technology layer above (that is facilitated by) IT and KM is genetic manipulation technologies that can alter life itself. In this chapter, we focus only on the top-layer KM technologies; indeed, numerous lower level technologies cited in the table (e.g., quantum computing and optical storage) have the potential to revolutionize the speed, scale, and complexity of IT capabilities—enabling significant improvements in KM for intelligence applications. Nanotechnology will enable ubiquitous sensing, quantum, micro, and optoelectronic technology will enable ubiquitous communications and computation, while IT and KM technologies will allow distributed teams of collaborating analysts to translate the global breadth and depth of information on complex situations to actionable intelligence.

Before exploring the implications of future KM technologies, it is valuable to briefly review the rapid growth of IT to date and its impact on KM. Over the past 3 decades, computing (hardware) and software technologies have rapidly transitioned from the centralized mainframe to the distributed network (Figure 10.1). As computing hardware has moved homogeneous computing toward networked heterogeneous computing nodes, software has likewise moved from a single operating program to distributed services, assembled from component objects at *run time* to provide the requested service. The granularity of operations has moved from the dedicated batch run time to a *session* on the network. Administration of the single mainframe has been replaced by a distributed network administration that brings complexity, redundancy, flexibility, and greater challenges to security.

The ubiquity of networked computing, propelled by increasing computing power, storage capacity, ubiquitous communications, broadband connectivity, and open protocols set the stage for widespread explicit knowledge capture in databases, sharing, and analysis by OLAP. The future of IT (at the right of the figure) will be characterized by:

- Fully distributed networks of heterogeneous knowledge processes and stores; these networks will be organized in logical *tiers* of greater depth than today's two-tier client-server architectures.

- The structures of these future networks will be self-organizing—with agents detecting new capabilities as they come online and organizing data and processes to deliver services based on the best available current capabilities.

- Autonomous agents will traverse the network to locate and apply services required to solve complex problems (not just queries) tasked by human analysts. These dynamic agents will understand purpose and

Era:	Mainframe ⟶	PC ⟶	Client-server ➜	Networked
Organizing form and control	Hierarchy			Network
	Centralized	⟶		Distributed
Computing	Structure	Mainframe homogeneity	Personal computing	Client-server	N-tier networked heterogeneity
and	Organizing principle	Structured	Object oriented	Component service	Secure self-organizing services
Software technology	Language generation	1GL (Fortran, Cobol)	2GL (C, Basic, Pascal)	3GL (C++)	4GL (Java, platform independent)
	Function unit granularity (organizer) process granularity	Subroutines (op system) a batch "run"	Modules (op system) application "run-time"	Objects (obj brokers) "session"	Object, component, service (dynamic agents) "domain (of services) life"
System control and behavior	Organizing control	Single administrator	Single user	Multiple users organize by collaborative work tools	Self-organizing by automated agency
Implications for knowledge creation	Independent analysis, product sharing face-face, ⟶ telephone collaboration			User-to-user teleconference ➜ collaboration	User-to-agent data collaboration

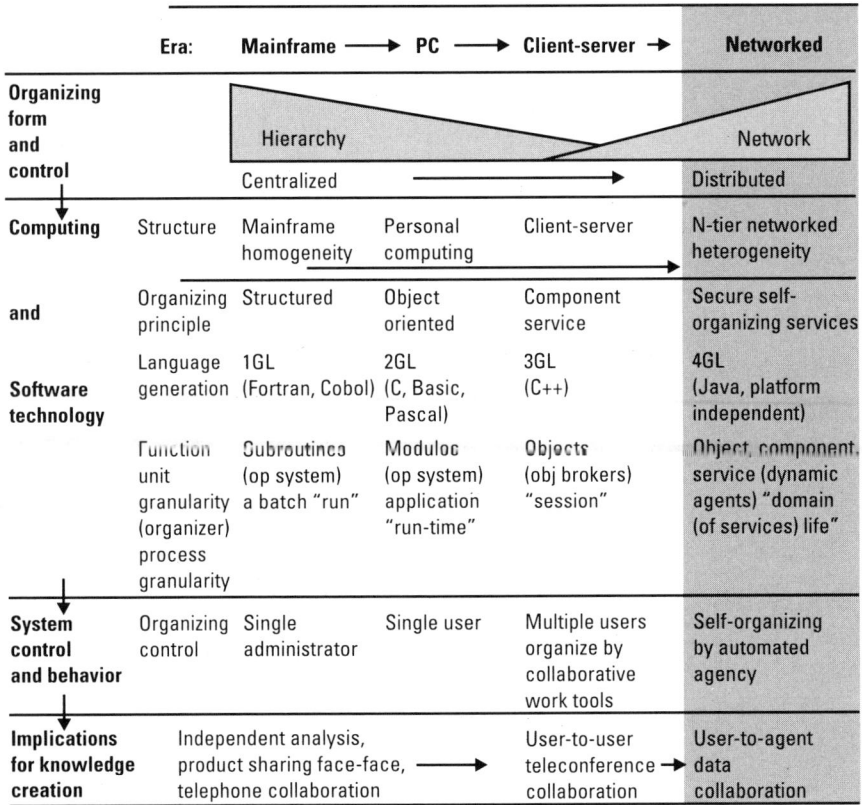

Figure 10.1 Four generations of IT.

context, and they will guide the organizing properties of the network to optimize service delivery.

- Security services will provide strong, biometric-based protection of service and information.

Of course, these are the technologies that have enabled organizations to move from a focus on lower level data processing to the higher level attention to KM. The implications for intelligence analysis-synthesis are profound. The analytic processes of the past were founded on individual SMEs with limited-text soft-copy products, and they were restricted to telephone or face-to-face collaboration with others. Today's KM technologies are enabling online collaboration and problem solving with dispersed analytic teams, using the support of analytic tools to explicitly model problems and solutions. Tomorrow, IT will enable highly effective collaboration and problem solving among teams of humans and their autonomous agents across secure networks with access to vast arrays of heterogeneous data stores of structured and unstructured data. These agents will empower analysts to query directly for knowledge (i.e., answers, rather than querying for data or information) across multiple languages and media. They will also support analysts to understand the results of queries—representing and explaining multidimensional information for human understanding. This human-agent collaboration will, however, introduce new cultural and social issues yet to be appreciated.

10.2 KM Research for National Security Applications

The quest to develop these future IT capabilities in the United States for national security applications, including intelligence, is performed by a wide range of components within the DoD and IC as depicted in Figure 10.2.

The DCI Strategic Intent specifically identifies KM and supporting technologies as key elements of the processes of unifying the community through collaborative processes and investing in people and knowledge (these are the first two of the Strategic Intent objectives) [2]. KM technology applications are categorized by the IC CIO in the following areas:

Joint Services			
JCS- Info superiority/net-centric war leadership			
OSD/C3I—architectures, C4ISR systems			
DISA/D2—C4I & intelligence / D7—rqmts analysis DMSO—modeling and simulation	• Nat'l defense univ—NCW and IW academic theory	DARPA—info technology and Systems (KM) research	
Air Force	**Army**	**Navy/Marines**	**Intel Community**
AF Office of Scientific Research–artificial intelligence, perception and cognition AF Science Advisory Board—battlespace infosphere AF Research Lab —Info Directorate C4ISR (Rome), sensor Directorate (Dayton) AF Battle Labs—info Ops experiments	Army Research Office —Knowledge-based systems, sensor webs Army Research Lab —battlefield KM Nets and decision making Army Digitization Office —experiment, demo integration Army Space Pgm Office	Office of Naval Research —Cognitive augmentation, training, intelligent systems SPAWAR System Center San Diego —NCW systems R&D Naval Research Lab Navy Ctr AI Research—KM and intelligence research NWDC—NCW experimentation	ARDA-Adv R&D Agency In-Q-Tel—intel community CIA/DIA—all source KM, analysis and digital production R&D NRO, NSA, NIMA, CMO —Tasking, processing, exploitation, and dissemination R&D

Figure 10.2 U.S. DoD and IC KM technology developers.

- *Intelligence business processes*—governance, applying resources, IT competency, and IT service delivery;

- *KM*—collaboration, intelligence applications, directory services, information storage and management, subscription and delivery services, search and access facilitation, administration applications, and messaging (and e-mail);

- *Infrastructure*—networking, information assurance, computing platforms, and infrastructure management.

Emerging KM enterprise concepts and supporting technologies have been studied in numerous DoD and IC studies, including:

- *Advanced battlespace information system.* This U.S. DoD Director Defense Research and Engineering (DDR&E) study in 1998 evaluated the application of emerging ITs to provide military KM capabilities to support capabilities sought in Joint Vision 2010.

- *Building the joint battlespace infosphere.* This is a 2000 Air Force Science Advisory Board study of Air Force implementations of KM concepts.

- *Navy command center of the future.* This is a 2000 Naval Research Advisory Committee study of Naval KM for command and control.

- *The global technology revolution bio/nano/material trends and their synergies with information technology by 2015.* This is a 2001 RAND National Defense Research Institute study for the National Intelligence Council examining the implications of potential discontinuities in IT.

- *Global Trends 2015: A Dialogue About the Future with Nongovernment Experts.* This is a National Intelligence Council study of world futures, including effects of global IT, published in 2000.

In addition to government studies, it is important to note that IT consulting organizations routinely publish projections of commercial KM technologies for business applications [3].

10.3 A KM Technology Roadmap

The sequence of integration and implementation of ITs into KM capabilities can be illustrated in a technology roadmap (Figure 10.3). The trend of KM capabilities includes a sequence of major generations, each characterized and enabled by a unique integration of available IT.

First generation KM   Second generation KM   Future KM

Emphasis on capture, store knowledge (like physical assets)	Emphasis on multi-media, sharing culture, collaboration process, and networks	Emphasis on integration of knowledge into all aspects of business culture and operations to revolutionize business and processes

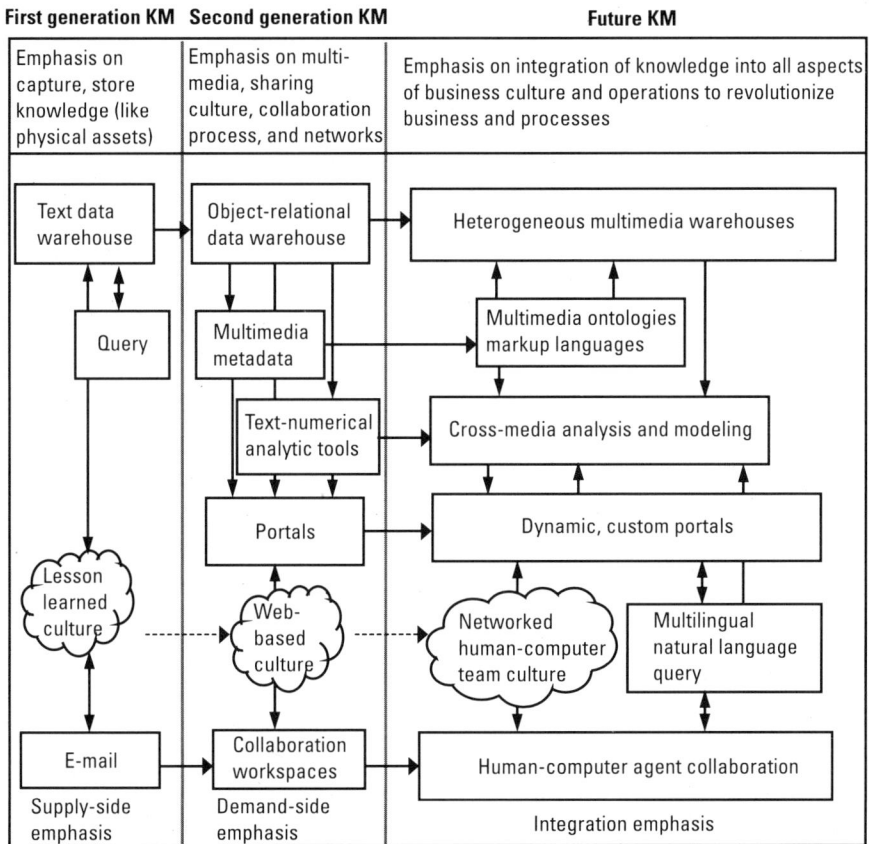

Figure 10.3 A roadmap of three KM technology generations.

The first generation of KM (1985–1995) was enabled by early networked computing technologies: shared databases, email, and service directories. This generation emphasized the capture and storage of corporate explicit knowledge. Often called a *supply-side* emphasis, this generation focused on the capture of organizational knowledge in the form of text data in relational databases. E-mail, reports, business process data, and scanned paper documents were indexed and entered into databases accessible by search engines. These capabilities provided unprecedented access to organizational data by key word queries. The introduction of email in this period also encouraged asynchronous collaboration and the building of communities of practice across time and space. These capabilities encouraged a first generation KM culture that captured and shared lessons learned and developed best practices to be shared across the organization.

The second generation of KM enterprise technology now being implemented emphasizes the *demand-side* by focusing on the delivery of products to users. To effectively deliver accurate, timely, and relevant knowledge to users, technologies are being developed to support four demand-side goals:

1. *Complete use of all data types.* IT is being applied to capture, store, search, and manipulate multimedia information (multilingual audio, video, and text; imagery and geospatial data). Object-relational databases and warehouses store structured and unstructured data; metadata tagging (e.g., XML) is used to increase the content description of data sets, while content-analysis technology is increasing the ability to search, categorize, and tag unstructured data.

2. *Organizationwide collaboration.* Integrated synchronous-asynchronous collaboration tools enable secure multiple-mode text, graphics, audio-video conferencing, and data sharing.

3. *Comprehensive analysis of all sources.* Data-mining and data-fusion tools process a wide variety of numerical and textual data sets to correlate content, detect patterns of known behavior, and discover relationships across diverse data types.

4. *Organizationalwide tailored dissemination of knowledge.* Corporate portals and information-assurance technologies are providing secure access to knowledge across the organization. User profiling (manually entered by the user and automatically learned profiles derived from user behavior) tailors the information presented to the user.

The effect of these current second-generation technologies on the culture has been to enable a Web-based culture that accepts larger distributed communities of practice comprised of people who perform collaborative knowledge sharing and problem solving.

Future generations of KM will emphasize the creation of knowledge by completely integrating the supply and demand sides of the enterprise with greater machine intelligence to support the human users of the enterprise. These future technologies will add to the previous generations by adding:

- *Deep content understanding.* Ontologies of general (common-sense) and domain-specific structured knowledge will enable greater access to, and reasoning about, heterogeneous data sources to provide deeper understanding of context and content of captured and stored *explicit* knowledge.

- *Intelligent support to users.* Autonomous intelligent agents will contribute to collaborative analytic teams, first as supporting contributors with powerful, yet narrow, capabilities (e.g., search) and ultimately as near equals to their human leaders in more general reasoning. This capability will contribute to a deeper exchange of *tacit* and explicit knowledge between humans and agents, respectively.

- *Dynamic human-agent collaboration.* Agents will collaborate with human users at high levels of abstract reasoning to collaboratively solve problems.

These future generations will move toward a more collaborative human-machine culture that teams humans and their agents to create knowledge to achieve organizational goals. Trusted autonomous agents, almost indistinguishable from human analysts in analytic behavior, yet distinctively more powerful in certain areas of reasoning and search, will support human teams in ways not yet fully understood; this will certainly require new cultural adjustments [4].

10.4 Key KM Technologies

The technologies critical to moving toward future generations of KM capabilities are enumerated in the technology matrix (Table 10.2), which distinguishes technologies as core (the basis of today's first and second generation KM implementations and capabilities), enabling (the technology base for the next generation), and emerging (beyond next generation, a revolutionary departure from current KM practices). The matrix further distinguishes technologies by their contribution to explicit or tacit knowledge creation or to explicit-tacit knowledge exchange.

The following subsections highlight several of the most pertinent enabling and emerging KM technology capabilities that are identified on the technology matrix. The paragraphs describe the projected *capabilities* (what), which will be the result of current basic and applied research that is developing the *methods* (how) of the technologies. Endnotes in the paragraphs cite several of the current research and development programs supporting the U.S. IC in many of these areas.

10.4.1 Explicit Knowledge Combination Technologies

Future explicit knowledge combination technologies include those that transform explicit knowledge into useable forms and those that perform combination processes to create new knowledge.

- *Multimedia content-context tagged knowledge bases.* Knowledge-base technology will support the storage of multimedia data (structured and

Table 10.2
KM Technology Matrix

	Knowledge Exchange Category		
	Section 10.4.1 **Explicit ⟷ Explicit**	**Section 10.4.2** **Explicit ⟷ Tacit**	**Section 10.4.3** **Tacit ⟷ Tacit**
Core: Required to Maintain and Sustain Current Capabilities	Text and measurement (sensor) data capture, indexing, mapping, and warehousing; HTML and XML IIDR Semiautomated data fusion and mining of structured data	Push/pull (subscribe/query) dissemination 3D and synthetic n-dimensional visualization—virtual reality Object- and text-based tagging, indexing, search, and retrieval	Distance e-learning Synchronous/asynchronous collaboration tools Creativity supportive tools Visual and strategy simulation tools (experience)
Enabling: Tech Base for Next Generation Capabilities	Global-scale multilingual natural language Multimedia content-context tagged knowledge bases Integrated deduction-induction	Collaborative agent-teams Rapid expert knowledge acquisition Situation immersion Natural language query and conversation Multilingual speech interaction	Tailored naturalistic collaboration tools Human-agent problem-solving collaboration Intimate tacit simulations Combined human-agent learning; personal agent tutors, mentors, and models
Emerging: Beyond Next Generation, a Departure From Current Practices	Automated deductive-inductive reasoning and learning Automated ontology creation	Purposeful, aware agents Human cognition augmentation Direct brain interaction Pervasive personal networked computers	Indistinguishable human-like agent partners, communities of practice, and teams Direct brain tacit knowledge awareness, tracking, articulation, and capture

unstructured) with tagging of both content and context to allow comprehensive searches for knowledge across heterogeneous sources.

- *Multilingual natural language.* Global natural language technologies will allow accurate indexing, tagging, search, linking, and reasoning about multilingual text (and recognized human speech at both the content level and the concept level. This technology will allow analysts to conduct multilingual searches by topic and concept at a global scale; it will also allow linkage of concepts across cultures and continents. The technology will also allow context and content summarization, understanding, and explanation [5].

- *Integrated deductive-inductive reasoning.* Data-fusion and-data mining technologies will become integrated to allow interactive deductive and inductive reasoning for structured and unstructured (text) data sources. Data-fusion technology will develop level 2 (situation) and level 3 (impact, or explanation) capabilities using simulations to represent complex and dynamic situations for comparison with observed situations. Data-mining technology will enable knowledge discovery to be performed in unstructured data; discovery will be performed at the content level (correlation of data) and at the context level (explanations of the reasons for correlations) to support the discrimination of meaningful high-level patterns and relationships [6].

- *Purposeful deductive-inductive reasoning.* Agent-based intelligence will coordinate inductive (learning and generalization) and deductive (decision and detection) reasoning processes (as well as abductive explanatory reasoning) across unstructured multilingual natural language, common sense, and structured knowledge bases. This reasoning will be goal-directed based upon agent awareness of purpose, values, goals, and beliefs.

- *Automated ontology creation.* Agent-based intelligence will learn the structure of content and context, automatically populating knowledge bases under configuration management by humans.

10.4.2 Human-Computer Tacit-Explicit Exchange Technologies

Human-computer technologies include all capabilities to interface and integrate human tacit (and articulable explicit) knowledge with computing machines and communication.

- *Collaborative agent teams.* Agent technology will provide agents as collaborative team members, first as supporting helpers, then as intelligence participants and peers who can become expert contributors. The agents will operate with autonomy, purpose, intelligence, and the capability to explain the rationale for their contributions to the team.

- *Rapid expert knowledge acquisition.* Distributed, continually updating, and reusable knowledge bases will be maintained for a wide range of intelligence domain topics; these knowledge bases will enable rapid and continual creation, refinement, and validation of knowledge by distributed contributors. These knowledge bases will be integrated with deductive-inductive reasoning engines (see prior paragraph) to support analysis, situation assessment, and decision making.

- *Situation immersion.* Virtual-reality technologies will ingrate visual, tactile, and agent storytelling capabilities to provide integrated explicit and tacit (respectively) experiences of data, information, and finished intelligence to humans. These immersive technologies will provide *mixed reality* experiences—integrating synthetic and actual data to allow analysts to be immersed in complex high-dimensionality situations, including the real and the possible. These technologies will be automatically populated by data.

- *Natural language query and conversation.* General question-answer queries at high levels of abstraction will be able to be posed to, understood by, and answered by intelligent systems [7].

- *Multilingual speech interaction.* Natural, bidirectional speech across the major global languages will not only enhance the capture of explicit conversation, it will enable more natural and accurate collaboration between analysts and foreign intelligence partners and contributors.

- *Purposeful, aware agents.* Agent technology will provide a degree of self-awareness, purpose, and understanding to agents that will provide an appreciation of the overall intelligence context, enabling them to adapt, explain, and interact with analysts at a high level of abstraction.

- *Human cognition augmentation.* Technologies that will monitor human attention and focus will be used track performance and guide the flow of information to augment cognition; these technologies will also counteract human limitations in memory, attention, learning, and sensing (visualization) [8]. Instinctual systems will detect human sense and autonomic reactions, as well as recognition of subliminal cognitive cues (e.g., the unspoken "huh?"; "hmm"; "oh!"; and "ah ha!") to cue tight interactive analysis collaboration between human and machine.

- *Direct brain interaction.* This is direct multidimensional presentation to human brain—computing to humans (multilingual speech and context recognition) and computing to physical world (robotics). Current direct brain-to-computer research is focused on the control of physiological functions to enable disabled individuals to control robotic devices by thought. The more complex emerging technologies will go beyond such binary or discrete exchanges to higher level cognitive *conversations.*

- *Pervasive personal networked computers.* While agents will provide personalized intelligent support, wearable or body-augmented networked (wireless) computations will enable their continuous presence to monitor support and anticipate human activity. Analysis, problem solving, and collaboration will not be constrained to the desktop.

10.4.3 Knowledge-Based Organization Technologies

Technologies that support the socialization processes of tacit knowledge exchange will enhance the performance and effectiveness of organizations; these technologies will increasingly integrate intelligence agents into the organization as aids, mentors, and ultimately as collaborating peers.

- *Tailored naturalistic collaboration.* Collaboration technologies will provide environments with automated capabilities that will track the context of activities (speech, text, graphics) and manage the activity toward defined goals. These environments will also recognize and adapt to individual personality styles, tailoring the collaborative process (and the mix of agents-humans) to the diversity of the human-team composition.

- *Intimate tacit simulations.* Simulation and game technologies will enable human analysts to be immersed in the virtual physical, symbolic, and cognitive environments they are tasked to understand. These technologies will allow users to explore data, information, and complex situations in all three domains of reality to gain tacit experience and to be able to share the experience with others.

- *Human-like agent partners.* Multiagent system technologies will enable the formation of agent communities of practice and teams—and the creation of human-agent organizations. Such hybrid organizations will enable new analytic cultures and communities of problem-solving.

- *Combined human-agent learning.* Personal agent tutors, mentors, and models will *shadow* their human partners, *share* experiences and observations, and *show* what they are learning. These agents will learn monitor subtle human cues about the capture and use of tacit knowledge in collaborative analytic processes [9].

- *Direct brain tacit knowledge.* Direct brain biological-to-machine connections will allow monitors to provide awareness, tracking, articulation, and capture of tacit experiences to augment human cognitive performance.

10.5 Summary

KM technologies are built upon materials and ITs that enable the complex social (organizational) and cognitive processes of collaborative knowledge creation and dissemination to occur over large organizations, over massive scales of knowledge. Technologists, analysts, and developers of intelligence enterprises must monitor these fast-paced technology developments to continually reinvent the enterprise to remain competitive in the global competition for knowledge. This continual reinvention process requires a wise application of technology in three modes. The first mode is the direct *adoption* of technologies by upgrade and integration of COTS and GOTS products. This process requires the continual monitoring of industry standards, technologies, and the marketplace to project the lifecycle of products and forecast adoption transitions. The second application mode is *adaptation,* in which a commercial product component may be *adapted* for use by wrapping, modifying, and integrating with commercial or custom components to achieve a desired capability. The final mode is custom *development* of a technology unique to the intelligence application. Often, such technologies may be classified to protect the unique investment in, the capability of, and in some cases even the existence of the technology.

Intelligence organizations have a mandate to remain competitive and a mandate to leverage the significant commercial investment in information and KM technologies. The Government Electronics and Information Technology Association reported that in 1997 global commercial IT research investments exceeded the entire (not just IT-related) U.S. DoD research development, test, and evaluation budget and was increasing at a rate significantly higher than DoD investments [10]. This commercial technology is available to all intelligence competitors and the intelligence technologist requires the wisdom to know *how and what technologies to adopt, adapt, and develop.* Technology is enabling, but it is not sufficient; intelligence organizations must also have the vision to apply these technologies while transforming the intelligence business in a rapidly changing world.

Endnotes

[1]   "The Global Technology Revolution Bio/Nano/Material Trends and Their Synergies with Information Technology by 2015," Study for the National Intelligence Council, RAND National Defense Research Institute, 2001. In addition, the projected global societal effects of IT were reported in Anderson, R., et al., *The Global Course of the Information Revolution: Technological Trends: Proceedings of an International Conference,* RAND CF-157-NIC, 2000. In the terminology of that report, KM capabilities would be considered artifacts or services, rather than a technology.

[2]   The Strategic Intent is classified document, but essential KM activities are described in unclassified documents. See IC CIO, *Advancing Information Systems to Improve the Business of Intelligence*, DCI/CMS/IC CIO, 2000.

[3]   For example, see "Enterprise Portals: Connecting People, Information, and Applications," META Group Publications, 2000; "Advanced Technologies and Applications Study," The Gartner Group, 2001; and "Corporate Portals," The Delphi Group, 2001.

[4]   The reader should be aware that this future possibility is envisioned by some serious thinkers as a grave danger. In the context of this book, we deal only with the future certainty of increasing machine intelligence and the potential benefits to knowledge creation. For the concerns on this issue, see Kurtzweil, R., *The Age of Spiritual Machines: When Computers Exceed Human Intelligence,* New York: Penguin Books, 1999; and Joy, Bill, "Why the Future Doesn't Need Us," *Wired*, 8.04, April 2000, accessed online on August 6, 2002, at http://www.wired.com/wired/archive/8.04/joy.html.

[5]   The DARPA Translingual Information Detection, Extraction, and Summarization Program is conducting research in relevant topic detection, extraction, and tracking to enable English-speaking users to access, correlate, and interpret multilingual sources of real-time information.

[6]   The DARPA Evidence Extraction, Linking, and Discovery program is developing knowledge discovery technology to overcome the domain dependent limitations of current technologies to provide extraction of evidence from unstructured data, automated detection of relevant links, and learning of a wide range of pattern types (e.g., temporal, organizational, and transactional.)

[7]   The Advanced Research and Development Agency Advanced Question and Answer Program (Acquaint) is conducting research to: question understanding and interpretation (including contextual interpretation, query expansion, and query taxonomy), answer derivation (including information retrieval and extraction from multiple media/languages and data types, interpretation, synthesis, resolving conflicting information, and justification), and answer formulation and presentation (including summarization, synthesis, and generation).

[8]   The DARPA Augmented Cognition program is conducting research methods to measure human cognitive load and capacity to optimize the flow of information to the human sensor channels in an effort to overcome cognitive limitations. The DARPA-sponsored Info-Cockpit research conducted by Carnegie-Melon/University of Virginia is an example of an immersive visual environment constructed to enhance human memory of complex situations.

[9]   The ARDA Novel Intelligence from Massive Data program is researching tacit knowledge capture and use technologies.

[10] "Federal Information Technology Forecast," Government Electronics and Information Technology Association, Washington D.C., 1998.

About the Author

Ed Waltz is the technical director for intelligence systems with the Veridian Systems Division, Applied Science and Technology Sector. He leads research and development for advanced KM technologies applied to difficult intelligence problems. His more than 30 years of engineering experience have encompassed a wide range of signal and data processing applications, with emphasis on non-cooperative and automatic target recognition, data fusion, modeling and simulation, and KM systems. He has broad experience working with U.S. intelligence, surveillance, and reconnaissance and C2 systems.

He is internationally recognized for his expertise on the subjects of data fusion, information warfare, and knowledge management; he has lectured on these subjects throughout the United States and Europe, and in Canada and the Middle East. Mr. Waltz is also the author of *Information Warfare Principles and Operations* and coauthor of *Multisensor Data Fusion* (published by Artech House, 1998 and 1990, respectively). He has authored or coauthored more than 35 other technical publications on advanced sensors, information warfare, and intelligence analysis.

Mr. Waltz received a B.S. in electrical engineering from the Case Institute of Technology, Cleveland, Ohio, (1968), and an M.S. in computer, information, and control engineering from the University of Michigan, Ann Arbor, Michigan (1971).

Index

Abductive reasoning, 173–75
    creating/testing hypotheses, 174–75
    critical stage of, 174
    defined, 173
    hypothesis selection, 175
    in intelligence example, 178–79
    Peirce's abductive method, 174
    process, 177–78
    process form, 173
    See also Reasoning
Abstraction, 68
Acquisition, 69
Agent-based simulation (ABS), 294
Agents
    aware, 338
    brokering, 266–67
    defined, 265–66
    functions, 267
    human-like partners, 339
    interactive, 267
    personal information (PIMs), 266
    planning, 267
Alternative analysis, 146–47
American Intelligence Journal, 21
American National Standards Institute
    (ANSI), 322
American Productivity and Quality Center
    (APQC), 22
Analysis, 159–91, 225
    alternative, 146–47

availability, 48
capability maturity levels of, 124
collaborative, 143, 173
collaborative tools, 44
data, 5
decision, 147
defined, 160, 164–65
hypotheses, 226
impact refinement, 202
matrix, 225
MI objects of, 32
new, 49
situation refinement, 202
subject of, 209
user availability, 47
See also Analysis-synthesis
Analysis of competing hypotheses (ACH),
    144, 224–27
    for collaborative analytic reasoning, 225
    matrix, 255
    process flow illustration, 226
    structure, 225
Analysis-synthesis, 161–62
    collaborative, 235
    distinctions, 163
    intelligence consumer expectations and,
        196–98
    with intelligence cycle, 161–62
    in intelligence workflow, 198–202
    as modeling process, 180–86

The Artech House Information Warfare Library

Electronic Intelligence: The Analysis of Radar Signals, Second Edition, Richard G. Wiley

Electronic Warfare for the Digitized Battlefield, Michael R. Frater and Michael Ryan

Electronic Warfare in the Information Age, D. Curtis Schleher

EW 101: A First Course in Electronic Warfare, David Adamy

Information Warfare Principles and Operations, Edward Waltz

Introduction to Communication Electronic Warfare Systems, Richard Poisel

Knowledge Management in the Intelligence Enterprise, Edward Waltz

Principles of Data Fusion Automation, Richard T. Antony

Tactical Communications for the Digitized Battlefield, Michael Ryan and Michael R. Frater

For further information on these and other Artech House titles, including previously considered out-of-print books now available through our In-Print-Forever® (IPF®) program, contact:

Artech House
685 Canton Street
Norwood, MA 02062
Phone: 781-769-9750
Fax: 781-769-6334
e-mail: artech@artechhouse.com

Artech House
46 Gillingham Street
London SW1V 1AH UK
Phone: +44 (0)20-7596-8750
Fax: +44 (0)20-7630-0166
e-mail: artech-uk@artechhouse.com

Find us on the World Wide Web at:
www.artechhouse.com